Progress in Mathematics
Volume 82

M. Duflo N.V. Pedersen M. Vergne

The Orbit Method in Representation Theory

Proceedings of a Conference Held in Copenhagen, August to September 1988

With 23 Figures

1990 Birkhäuser
Boston · Basel · Berlin

M. Duflo
University of Paris-VII
75251 Paris Cedex 05
France

N.V. Pedersen
Mathematics Department
University of Copenhagen
2100 Copenhagen
Denmark

M. Vergne
CNRS
DMI
45 rue d'Ulm
75005 Paris
France

Library of Congress Cataloging-in-Publication Data
The Orbit method in representation theory: proceedings of a conference held in Copenhagen, August to September 1988/M. Duflo, N.V. Pedersen, M. Vergne, editors.
p. cm.—(Progress in mathematics; v. 82)
"Held at the University of Copenhagen from August 29 to September 2, 1988 . . . in honor of L. Pukanszky"—Pref.
Includes bibliographical references.
ISBN 0-8176-3474-6 (alk. paper)
1. Orbit method—Congresses. 2. Lie groups—Congresses. 3. Representations of groups—Congresses. 4. Lie algebras—Congresses. 5. Representations of algebras—Congresses. 6. Pukanszky, L.—Congresses. I. Duflo, Michel. II. Pedersen, N.V. (Niels Vigand) III. Vergne, Michèle. IV. Pukanszky, L.V. Københavns Universitet. VI. Series: Progress in mathematics (Boston, Mass.); vol. 82.
QA387.073 1990
512'.55—dc20

89-18439

Printed on acid-free paper.

ISBN 0-8176-3474-6
ISBN 3-7643-3474-6

Camera-ready copy supplied by the editors using TEX.
Printed and bound by Edwards Brothers, Inc., Ann Arbor, Michigan.

9 8 7 6 5 4 3 2 1

Preface

The present volume contains the proceedings of the conference "The Orbit Method in Representation Theory" held at the University of Copenhagen from August 29 to September 2, 1988. Ever since its introduction around 1960 by Kirillov, the Orbit Method has played a major role in representation theory of Lie groups and Lie algebras. We therefore felt it was desirable to devote a conference fully to this topic. As one of the main contributors to the orbit method, L. Pukanszky, celebrated his sixtieth birthday in November 1988, we decided to hold the conference in honor of him.

At the conference were given 22 invited lectures of which 9 are published in these proceedings. Furthermore, L. Pukanszky, who unfortunately became ill just prior to the conference, has submitted a paper for the proceedings. The conference was organized by M. Duflo, M. Flensted-Jensen, H. Plesner Jakobsen, N. Vigand Pedersen, and M. Vergne. These proceedings are edited by M. Duflo, N. Vigand Pedersen, and M. Vergne.

Most of the manuscripts were typeset (in TEX) at the Mathematics Department, University of Copenhagen, and the final manuscript was prepared there.

We heartily thank the following institutions for generous support: Carlsberg Fondet, The Danish Natural Science Research Council, Den Danske Bank, The Danish Mathematical Society, Julius Skrikes Stiftelse, Knud Højgårds Fond, The Danish Ministry of Education, Otto Mønsteds Fond and Tuborg Fondet.

Copenhagen
September 1989

NIELS VIGAND PEDERSEN

Contents

List of Participants

G. Almquist, Lund
H. Haahr Andersen, Århus
M. Andler, Paris
E. van den Ban, Utrecht
J. Bang-Jensen, Odense
Y. Benoist, Paris
C. Berg, Copenhagen
T. Branson, Copenhagen and Iowa City
C.J.B. Brookes, Cambridge (England)
J.-Y. Charbonnel, Paris
E. Christensen, Copenhagen
L. Corwin, New Brunswick
B. Currey, Saint Louis
V.K. Dobrev, Trieste
F. Du Cloux, Palaiseau
J.-Y. Ducloux, Paris
M. Duflo, Paris
E.G. Dunne, Oxford
G. Elliott, Copenhagen
R. Felix, Eichstätt
A. Fialovsky, Bonn
M. Flensted-Jensen, Copenhagen
H. Fujiwara, Kinki
R. Goodman, New Brunswick
F. Greenleaf, New York
U. Haagerup, Odense
G. Heckman, Leiden
J. Hilgert, Erlangen
J. Jacobsen, Århus
H. Plesner Jakobsen, Copenhagen
J.C. Jantzen, Hamburg
S. Jøndrup, Copenhagen
A. Joseph, Paris and Rehovot
E. Kehlet, Copenhagen
M.S. Khalgui, Tunis
A.A. Kirillov, Moscow
S. Kleiman, Cambridge (USA)
N.J. Kokholm, Copenhagen
T. Koornwinder, Amsterdam
A. Koranyi, New York
B. Kostant, Cambridge (USA)
K. Kumahara, Tottori
H. Leptin, Bielefeld
R.L. Lipsman, College Park
W. Lisiecki, Warsaw
J. Ludwig, Luxemburg
L.-E. Lundberg, Copenhagen
B. Magneron, Paris
A. Melin, Lund
I. Mladenov, Sofia
T. Moons, Diepenbeek
O.A. Nielsen, Kingston
A. Ocneanu, University Park
G. 'Olafsson, Göttingen
D. Olesen, Roskilde
J. Børling Olsson, Copenhagen
B. Ørsted, Odense
T. Oshima, Tokyo
G. Kjærgård Pedersen, Copenhagen
N. Vigand Pedersen, Copenhagen
R. Penney, Purdue
M. Poel, Groningen
D. Poguntke, Bielefeld
H. Prado, Iowa City
M. Reimann, Bern
R. Rentschler, Orsay
H. Schlichtkrull, Copenhagen

W. Soergel, Hamburg
H. Stetkær, Århus
T. Sund, Oslo
E. Thieleker, Tampa
P. Torasso, Poitiers
M. Vergne, Cambridge (USA) and Paris
N. Wildberger, Toronto
H. Yamada, Japan
S. Yamagami, Tohoku

Towards Harmonic Analysis on Homogeneous Spaces of Nilpotent Lie Groups

LAWRENCE CORWIN[1]

The work described here is a joint project with Fred Greenleaf.

Let G be a simply connected nilpotent Lie group, with Lie algebra $\mathfrak{G}$, and let K be a connected Lie subgroup, with Lie algebra $\mathfrak{K}$. We would like to do harmonic analysis on $K\backslash G$. More generally, let χ be a character of K; we would like to consider questions of harmonic analysis for the unitary representation τ induced from χ. The first task facing us is to understand what that means.

In order to begin to do harmonic analysis, we must solve a problem in group representation theory: determining the direct integral decomposition of the representation τ on $\mathcal{H}_\tau = \{f : G \to \mathbf{C} \mid f(kx) = \chi(k)f(x)$ if $k \in K$, f measurable, and $|f| \in \mathcal{L}^2(K \setminus G)\}$. That problem was answered in [4] (see also [6]). Let $\ell' \in \mathfrak{G}^*$ be such that for $Y \in \mathfrak{K}$, $\chi(\exp Y) = e^{2\pi i \ell'(Y)}$, and set $\Omega_\tau = \ell' + \mathfrak{K}^\perp \subset \mathfrak{G}^*$. The Kirillov correspondence gives a bijection between $G^\wedge$ and $\mathrm{Ad}^*(G)$-orbits in $\mathfrak{G}^*$. The decomposition of τ can be described as follows:

$$\tau \cong \int_{G^\wedge}^{\oplus} n_\pi \pi d\mu(\pi)\,, \tag{1}$$

where μ is the push-forward of (a finite measure equivalent to) Lebesgue measure on Ω_τ to $\mathfrak{G}^*/\,\mathrm{Ad}^*(G)$ and thence to $G^\wedge$, and n_π is the number of $\mathrm{Ad}^*(K)$-orbits in $\mathcal{O}_\pi \cap \Omega_\tau$. (Here and in the future, $\mathcal{O}_\pi$ is the Ad^*-orbit in $\mathfrak{G}^*$ corresponding to π.) Furthermore, n_π is either μ-a.e. finite or μ-a.e. infinite; the former occurs iff for generic $\ell \in \mathfrak{K}^\perp$,

$$\dim G \cdot \ell = 2 \dim K \cdot \ell\,.$$

(In fact, n_π is then either μ-a.e. even or μ-a.e. odd; see [1].) We can also decompose ρ as

$$\tau \cong \int_{\Omega_\tau/\,\mathrm{Ad}^* K}^{\oplus} \pi_O d\nu(O)\,, \tag{2}$$

[1]Supported by NSF grant DMS-86-03169

where, for an $\mathrm{Ad}^*(K)$-orbit O, $\pi_O = \pi$ is the representation of G such that $O \subseteq \mathcal{O}_\pi$, and ν is the push-forward of (a finite measure equivalent to) Lebesgue measure on Ω_τ. Direct proofs of (2) are given in [2] and [7].

Harmonic analysis should mean more than this, of course. For example, one might want to have an explicit intertwining operator giving the direct integral decomposition. In the case where n_π is not 1 a.e., there is a great deal of choice possible in describing this operator. In order to make the situation more rigid, we restrict to the case where n_π is a.e. finite. There are three reasons for this restriction. The first two are practical, the third conjectural. First of all, experience shows that the case of finite multiplicity is more tractable; the main cases where progress has been made in nonabelian harmonic analysis (compact homogeneous spaces; symmetric spaces; the regular representation, usually regarded as a two-sided representation of $G \times G$) involve finite multiplicity. Second, the proof of the direct integral formula (2) gives an (inductively defined) intertwining operator between τ and the right hand side of (2) in the finite multiplicity case; in the case of infinite multiplicity, no such operator is produced. This also makes the analysis somewhat easier if we assume finite multiplicity.

The third reason requires some further explanation and notation. We denote by $D_\tau(K \setminus G)$ the algebra of differential operators D on $\mathcal{S}_\tau^\infty(K \setminus G) = \{f \in \mathcal{H}_\tau | f \text{ is } C^\infty \text{ and rapidly vanishing transversally to } K\}$ commuting with τ. There is a different, and useful, description of this algebra. We may regard $\mathcal{S}_\tau^\infty(K \setminus G)$ as a subspace of $C^\infty(G)$ in the obvious way, and therefore every element of $\mathfrak{U}(\mathfrak{G})$ acts on $\mathcal{S}_\tau^\infty(K \setminus G)$ as a right invariant operator. Let $\mathfrak{a}_\tau$ be spanned by the elements $Y + 2\pi i \ell_0(Y)$, $Y \in \mathfrak{K}$, let $\mathfrak{I}_\tau = \mathfrak{U}(\mathfrak{G})\mathfrak{a}_\tau$ be the left ideal generated by $\mathfrak{a}_\tau$, and let $\mathfrak{S}(\mathfrak{G}, \mathfrak{K}; \tau)$ denote the subspace of elements $A \in \mathfrak{U}(\mathfrak{G})$ such that $[A, Y] \in \mathfrak{I}_\tau$ for all $Y \in \mathfrak{K}$; it is not hard to check that elements of $\mathfrak{S}(\mathfrak{G}; \mathfrak{K})$ map $\mathcal{S}_\tau^\infty(K \setminus G)$ to itself, and that elements of $\mathfrak{I}_\tau$ map $\mathcal{S}_\tau^\infty(K \setminus G)$ to 0. In fact, we have the following result (which is well-known in the case $\chi \equiv 1$):

PROPOSITION 1. *With notation as above,* $D_\tau(K \setminus G) \cong \mathfrak{S}(\mathfrak{G}, \mathfrak{K}; \tau)/\mathfrak{I}_\tau$.

For future convenience, we denote by Φ the canonical map from $\mathfrak{S}(\mathfrak{G}, \mathfrak{K}; \tau)$ to $D_\tau(K \setminus G)$.

We can now give our first result:

THEOREM 1. *(a) In the finite multiplicity case,* $D_\tau(K \setminus G)$ *is commutative. (b) The spectral decomposition of* $D_\tau(K \setminus G)$ *also gives a direct integral decomposition of* τ *into irreducibles.*

The meaning of part (b) of the theorem is this: the elements of the algebra $D_\tau(K\backslash G)$ are (essentially) normal operators, and there is a minimal common spectral decomposition for all of them. This spectral decomposition also decomposes τ, and thus gives a direct integral decomposition of τ. It turns out that the components are a.e. irreducible, even though τ may have (finite) multiplicity. A sketch of the proof is given below.

The description of $D_\tau(K\backslash G)$ that (3) gives is not always easy to work with. Define $\mathfrak{S}^K(G)$ to be the subalgebra of elements $A \in \mathfrak{U}(\mathfrak{G})$ such that $[A, Y] = 0$ for all $Y \in \mathfrak{K}$; say that (G, K, τ) is *generic* if Ω_τ meets the set of generic orbits in $\mathfrak{G}^*$.

"CONJECTURE" 1. *In the finite multiplicity case, and for generic* (G, K), $\mathfrak{S}(\mathfrak{G}, \mathfrak{K}; \tau) = \mathfrak{S}^K(\mathfrak{G}) + \mathfrak{I}_\tau$.

I put quotation marks around the word "conjecture" because I do not think that the evidence for it is as strong as it should be for a true conjecture.

It may be worth noting that the hypothesis of finite multiplicity is necessary. For example, let $\mathfrak{G}$ be the algebra of strictly upper triangular 4×4 matrices, and let $E_{i,j} (1 \le i < j \le 4)$ be the matrix with a 1 in the (i, j) entry and 0's elsewhere; let $\mathfrak{K}$ be spanned by $E_{2,3}$ and $E_{2,4}$. Then $E_{3,4} \in \mathfrak{S}(\mathfrak{G}, \mathfrak{K}; \tau)$, but $\notin \mathfrak{S}^K(\mathfrak{G}) + \mathfrak{I}_\tau$. Some sort of genericity condition is also necessary; for instance, if $\mathfrak{K}$ is allowed to meet the center of $\mathfrak{G}$, then one can easily construct counterexamples.

Assume now that we are in the generic case and that the above "Conjecture" 1 holds. What else could we ask for? In the case of finite multiplicity, $\dim(\mathcal{O}_\pi \cap \Omega_\tau) = \dim(\mathcal{O}_\pi)/2$ for generic orbits $\mathcal{O}_\pi$ meeting Ω_τ. Since π is naturally defined on $\mathbf{R}^k$, $k = \dim(\mathcal{O}_\pi)/2$, the following makes sense:

CONJECTURE 2. *There are "natural" realizations of* π *on* $\mathcal{L}^2(O)$ $(O = \mathcal{O}_\pi \cap \Omega_\tau)$ *and a "natural" unitary map* $U : \mathcal{L}^2(K \setminus G) \to \mathcal{L}^2(\mathfrak{K}^\perp)$ *such that if one regards* $\mathcal{L}^2(\Omega_\tau)$ *as* $\int^\oplus_{\Omega_\tau/\,\mathrm{Ad}^* K} \mathcal{L}^2(O) d\nu(O)$, *then* U *gives an explicit intertwining map for (2).*

For the case of $G \times G$ acting on $\mathcal{L}^2(G)$, this says that there is a "natural" unitary map of $\mathcal{L}^2(G)$ to $\mathcal{L}^2(\mathfrak{G}^*)$ such that $\pi \times \bar{\pi}$ is realized on $\mathcal{L}^2(\mathcal{O}_\pi)$. (Here, $\bar{\pi}$ is the contragredient of π.) Constructions like this are found in [10]. We also have:

THEOREM 2. *Let* G, K, χ, τ *be as above. Suppose that:*

(1) (G, K, τ) *is generic;*

(2) *There is a subalgebra* $\mathfrak{b}$ *normalized by* $\mathfrak{K}$ *that is polarizing for generic elements in* Ω_τ.

Then Conjecture 2 holds (with the natural map $\mathcal{F}$ being given essentially by a Fourier transform).

Even without Conjecture 2, we can try to associate the spectral decomposition of $D(K \setminus G)$ more closely with the direct integral decomposition of τ. In the case of $G \times G$ acting on $\mathcal{L}^2(G)$, $D_\tau(K \setminus G) = \mathfrak{Z}\mathfrak{U}(\mathfrak{G})$, the center of the universal enveloping algebra; it is known (see [5]) that for $A \in \mathfrak{Z}\mathfrak{U}(\mathfrak{G})$, $\pi(A) = P_A(2\pi i \ell)I$, where $\ell \in \mathcal{O}_\pi$ and $P_A \in S(\mathfrak{G})$ (the symmetric algebra of $\mathfrak{G}$, regarded here as the polynomial algebra on $\mathfrak{G}^\perp$) is the ($\mathrm{Ad}^*(G)$-invariant) polynomial mapped to A by the symmetrization map σ. (The factor of 2π is the result of our normalization of the Kirillov map.) These facts have been generalized to nilpotent symmetric spaces $K \setminus G$ by Y. Benoist [0]. In our case, there is generally no way to associate $\mathrm{Ad}^*(K)$-invariant polynomials on $\mathfrak{K}^\perp$ bijectively with elements of $D(K \setminus G)$.
However, if "Conjecture" 1 and Conjecture 2 both hold, then:

"CONJECTURE" 3. *In the finite multiplicity case, with (G, K, τ) generic, for every $D \in D_\tau(K \setminus G)$ there is an $\mathrm{Ad}^*(K)$-invariant polynomial P_D such that*

(i) $\Phi \circ \sigma(P_D) = D$;

(ii) *under the direct integral decomposition (3), D maps to the scalar field of operators whose value on an orbit O is $P_D(\ell)$ for any $\ell \in O$.*

We'll give an example soon. For the moment, notice that the above statement is complicated partly because the symmetrization map operates on polynomials on $\mathfrak{G}^*$, while we really ought to be concerned with polynomials on Ω_τ. Any polynomial on Ω_τ extends to $\mathfrak{G}^*$ in many ways, and any two extensions differ by an element in the ideal (in $S(\mathfrak{G})$) generated by $\mathfrak{a}_\tau$. Under symmetrization, polynomials in this ideal do not necessarily map to $\mathfrak{I}_\tau$, but their leading terms do. This suggests the following:

"CONJECTURE" 4. *There is a linear bijection between the space of $\mathrm{Ad}^*(K)$-invariant polynomials on Ω_τ and $D_\tau(K \setminus G)$ given by mapping P to $\Phi \circ \sigma(P^\sim)$, where $P^\sim$ is some Ad^*-invariant extension of P to $\mathfrak{G}^*$.*

One could even be greedy and hope for an algebra isomorphism.

Here is an example of a symmetric space that may make things clearer. Let $\mathfrak{G}$ be spanned by W, X, Y, and Z, with nontrivial brackets $[W, X] = Y$ and $[W, Y] = Z$; let $\mathfrak{K}$ be spanned by W, and let χ be trivial. Write a typical element of G as $(w, z, y, x) = \exp wW \exp zZ \exp yY \exp xX$, and use the elements $(x, y, z) = (0, x, y, z)$ as a cross-section for $K \setminus G$.

The irreducible representations $\pi_{\alpha,\beta}$ ($\alpha \neq 0$) of G in general position act on $\mathcal{L}^2(\mathbf{R})$ and are given infinitesimally by

$$\pi_{\alpha,\beta}(Z) = 2\pi i\alpha I\,, \quad \pi_{\alpha,\beta}(Y) = 2\pi it\,,$$

$$\pi_{a,\beta}(X) = 2\pi i\left(\frac{\alpha t^2}{2} + \beta\right)\,, \quad \pi_{\alpha,\beta}(W) = \frac{d}{dt}\,.$$

On $\mathcal{L}^2(K \setminus G)$, τ is given by

$$\begin{aligned}&(\tau(w,x,y,z)f)(x_0,y_0,z_0)\\ &\quad = f\left(z_0 + z - wy_0 + \frac{w^2 x_0}{2}, y + y_0 - wx_0, x + x_0\right)\,.\end{aligned}$$

To decompose this, write a typical element of $\Omega_\tau = \mathfrak{K}^\perp$ as (α,t,β), where (α,t,β) maps $wW + zZ + yY + xX$ to $\alpha z + ty + \beta x \in \mathbf{R}$. Then consider the Fourier transform map U: $\mathcal{L}^2(K \setminus G) \to \mathcal{L}^2(\mathfrak{K}^\perp)$ given by

$$Uf(\alpha,t,\beta) = |\alpha|^{1/2}\int_{\mathbf{R}^3} f(x_0,y_0,z_0)e^{-2\pi i(\alpha z_0+\alpha t y_0+\beta x_0)}dx_0dy_0dz_0\,.$$

Let $\tau_1 = U\tau U^{-1}$. Some calculation shows that for $\varphi \in \mathcal{L}^2(\mathfrak{K}^\perp)$,

$$\begin{aligned}&\tau_1(w,z,y,x)\varphi(\alpha,t,\beta)\\ &= \varphi\left(\alpha, t+w, \beta + \alpha\frac{(w+t)^2 - w^2}{2}\right)e^{2\pi i\alpha(z+(w+t)y+wt-wx^2/2)}e^{2\pi i\beta x}\,.\end{aligned}$$

This means that for $g \in G$ and $\ell \in \mathfrak{K}^\perp$, $\tau_1(g)\varphi(\ell)$ depends only on the values of φ on the $\mathrm{Ad}^*(K)$-orbit of ℓ. (The $\mathrm{Ad}^*(K)$-orbit of $(\alpha,0,\beta)$ is $\{(\alpha,t,\beta+\alpha t^2/2) : t \in \mathbf{R}\}$, for generic ℓ.) Some further calculation shows that restricting to the orbit of $(\alpha,0,\beta)$ gives the representation $\pi_{\alpha,\beta}$. Thus Conjecture 2 holds in this example. The other "conjectures" also hold. One determines $D_\tau(K\backslash G)$ as follows: $\mathfrak{Z}\mathfrak{U}(\mathfrak{G})$ is generated by Z and $U = Y^2 - 2ZX$. Modulo elements of $\mathfrak{J}_\tau = \mathfrak{U}(\mathfrak{G})\mathfrak{K}$, every element of $\mathfrak{U}(\mathfrak{G})$ can be written uniquely as $A = \sum_{j=0}^n P_j(U,Z)X^j + Q_j(U,Z)X^jY$, P_j and Q_j polynomials . Then

$$[W,A] = \sum_{j=0}^n jP_j(U,Z)X^{j-1}Y + (2j+1)ZQ_j(U,Z)X^j + jUQ_j(U,Z)X^{j-1}\,;$$

this is in $\mathfrak{U}(\mathfrak{G})\mathfrak{K}$ iff it is 0, and it is easy to see that $[W,A] = 0$ only if every $Q_j \equiv 0$ and $P_j \equiv 0$ for $j \geq 1$.

Another example, more interesting as an example of the theorem, has $\mathfrak{G}$ as above,but $\mathfrak{K}$ spanned by X (and χ trivial). Now τ has multiplicity 2, and $D_\tau(K \setminus G)$ is generated by (the images under Φ of) Y and Z. Here, $\mathfrak{J}_\tau = \mathfrak{Z}\mathfrak{U}(\mathfrak{G})$ corresponds to the subalgebra generated by Y^2 and Z, the conjecture holds, and the primary decomposition (1) corresponds to the spectral decomposition of the subalgebra while the decomposition (2) corresponds to the spectral decomposition of $D_\tau(K \setminus G)$; multiplicity 2 is explained by the change from Y to Y^2.

We have one result along the lines of "Conjecture" 3.

Theorem 3. *Let G, K, χ satisfy the hypotheses of Theorem 2. Suppose that $A \in \mathfrak{U}(\mathfrak{G})$ is such that A (as a right invariant operator) commutes with the map $Q_\tau : \mathcal{S}(G) \to \mathcal{S}_\tau^\infty(K \setminus G)$ given by*

$$(Q_\tau \omega)(x) = \int_K \omega(kx)\chi(k^{-1})dk\,.$$

Then $A \in \mathfrak{S}(\mathfrak{G}, \mathfrak{K}; \tau)$, and $\mathcal{F}\Phi(A)\mathcal{F}^{-1}$ (where $\mathcal{F}$ is the map of Theorem 2) is multiplication by a K-invariant polynomial on Ω_τ.

We turn now to a description of the proofs. Theorem 1 is proved by induction on $\dim(K \setminus G)$, there being nothing to prove when $K = G$. Construct a chain of subalgebras $\mathfrak{K} = \mathfrak{K}_0 \subset \mathfrak{K}_1 \subset \cdots \subset \mathfrak{K}_n = \mathfrak{G}$, each of codimension 1 in the next, and assume the theorem for $K_{n-1} = G_0$ (of course, $K_i = \exp \mathfrak{K}_i$). We have $\tau_0 = \mathrm{Ind}_K^{G_0}$ operating on $\mathcal{H}_0$, and by hypothesis we have a map $U_0 : \mathcal{H}_0 \to \mathcal{L}^2(E_0 \times \mathbf{R}^r, \nu_0 \times m)$, where E_0 is a cross-section for the generic K-orbits in Ω_{τ_0} (the preimage of ℓ' under the obvious projection), r is the dimension of these generic orbits, and m is Lebesgue measure; U intertwines τ_0 with $\int_{E_0}^\oplus \pi_{\ell_0} d\nu_0(\ell_0)$. Let $\pi_0 = \pi_{\ell_0}$ be a typical representation in this direct integral. When we induce to G, either π_0 generically induces to an irreducible or π_0 generically extends to an irreducible. We call these two possibilities Case 1 and Case 2 respectively.

In Case 1, the dimensions of generic K-orbits also increase, so that $\tau \cong \int_E^\oplus \pi_\ell d\nu_0(\ell)$, where $E \approx E_0$. In this case, we want to show that the obvious injection of $D_{\tau_0}(K \setminus G_0)$ into $D_\tau(K \setminus G)$ is an isomorphism. Let $X \in \mathfrak{G} \setminus \mathfrak{G}_0$, so that any element of $\mathfrak{U}(\mathfrak{G})$ can be written as $A = \sum_{j=0}^r B_j X^j$ with the $B_i \in \mathfrak{U}(\mathfrak{G}_0)$. We show that if $A \in \mathfrak{S}(\mathfrak{G}, \mathfrak{K}; \tau)$ then (modulo $\mathfrak{S}^K(\mathfrak{G})$) $A \in \mathfrak{S}(\mathfrak{G}_0, \mathfrak{K}; \tau_0)$.

The idea is this: first of all, there is no loss of generality in assuming that $B_r \notin \mathfrak{U}^K(\mathfrak{G})$, since we can always make this so by reducing r. We'll assume that $r \geq 1$ and derive a contradiction. We may also assume that $\Phi(A)$ is self-adjoint, essentially because $\Phi(A)^*\Phi(A) = \Phi(A_1)$, where A_1

has the same form. We have

$$\tau \cong \int_{G^\wedge}^{\oplus} n_\pi \pi d\mu(\pi),$$

where the n_π are finite. Let $t \mapsto U_t$ be the one-parameter group associated with $\Phi(A)$.Then the U_t commute with τ, and under the direct integral they decompose as well:

$$U_t \cong \int_{G^\wedge}^{\oplus} V_\pi(t) d\mu(\pi),$$

where $V_\pi(t)$ commutes with $n_\pi \pi$ a.e. Therefore $V_\pi(t)$ is an $n_\pi \times n_\pi$ block matrix. Taking the infinitesimal generator of this transformed one-parameter group, we have

(3) $$A \cong \int_{G^\wedge}^{\oplus} A_\pi d\mu(\pi),$$

where the A_π are finite-dimensional operators. There is a number M, therefore, such that on almost every fiber the operators $A_\pi, A_\pi^2, \ldots, A_\pi^M$ are linearly dependent. However,we also have

(4) $$\tau \cong \int_{E_0}^{\oplus} (\mathrm{Ind}_{G_0}^{G} \pi_{\ell_0}) d\nu_0(\ell_0),$$

and from this we can show that on each fiber of (3) A_π is given by an operator of infinite order. This is the desired contradiction.

In Case 2, (4) still holds, but now $\mathrm{Ind}_{G_0}^{G} \pi_{\ell_0}$ is itself a direct integral: if ℓ_n is trivial on $\mathfrak{G}_0$ but not on $\mathfrak{G}$, then $\mathrm{Ind}_{G_0}^{G} \pi_{\ell_0} = \int_{\mathbf{R}}^{\oplus} \pi_{\ell + t\ell_n} dt$, where $\ell|_{\mathfrak{G}_0} = \ell_0$. We now need an element of $D_\tau(K \setminus G)$ that separates the representations in this direct integral. Let $\dim G = N$, and notice that N is a non-jump index for any strong Malcev basis through $\mathfrak{G}_0$ (see [3] or [9] for terminology). If the π_ℓ are generic in $G^\wedge$, then there is a central element in $\mathfrak{U}(\mathfrak{G})$ of the form $A = BX + C$, with $A, B \in \mathfrak{U}(\mathfrak{G}_0)$ and A central in $\mathfrak{U}(\mathfrak{G}_0)$, and this central element has the required properties. In fact, if A_1 is any other element of $\mathfrak{S}(\mathfrak{G}, \mathfrak{K}; \tau)$, then we can find $B_1 \in \mathfrak{S}(\mathfrak{G}_0, \mathfrak{K}; \tau_0)$ such that $\Phi(B_1) \neq 0$ and $A_1 B_1$ is equal (modulo an element of $\mathfrak{J}_\tau$) to a polynomial in A with coefficients in $\mathfrak{S}(\mathfrak{G}_0, \mathfrak{K}; \tau_0)$. If the representations π_ℓ are not generic, we need to manufacture an element A such that $\pi_\ell(A)$ is scalar if π_ℓ appears in the direct integral (1) and such that A distinguishes the representations in (4). We produce A by using results in [8], together with a little algebra. This completes the outline of the proof of Theorem 1.

For Theorem 2, the idea is this: the map $Q_\tau : \mathcal{S}(G) \to \mathcal{H}_\tau^\infty$, described in Theorem 3, intertwines the right action of G on $\mathcal{S}(G)$ with τ, and the map $Q_\pi : \mathcal{S}(G) \to \mathcal{H}_\pi^\infty$,

$$(Q_\pi \omega)(x) = \int_K \int_{K\cap B\backslash K} \chi(k^{-1})\chi_\ell(b^{-1})\omega(kbx)dkdb$$

($\pi = \pi_\ell, \ell \in \Omega_\tau \cap \mathcal{O}_\pi, B$ as in Theorem 2, $\chi_\ell(\exp Y) = e^{2\pi i \ell(Y)}$) intertwines the right action of G on $\mathcal{S}(G)$ with π_ℓ. We also define $\mathcal{F} : \mathcal{S}(G) \to \mathcal{S}(\mathfrak{G}^*)$ by using a weak Malcev basis through $\mathfrak{K}$ and $\mathfrak{K} + \mathfrak{B}$. Then there is a map Ψ such that $\Psi \circ Q_\tau = \mathcal{F}|_{\Omega_\tau}$, and there is a map Ψ_π such that $\Psi_\pi \circ Q_\pi = \mathcal{F}|_{K\cdot\ell}$. One then has an explicit decomposition of τ as a direct integral. If A is as in Theorem 3, then one can show that A decomposes under the direct integral and that it must be a differential operator with polynomial coefficients. Since A must be scalar on generic K-orbits, Theorem 3 follows.

We conclude with an example concerning exponential solvable groups. Let $\mathfrak{G}$ be spanned by A, X, Y, with $[A, X] = Y - X$ and $[A, Y] = X + Y$; let X span $\mathfrak{K}$, and let χ be trivial. As was noted in [6], $\tau = \mathrm{Ind}_K^G 1$ has infinite multiplicity although $2 \dim K \cdot \ell = \dim G \cdot \ell$ for generic $\ell \in \mathfrak{K}^\perp$. A calculation shows that $D_\tau(K \setminus G)$ is generated by $\Phi(Y)$ and is hence commutative; furthermore, the spectral decomposition for $\Phi(Y)$ decomposes τ into irreducibles. This suggests that the above theorems may hold for exponential solvable groups when the dimension condition is met.

References

[0] Y. Benoist, *Analyse harmonique sur les espaces symétriques nilpotents*, J. Functional Analysis **59** (1984), 211–253.

[1] Corwin, L., and Greenleaf, F.P., *Complex algebraic geometry and calculation of multiplicities for induced representations of nilpotent Lie groups*, Trans. Amer. Math. Soc. **305** (1988), 601–622.

[2] Corwin, L. and Greenleaf, F.P., *A canonical approach to induced and restricted representations of nilpotent Lie groups*, Comm. Pure Appl. Math. **41** (1988), 1051–1088.

[3] Corwin, L. and Greenleaf, F.P., "Representations of Nilpotent Lie Groups and their Applications," Cambridge Univ. Press, Cambridge (to appear).

[4] Corwin, L., Greenleaf, F.P. and Grélaud, G., *Direct integral decompositions and multiplicities for induced representations of nilpotent Lie groups*, Trans. Amer. Math. Soc. **304** (1987), 549–583.

[5] Dixmier, J., "Algèbres Enveloppantes," Gauthier-Villars, Paris, 1976.

[6] Grélaud, G., *Désintégration centrale des représentations induites d'un groupe résoluble exponentiel*, Thèse, Univ. de Poitiers (1984).

[7] Lipsman, R., *Orbital parameters for induced and restricted representations*, Trans. Amer. Math. Soc. (to appear).

[8] Pedersen N.V., *On the infinitesimal kernel of irreducible representations of nilpotent Lie groups*, Bull. Soc. Math. France **112** (1984), 423–467.
[9] Pukanszky, L., "Leçons sur les représentations des groupes," Paris, Dunod, 1967.
[10] Wildberger, N., *Quantization and harmonic analysis on nilpotent Lie groups*, Thesis, Yale University (1983).

Department of Mathematics
Rutgers University
New Brunswick, N.J. 08904
USA

Orbites Coadjointes et Cohomologie Équivariante

MICHEL DUFLO ET MICHÈLE VERGNE

INTRODUCTION

Soient G un groupe de Lie et $\mathfrak{g}$ son algèbre de Lie. Pour étudier les fonctions généralisées G-invariantes sur $\mathfrak{g}$, il est utile de considérer leur restriction à certaines sous-algèbres de Lie $\mathfrak{k}$. Les fonctions généralisées G-invariantes que nous considérons sont transformées de Fourier d'orbites de la représentation coadjointe. Si $\mathfrak{k}$ est l'algèbre de Lie d'un sous-groupe fermé K de G, c'est l'analogue "classique" de la restriction à K d'une représentation unitaire irréductible de G. Le but de cet article est de montrer que lorsque K est compact les *formes différentielles équivariantes* sont particulièrement bien adaptées à cette étude. On peut considérer cet article comme une suite à **[5]** (où la cohomologie équivariante est utilisée pour calculer les transformées de Fourier d'orbites coadjointes d'éléments elliptiques d'un groupe semi-simple connexe de centre fini G) et à **[11]** (où on calcule leur restriction à l'algèbre de Lie $\mathfrak{k}$ d'un sous-groupe compact maximal de G).

La première partie de cet article contient de brefs rappels sur ces fonctions généralisées et leurs restrictions. La seconde partie est un exposé de la théorie des formes différentielles équivariantes. Nous redémontrons les résultats qui sont utilisés dans la troisième partie, en particulier la formule de localisation de **[4]**, sa généralisation par Bismut **[6]** au cas de deux champs de vecteurs qui commutent et quelques faits concernant la classe d'Euler et la classe de Thom équivariantes d'un fibré équivariant inspirés par la construction de Mathai et Quillen **[18]**. Dans la troisième et dernière partie G est un groupe semi-simple connexe de centre fini, K un sous-groupe compact maximal et $F_M(X)$ la transformée de Fourier d'une orbite coadjointe semi-simple M. C'est une fonction généralisée sur $\mathfrak{g}$ qui admet une restriction $F_M|_{\mathfrak{k}}$ à $\mathfrak{k}$. Tout d'abord nous définissons certaines fonctions analytiques K-invariantes sur $\mathfrak{k}$ (ce sont très souvent des transformées de Fourier d'orbites de la représentation coadjointe de K dans le dual $\mathfrak{k}^*$ de $\mathfrak{k}$). Au passage, nous donnons une formule pour la fonction K-invariante P sur $\mathfrak{k}$ qui prolonge une fonction p sur une sous-algèbre de Cartan invariante par le groupe de Weyl. Le cas d'un polynôme p mérite d'être souligné puisqu'on obtient ainsi une version "explicite" du théorème de Chevalley. Le premier résultat sur F_M est que la fonction $\det^{1/2}_{\mathfrak{g}/\mathfrak{k}}(X)F_M|_{\mathfrak{k}}(X)$ est analytique sur $\mathfrak{k}$. Supposons de plus M elliptique, considérons la K-orbite $N = M \cap \mathfrak{k}^*$ et le fibré normal

E à N dans M. Soient $\sigma_{\mathfrak{k}}^{0}(X)$, $X \in \mathfrak{k}$, la structure symplectique K-équivariante de N et $\mathrm{Eul}_E(X)$ la classe d'Euler équivariante du fibré E. Nous construisons un inverse (au sens distribution) $\underline{\nu}(X)$ de $\mathrm{Eul}_E(X)$ et nous démontrons l'égalité de fonctions généralisées sur $\mathfrak{k}$

$$F_M(X) = (2i\pi)^{-\dim(M)/2} \int_N e^{i\sigma_{\mathfrak{k}}^{0}(X)} \underline{\nu}(X).$$

Nous montrons (grâce à un résultat sur l'intégration sur N d'une forme équivariante à coefficients fonctions généralisées) comment cette formule redonne les résultats de **[11]**.

1. Restriction de fonctions généralisées

Nous faisons quelques rappels sur les fonctions généralisées, leurs restrictions à des sous-variétés, et sur les transformées de Fourier d'orbites de la représentation coadjointe.

1.1 Front d'onde. Soit M une variété C^∞. On note $\mathcal{M}_c^\infty(M)$ l'espace des densités (à valeurs complexes) C^∞ à support compact sur M. Rappelons qu'une *fonction généralisée* sur M est une forme linéaire continue sur $\mathcal{M}_c^\infty(M)$. Par exemple, une fonction localement sommable F sur M définit canoniquement une telle fonction généralisée. Nous notons $C^{-\infty}(M)$ l'espace de ces fonctions généralisées. Soit θ une fonction généralisée sur M. Nous notons $\mathrm{WF}(\theta)$ son front d'onde C^∞ (voir **[15]** ch. VIII). C'est un cône fermé de $T^*(M)\backslash 0$ (l'espace cotangent privé de la section nulle). Soit $m \in M$. On pose $\mathrm{WF}(\theta)_m = T_m^*(M) \cap \mathrm{WF}(\theta)$. L'ensemble des $m \in M$ tels que $\mathrm{WF}(\theta)_m = \emptyset$ est égal à l'ouvert des points au voisinage desquels θ est une fonction C^∞ (**[15]** prop. 8.1.3).

Soient N une variété C^∞ et $f : N \to M$ une application différentiable. Soit $\theta \in C^{-\infty}(M)$. On dit que $\mathrm{WF}(\theta)$ est transverse à f si pour tout $n \in N$ et tout $\xi \in \mathrm{WF}(\theta)_{f(n)}$, on a $\xi(df(T_n(N)) \neq 0$. Dans ce cas, on peut définir $f^*(\theta)$. C'est un élément de $C^{-\infty}(N)$. De plus, pour tout $n \in N$, on a, en notant f^* le transposé de l'application $df : T_n(N) \mapsto T_{f(n)}(M)$,

$$\mathrm{WF}(f^*(\theta))_n \subset f^*(\mathrm{WF}(\theta)_{f(n)}) \tag{1}$$

(**[15]** theorem 8.2.4).

Lorsque θ est différentiable $\mathrm{WF}(\theta)$ est vide. La condition de transversalité est trivialement vérifiée et $f^*(\theta)$ est simplement égale à la fonction composée $\theta \circ f$.

De même, si f est submersive, la condition de transversalité est vérifiée. Dans ce cas, si β est un élément de $\mathcal{M}_c^\infty(N)$, par intégration sur les fibres de f on obtient la densité image $f_*(\beta) \in \mathcal{M}_c^\infty(M)$ et l'on a $f^*(\theta)(\beta) = \theta(f_*(\beta))$.

Lorsque N est une sous-variété localement fermée de M et lorsque $\mathrm{WF}(\theta)$ est transverse à l'injection canonique, nous dirons que $\mathrm{WF}(\theta)$ est transverse à N et que $f^*(\theta)$ est la restriction de θ à N. Nous la noterons $\theta|_N$.

1.2 Représentation adjointe. Soit G un groupe de Lie réel d'algèbre de Lie $\mathfrak{g}$. Nous notons $C^{-\infty}(\mathfrak{g})^G$ l'espace des fonctions généralisées sur $\mathfrak{g}$ invariantes par l'action adjointe.

Introduisons quelques notations. Si V est un espace vectoriel de dimension finie sur $\mathbf{R}$, nous notons V^* l'espace vectoriel dual, $V_{\mathbf{C}}$ l'espace vectoriel complexifié. L'algèbre symétrique $S(V_{\mathbf{C}})$ s'identifie à l'algèbre des fonctions polynomiales sur V^* et à l'algèbre des opérateurs différentiels à coefficients constants sur V. Nous noterons ∂_p l'opérateur différentiel correspondant à un élément $p \in S(V_{\mathbf{C}})$. Si W est un sous-espace vectoriel de V, nous notons $W^{\perp}$ son orthogonal dans V^*.

Soit $X \in \mathfrak{g}$. Nous noterons $\mathrm{ad}^* X$ l'opérateur $-{}^t \mathrm{ad}\, X$ de $\mathfrak{g}^*$. Si $f \in \mathfrak{g}^*$, nous écrirons Xf pour $\mathrm{ad}^* X(f)$.

Si G opère à gauche dans une variété différentiable M et si $X \in \mathfrak{g}$, on note X_M le champ de vecteurs sur M donné par

$$(X_M \cdot \varphi)(x) = \frac{d}{d\varepsilon}\varphi((\exp -\varepsilon X)x)|_{\varepsilon=0}. \tag{2}$$

Le lemme suivant permet de majorer le front d'onde d'une fonction généralisée.

LEMME 1. *(a) Soit $\theta \in C^{-\infty}(\mathfrak{g})^G$. Soit $X \in \mathfrak{g}$. On a*

$$\mathrm{WF}(\theta)_X \subset \ker(\mathrm{ad}^*(X))$$

(b) Soit I un idéal de $S(\mathfrak{g}_{\mathbf{C}})$. Soit J l'idéal gradué associé de $S(\mathfrak{g}_{\mathbf{C}})$ et soit $\mathcal{N}$ le cône de $\mathfrak{g}^*$ formé des zéros de J. Soit $\theta \in C^{-\infty}(\mathfrak{g})$ un élément annulé par I, c'est-à-dire vérifiant les équations différentielles $\partial_p(\theta) = 0$ pour tout $p \in I$. Alors on a $\mathrm{WF}(\theta)_X \subset \mathcal{N}$ pour tout $X \in \mathfrak{g}$.

DÉMONSTRATION: (a) Soit $Y \in \mathfrak{g}$. On note $Y_{\mathfrak{g}}$ le champ de vecteurs sur $\mathfrak{g}$ correspondant à l'action adjointe. Sa valeur $Y_{\mathfrak{g}}(X)$ en un point $X \in \mathfrak{g}$ est égale à $-[Y, X]$. Comme θ est G-invariante, elle vérifie l'équation différentielle $Y_{\mathfrak{g}}(\theta) = 0$. En appliquant le théorème 8.3.1 de [**15**], on voit que si f est un élément de $\mathrm{WF}(\theta)_X$, on a $f([X, Y]) = 0$. En appliquant ceci à tout $Y \in \mathfrak{g}$, on voit que l'on a $Xf = 0$ pour tout $f \in \mathrm{WF}(\theta)_X$.

(b) Soit $p \in J$ un élément homogène non nul de degré $n > 0$. Il existe un polynôme $q \in I$ dont la composante homogène de plus haut degré non nulle est p. Le symbole de l'opérateur ∂_q est p, et donc, pour tout $X \in \mathfrak{g}$ et pour tout $f \in \mathrm{WF}(\theta)_X$, on a $p(f) = 0$, toujours d'après le théorème 8.3.1 de [**15**]. ∎

Le corollaire suivant résume les situations dans lesquelles nous étudierons les restrictions de fonctions généralisées.

COROLLAIRE 2. *Soient* $\mathfrak{h}$ *une sous-algèbre de Lie de* $\mathfrak{g}$, $\mathcal{N}$ *un cône fermé de* $\mathfrak{g}^*$ *et* $\theta \in C^{-\infty}(\mathfrak{g})^G$. *On suppose que, pour tout* $X \in \mathfrak{g}$, *on a* $\mathrm{WF}(\theta)_X \subset \mathcal{N}$. *Soit* $\mathfrak{h}''$ *l'ensemble des* $X \in \mathfrak{h}$ *tels que* $\mathfrak{h}^\perp \cap \mathcal{N} \cap \ker(\mathrm{ad}^*(X)) = \{0\}$. *Alors* $\mathfrak{h}''$ *est ouvert dans* $\mathfrak{h}$ *et transverse à* $\mathrm{WF}(\theta)$.

DÉMONSTRATION: Compte tenu du lemme qui précède, il reste à démontrer que $\mathfrak{h}''$ est ouvert dans $\mathfrak{h}$. Nous démontrons que le complémentaire est fermé. Soit X_n, $n = 1, 2, \ldots$ une suite de points de $\mathfrak{h}$ tendant vers une limite X_0, et telle que pour tout $n \geq 1$, il existe $f_n \neq 0$ dans $\mathfrak{h}^\perp \cap \mathcal{N} \cap \ker(\mathrm{ad}^*(X_n))$. On extrait de la suite $1, 2, \ldots$ une suite n_j, $j = 1, 2, \ldots$ telle que la droite $\mathbf{R} f_{n_j}$ tende vers une limite (c'est possible car l'espace projectif est compact). Soit f_0 un point non nul de la droite limite. On a $f_0 \in \mathfrak{h}^\perp \cap \mathcal{N} \cap \ker(\mathrm{ad}^*(X_0))$, et donc f_0 n'appartient pas à $\mathfrak{h}''$. ∎

Nous considérons quelques exemples.

EXEMPLE 1. Lorsque $\mathcal{N} = \mathfrak{g}^*$, l'ensemble $\mathfrak{h}''$ du lemme est égal à l'ensemble $\mathfrak{h}'$ des $X \in \mathfrak{h}$ tels que l'application naturelle de $G \times \mathfrak{h}$ dans $\mathfrak{g}$ soit submersive au point $(1, X)$, ou encore à l'ensemble des $X \in \mathfrak{h}$ tels que $ad_{\mathfrak{g}/\mathfrak{h}}(X)$ soit inversible.

EXEMPLE 2. Soit $S(\mathfrak{g}_{\mathbf{C}})^G$ l'ensemble des éléments de l'algèbre symétrique $S(\mathfrak{g}_{\mathbf{C}})$ qui sont invariants par l'action adjointe de G. Soit $\mathcal{N}$ le cône fermé de $\mathfrak{g}^*$ formé des éléments annulés par tous les polynômes de $S(\mathfrak{g}_{\mathbf{C}})^G$ sans terme constant. Soit $\theta \in C^{-\infty}(\mathfrak{g})^G$ une fonction généralisée annulée par un idéal de codimension finie I de $S(\mathfrak{g}_{\mathbf{C}})^G$, c'est-à-dire vérifiant les équations différentielles $\partial_p(\theta) = 0$ pour tout $p \in I$. Alors on a $\mathrm{WF}(\theta)_X \subset \mathcal{N}$.

EXEMPLE 3. Soit G un groupe de Lie connexe semi-simple d'algèbre de Lie $\mathfrak{g}$. Soit $\mathfrak{g}_r$ l'ensemble des éléments $X \in \mathfrak{g}$ qui sont semi-simples et réguliers, c'est-à-dire dont le centralisateur $\mathfrak{g}(X)$ dans $\mathfrak{g}$ est une algèbre de Cartan de $\mathfrak{g}$. On identifie $\mathfrak{g}$ et $\mathfrak{g}^*$ au moyen de la forme de Killing. Le cône $\mathcal{N}$ de l'exemple 2 est le cône des éléments nilpotents de $\mathfrak{g}$. Pout tout $X \in \mathfrak{g}$, l'espace $\ker(\mathrm{ad}^*(X))$ est égal à la sous-algèbre $\mathfrak{g}(X)$. Soit θ une fonction généralisée G–invariante sur $\mathfrak{g}$ et $S(\mathfrak{g}_{\mathbf{C}})^G$-finie. Il résulte de ce qui précède que θ est une fonction différentiable dans l'ouvert de $\mathfrak{g}$ formé des X tels que $\mathfrak{g}(X) \cap \mathcal{N} = 0$, et en particulier dans $\mathfrak{g}_r$. Ceci est un résultat d'Harish-Chandra (voir e.g. **[22]** prop. 13 p. 62 et th. 3 p. 93).

Soit s un automorphisme semi-simple de $\mathfrak{g}$. Soit $\mathfrak{h} = \ker(1 - s)$ la sous-algèbre des points fixes de s. Posons $\mathfrak{q} = \mathrm{Im}(1 - s)$. C'est un supplémentaire de $\mathfrak{h}$. Lorsqu'on identifie $\mathfrak{g}$ et $\mathfrak{g}^*$ grâce à la forme de Killing, $\mathfrak{h}^\perp$ s'identifie à $\mathfrak{q}$. Soit $X \in \mathfrak{g}$. Posons $\mathfrak{q}(X) = \mathfrak{g}(X) \cap \mathfrak{q}$. L'ensemble $\mathfrak{h}''$ du corollaire 2 est donc égal à l'ensemble des $X \in \mathfrak{h}$ tels que $\mathfrak{q}(X) \cap \mathcal{N} = 0$. Il contient l'ouvert de Zariski *non vide* $\mathfrak{h} \cap \mathfrak{g}_r$. Si l'on connaît la fonction θ dans $\mathfrak{g}_r$, étudier sa restriction à $\mathfrak{h} \cap \mathfrak{g}_r$ est trivial, mais l'ouvert $\mathfrak{h}''$ peut être sensiblement plus gros que l'ensemble $\mathfrak{h} \cap \mathfrak{g}_r$, (e.g. si de plus $\mathfrak{q} \cap \mathcal{N} = 0$ comme dans l'exemple 4 ci-dessous) et l'étude de la restriction de θ à $\mathfrak{h}''$ peut poser des problèmes réels même lorsque l'on connaît la fonction θ dans $\mathfrak{g}_r$.

EXEMPLE 4. Lorsque $\mathcal{N} \cap \mathfrak{h}^\perp = \{0\}$, on peut restreindre θ à $\mathfrak{h}$ tout entier. Décrivons deux situations intéressantes où cette relation est vérifiée.

Soit $\mathfrak{s}$ l'algèbre de Lie d'un groupe de Lie connexe S. Soit G le groupe produit $S \times S$. On note $\mathfrak{g} = \mathfrak{s} \times \mathfrak{s}$ son algèbre de Lie, et $\mathfrak{h}$ la sous-algèbre diagonale (qui est naturellement isomorphe à $\mathfrak{s}$). Soient θ_1 et θ_2 des éléments de $C^{-\infty}(\mathfrak{s})^S$, et soient $\mathcal{N}_1$ et $\mathcal{N}_2$ des cônes fermés de $\mathfrak{s}^*$ tels que l'on ait $\mathrm{WF}(\theta_1)_X \subset \mathcal{N}_1$ et $\mathrm{WF}(\theta_2)_X \subset \mathcal{N}_2$ pour tout $X \in \mathfrak{s}$. On suppose que l'on a $\mathcal{N}_1 \cap -\mathcal{N}_2 = \{0\}$. La restriction de $\theta = \theta_1 \otimes \theta_2$ à $\mathfrak{h}$ est alors définie et peut être adoptée comme définition du produit des deux fonctions généralisées θ_1 et θ_2 (voir **[15]** theorem 8.2.10).

Soit maintenant G un groupe de Lie connexe semi-simple comme dans l'exemple 3 dont nous conservons les notations. Considérons le cas particulier de la conjugaison de Cartan, étudié dans **[11]** et sur lequel nous reviendrons plus bas. Soit donc $\mathfrak{g} = \mathfrak{k} \oplus \mathfrak{p}$ une décomposition de Cartan de $\mathfrak{g}$. Comme $\mathfrak{p} \cap \mathcal{N} = 0$, on a $\mathfrak{k}'' = \mathfrak{k}$, et tout élément $S(\mathfrak{g}_{\mathbf{C}})^G$-fini de $C^{-\infty}(\mathfrak{g})^G$ admet une restriction à $\mathfrak{k}$ tout entier.

1.3 Transformées de Fourier d'orbites coadjointes. Soit G un groupe de Lie séparable d'algèbre de Lie $\mathfrak{g}$. Il opère dans $\mathfrak{g}^*$ par la représentation coadjointe. Si $f \in \mathfrak{g}^*$, on note $G(f)$ son stabilisateur dans G et $\mathfrak{g}(f)$ son stabilisateur dans $\mathfrak{g}$. Soit M une orbite de la représentation coadjointe. Les champs de vecteurs X_M, $X \in \mathfrak{g}$, engendrent l'espace tangent en chaque point f de M et on a $(X_M)_f = -Xf$, avec $-(Xf)(Y) = f([X,Y])$.

Rappelons que la forme $\sigma(X_M, Y_M)_f = -f([X,Y])$ définit sur M une forme symplectique G-invariante. La dimension $n = 2d$ de M est paire, et on note

$$\beta_M = (2\pi)^{-d} \frac{\sigma^d}{d!}$$

la forme de Liouville.

On munit M de l'orientation définie par β_M, de sorte que β_M définit sur M une mesure positive qui sera notée aussi β_M. On note encore β_M la mesure positive sur $\mathfrak{g}^*$ image de β_M par l'injection canonique (ce n'est pas nécessairement une mesure de Radon).

Rappelons qu'une mesure positive μ sur $\mathfrak{g}^*$ est dite tempérée si, étant donnée une norme $f \mapsto \|f\|$ sur $\mathfrak{g}^*$, il existe une constante $N > 0$ telle que l'on ait $\int_{\mathfrak{g}^*}(1+\|f\|)^{-N}d\mu(f) < \infty$. Dans ce cas, on peut définir sa transformée de Fourier. C'est la fonction généralisée θ sur $\mathfrak{g}$ définie par la formule

$$\theta(X) = \int_{\mathfrak{g}^*} e^{if(X)}d\mu(f).$$

Cela signifie que pour tout $\Phi \in \mathcal{M}_c^\infty(\mathfrak{g})$ on a

$$\theta(\Phi) = \int_{\mathfrak{g}^*}\Big(\int_{\mathfrak{g}} e^{if(X)}d\Phi(X)\Big)d\mu(f).$$

Lorsque β_M est tempérée, on notera F_M sa transformée de Fourier. C'est donc la fonction généralisée G-invariante sur $\mathfrak{g}$ définie par la formule

$$F_M(X) = \int_{\mathfrak{g}^*} e^{if(X)}d\beta_M(f). \tag{3}$$

Pour étudier F_M, nous commençons par quelques généralités sur les transformées de Fourier de mesures tempérées.

Si M est un sous-ensemble de $\mathfrak{g}^*$, on note $\mathrm{CA}(M)$ le cône asymptote de M. C'est l'ensemble des $f \in \mathfrak{g}^*\backslash 0$ tels que tout voisinage conique de f dans $\mathfrak{g}^*$ coupe M suivant un ensemble non borné. C'est un cône fermé dans $\mathfrak{g}^*\backslash 0$. Rappelons le lemme 8.1.7 de **[15]**, (valable en fait pour toute distribution tempérée μ).

LEMME 3. *Soit θ la transformée de Fourier d'une mesure positive tempérée μ sur $\mathfrak{g}^*$, et soit* $\mathrm{supp}(\mu)$ *le support de μ. Alors, pour tout $X \in \mathfrak{g}$, on a* $\mathrm{WF}(\theta)_X \subset \mathrm{CA}(\mathrm{supp}(\mu))$.

Soit μ une mesure positive tempérée sur $\mathfrak{g}^*$ et θ sa transformée de Fourier. Nous notons $\mathrm{CE}(\mu)$ (pour *cône essentiel*) l'ensemble des $f \in \mathfrak{g}^*\backslash 0$ tels que, pour tout voisinage conique C de f dans $\mathfrak{g}^*$ il existe une constante $N > 0$ telle que l'on ait $\int_C \|f\|^N d\mu(f) = \infty$. C'est un cône fermé dans $\mathfrak{g}^*\backslash 0$. Il est contenu dans $\mathrm{CA}(\mathrm{supp}(\mu))$. Le lemme suivant précise le lemme 3.

LEMME 4. *Soit θ la transformée de Fourier d'une mesure positive tempérée μ sur $\mathfrak{g}^*$.*

a) On a $\mathrm{WF}(\theta)_X \subset \mathrm{WF}(\theta)_0$ *pour tout $X \in \mathfrak{g}$.*

b)On a $\mathrm{WF}(\theta)_0 = \mathrm{CE}(\mu)$.

DÉMONSTRATION: Choisissons une mesure de Lebesgue dX sur $\mathfrak{g}$. Si $\Phi \in C_c^{-\infty}(\mathfrak{g})$ on pose

$$\hat{\Phi}(f) = \int_{\mathfrak{g}} e^{-if(X)}\Phi(X)dX.$$

La fonction $\hat{\Phi}$ est C^∞ sur $\mathfrak{g}^*$.

Soit $X_o \in \mathfrak{g}$. Par définition (voir [15] ch. VIII), un élément $f_o \in \mathfrak{g}^*\backslash 0$ est dans le complémentaire de $\mathrm{WF}(\theta)_{X_o}$ s'il existe $\Phi \in C_c^\infty(\mathfrak{g})$ tel que $\Phi(X_o) \neq 0$, et un voisinage conique C de f_o tels que, pour tout $N > 0$, il existe c_N tel que $|\widehat{\Phi\theta}|(f)| \leq c_N(1 + \|f\|)^{-N}$ pour tout $f \in C$. (Nous dirons que $|\widehat{\Phi\theta}|$ est à décroissance rapide dans C). D'après le lemme 8.1.1 et la formule (8.1.5) de [15], il existe alors un voisinage ouvert V de X_o et un voisinage conique C de f_o tels que $|\widehat{\Phi\theta}|$ soit à décroissance rapide dans C pour tout $\Phi \in C_c^\infty(V)$.

Comme μ est une mesure tempérée, la fonction $\widehat{\Phi\theta}$ est égale au produit de convolution $\hat{\Phi} * \mu$ défini par l'intégrale absolument convergente $\hat{\Phi} * \mu(f) = \int_{\mathfrak{g}^*} \hat{\Phi}(f - u) d\mu(u)$.

Démontrons a). Soit $f_o \notin \mathrm{WF}(\theta)_0$. Il résulte de ce qui précède que l'on peut trouver $\beta \in C_c^\infty(\mathfrak{g})$ telle que $\beta(0) \neq 0$ et $\hat{\beta} \geq 0$, et un voisinage conique C de f_o tels que $|\widehat{\beta\theta}|$ soit à décroissance rapide dans C (on choisit par exemple β de la forme $\gamma * \check{\gamma}$, où $\gamma \in C_c^\infty$ a son support suffisamment petit, et $\check{\gamma}(X) = \overline{\gamma(-X)}$).

On a donc $\widehat{\beta\theta} \geq 0$. Soit $X_o \in \mathfrak{g}$. Définissons Φ par la formule $\Phi(X) = \beta(X - X_o)$. On a $\Phi(X_o) \neq 0$ et $|\widehat{\Phi\theta}| = \widehat{\beta\theta}$. Donc $|\widehat{\Phi\theta}|$ est à décroissance rapide dans C et l'on a $f_o \notin \mathrm{WF}(\theta)_{X_o}$.

Démontrons b). Prouvons que l'on a $\mathrm{CE}(\mu) \subset \mathrm{WF}(\theta)_0$. Soit $f_o \notin \mathrm{WF}(\theta)_0$. On choisit comme ci-dessus $\beta \in C_c^\infty(\mathfrak{g})$ telle que $\beta(0) \neq 0$ et $\hat{\beta} \geq 0$, et un voisinage conique C de f_o tels que $|\widehat{\beta\theta}|$ soit à décroissance rapide dans C. Pour fixer les idées, nous supposons C ouvert, et notons χ_C la fonction caractéristique de C. Pour tout N, l'intégrale double $\int_{\mathfrak{g}^*}\int_{\mathfrak{g}^*} \chi_C(f)\|f\|^N \hat{\beta}(f - u) d\mu(u) df$ est donc convergente. Elle est égale à $\int\int \chi_C(f + u)\|f + u\|^N \hat{\beta}(f) d\mu(u) df$. Il existe donc $f \in \mathfrak{g}^*$ tel que l'intégrale $\int \chi_C(f + u)\|f + u\|^N d\mu(u)$ converge. Soit C' un voisinage conique fermé de f_o contenu dans C. Au voisinage de ∞ dans C', les fonctions $\chi_C(f+u)\|f+u\|^N$ et $\|u\|^N$ sont équivalentes. Donc l'intégrale $\int_C \|u\|^N d\mu(u)$ converge, et f_o n'appartient pas à $\mathrm{CE}(\mu)$.

Pour terminer, démontrons l'inclusion inverse $\mathrm{WF}(\theta)_0 \subset \mathrm{CE}(\mu)$. Soit $f_o \notin \mathrm{CE}(\mu)$. Choisissons un voisinage conique ouvert C de f_o tel que, pour tout N l'intégrale $\int_C \|f\|^N d\mu(f)$ converge. Ecrivons $\mu = \mu_1 + \mu_2$, où μ_1 est une mesure positive nulle dans C et μ_2 est une mesure positive nulle dans $\mathfrak{g}^*\backslash C$. La transformée de Fourier θ_2 de μ_2 est une fonction C^∞. Le front d'onde de θ est donc égal au front d'onde de la transformée de Fourier θ_1 de μ_1. Par construction f_o n'appartient pas au cône asymptote du support de μ_1. Le lemme 3 implique que f_o n'appartient pas à $\mathrm{WF}(\theta)_0$. ∎

Soit $\mathfrak{h}$ un sous-espace vectoriel de $\mathfrak{g}$. D'après le lemme 4, si l'intersection $\mathfrak{h}^\perp \cap \mathrm{CE}(\mu)$ est vide, $\mathfrak{h}$ est transverse à $\mathrm{WF}(\theta)$, et l'on peut restreindre θ à $\mathfrak{h}$. La proposition suivante donne une interprétation de la fonction généralisée $\theta|_\mathfrak{h}$. Notons p la projection naturelle de $\mathfrak{g}^*$ sur $\mathfrak{h}^*$, consistant à restreindre à $\mathfrak{h}$ une forme linéaire sur $\mathfrak{g}$.

PROPOSITION 5. *Soit μ une mesure tempérée positive sur $\mathfrak{g}^*$.*

a) Le sous-espace $\mathfrak{h}$ est transverse à $\mathrm{WF}(\theta)$ *si et seulement si l'intersection $\mathfrak{h}^\perp \cap \mathrm{CE}(\mu)$ est vide.*

b) Dans ce cas, la mesure image $p_*(\mu)$ est une mesure tempérée sur $\mathfrak{h}^*$, et $\theta|_\mathfrak{h}$ est la transformée de Fourier de $p_*(\mu)$.

La signification de b) est la suivante. Fixons une mesure de Lebesgue dY sur $\mathfrak{h}$. Soit $\beta \in C_c^\infty(\mathfrak{h})$, et posons, pour $f \in \mathfrak{g}^*$,

$$\tilde{\beta}(f) = \int_\mathfrak{h} \beta(Y) e^{if(Y)} dY.$$

Alors on a

$$\theta|_\mathfrak{h}(\beta dY) = \int_{\mathfrak{g}^*} \tilde{\beta}(f) d\mu(f), \tag{4}$$

où l'intégrale est absolument convergente.

DÉMONSTRATION: La première assertion résulte immédiatement du lemme 4.

Supposons que $\mathfrak{h}^\perp \cap \mathrm{CE}(\mu)$ soit vide. Choisissons un supplémentaire $\mathfrak{q}$ de $\mathfrak{h}$ dans $\mathfrak{g}$. Il lui correspond une décomposition en somme directe $\mathfrak{g}^* = \mathfrak{h}^* \oplus \mathfrak{q}^*$. On note $f = f_1 + f_2$ la décomposition correspondante d'un élément $f \in \mathfrak{g}^*$ (donc $f_1 = p(f)$). Il existe un nombre $c > 0$ tel que le cône fermé Γ de $\mathfrak{g}^* \backslash 0$ défini par l'équation $\|f_1\| \geq c\|f_2\|$ contienne un voisinage de $\mathrm{CE}(\mu)$. Ecrivons $\mu = \mu_1 + \mu_2$, où μ_1 est une mesure positive à support dans $\Gamma \cup \{0\}$, et μ_2 une mesure positive nulle dans $\Gamma \cup \{0\}$. Il suffit de prouver l'analogue de l'assertion b) pour chacune des mesures μ_1 et μ_2.

On voit facilement que, quel que soit N, l'intégrale $\int_{\mathfrak{g}^*} \|f\|^N d\mu_2(f)$ converge. La transformée de Fourier θ_2 de μ_2 est donc la fonction C^∞ définie par l'intégrale absolument convergente $\theta_2(X) = \int_{\mathfrak{g}^*} e^{if(X)} d\mu_2(f)$. Lorsque $X \in \mathfrak{h}$, cette intégrale est égale à $\int_{\mathfrak{h}^*} e^{if(X)} dp_*(\mu_2)(f)$, ce qui prouve notre assertion pour μ_2.

Pour simplifier les notations, nous supposons maintenant que μ_2 est nulle. C'est-à-dire que nous sommes ramenés à montrer la proposition lorsque de plus μ est à support dans Γ. Soit $N > 0$ tel que l'intégrale

$\int_{\mathfrak{g}^*}(1+\|f\|)^{-N}d\mu(f)$ converge. Dans Γ, la fonction $(1+\|f_1\|)^{-N}$ est majorée par un multiple de $(1+\|f\|)^{-N}$. On a

$$\int_{\mathfrak{g}^*}(1+\|f_1\|)^{-N}d\mu(f)=\int_{\mathfrak{h}^*}(1+\|f_1\|)^{-N}dp_*(\mu)(f_1)<\infty$$

et $p_*(\mu)$ est tempérée. Il reste à démontrer la formule (10). Soit $C_\Gamma^{-\infty}(\mathfrak{g})$ l'ensemble des fonctions généralisée sur $\mathfrak{g}$ dont le front d'onde est contenu dans $\mathfrak{g}\times\Gamma$. Pour tout entier $j>0$, notons μ_j la mesure $\chi_j\mu$, où χ_j est la fonction caractéristique de la boule de rayon j dans $\mathfrak{g}^*$. Soit θ_j la fonction C^∞ sur $\mathfrak{g}$ transformée de Fourier de μ_j. On vérifie facilement que la suite μ_j tend vers μ dans l'espace $C_\Gamma^{-\infty}(\mathfrak{g})$, c'est-à-dire que pour tout cône fermé C de $\mathfrak{g}^*$ disjoint de Γ, pour toute fonction $\phi\in\mathbf{C}_c^\infty(\mathfrak{g})$ et pour tout $N>0$, la suite $\|f\|^N\widehat{\phi(\theta-\theta_j)}(f)$ tend vers 0 quand j tend vers ∞, uniformément pour $f\in C$ (cf. [**15**], definition 8.2.1). Soit $\beta\in C_c^\infty(\mathfrak{h})$. Par définition (cf. [**15**], theorem 8.2.4), on a $\theta|_\mathfrak{h}(\beta dY)=\lim_{j\to\infty}\theta_j|_\mathfrak{h}(\beta dY)$. La formule (10) est évidente pour les fonctions θ_j, et l'on obtient (10) par passage à la limite. ∎

Remarque. Il peut arriver que la mesure $p_*(\mu)$ soit tempérée sans que $\mathfrak{h}^\perp\cap\mathrm{CE}(\mu)$ soit vide. En voici un exemple. Soit $\mathfrak{g}=\mathbf{R}^2$. Notons μ la mesure $\mu(\psi)=\int_{\mathfrak{g}^*}\psi(t,t^2/2)dt$. La fonction généralisée θ est donc définie par la formule $\theta(x,y)=\int_{-\infty}^{\infty}e^{i(xt+yt^2/2)}dt$. Le calcul en est bien connu: c'est la fonction localement intégrable définie par la formule $\theta(x,y)=\sqrt{2\pi}|y|^{-1/2}e^{i\,\mathrm{sgn}(y)\pi/4}e^{-ix^2/2y}$. Le cône $\mathrm{CE}(\mu)$ est égal à l'ensemble $\{(0,u)|u>0\}$. Si $\mathfrak{h}$ est le premier axe de coordonnée, c'est donc un ouvert non vide de $\mathfrak{h}^\perp$. La mesure $p_*(\mu)$ est la mesure de Lebesgue dt. Dans cette situation, bien que la condition de transversalité ne soit pas satisfaite, il est raisonnable de définir la restriction de θ à $\mathfrak{h}$ comme la fonction de Dirac $\delta(x)=\int e^{itx}dt$. On remarquera que la formule (1) n'est plus vraie –elle n'a d'ailleurs pas de sens.

Cet exemple apparaît naturellement lorsqu'on étudie les transformées de Fourier d'orbites coadjointes d'algèbres de Lie $\mathfrak{g}$ de dimension 4 contenant un idéal $\mathfrak{h}$ de dimension 3 qui est une algèbre d'Heisenberg. La proposition 5 ne couvre donc pas tous les cas où il est possible et intéressant de calculer des restrictions de transformées de Fourier d'orbites coadjointes.

Dans les applications, il est évidemment important de déterminer quand la mesure β_M est tempérée. Nous renvoyons à un article de Charbonnel [**10**] pour les résultats les plus complets. Nous nous contenterons de la proposition 8 ci-dessous, adaptée de Ginzburg [**14**] et Charbonnel [**8**]. Nous commençons par une définition.

Soit M une sous-variété différentiable régulièrement plongée dans $\mathfrak{g}^*$ (en particulier M est un sous-ensemble localement fermé de $\mathfrak{g}^*$). Soit ω une forme différentielle sur M. Nous dirons que ω est *régulière* s'il existe une famille finie de polynômes $q_j \in S(\mathfrak{g})$ telle que les ouverts $D_{q_j} = \{f \in \mathfrak{g}^* | q_j(f) \neq 0\}$ recouvrent M, et telle que chacune des formes $q_j\omega$ soit la restriction à M d'une forme différentielle sur $\mathfrak{g}^*$ à coefficients polynomiaux.

LEMME 5. *Soit M une orbite localement fermée de la représentation coadjointe. Alors la forme symplectique σ et la forme de Liouville β_M sont des formes régulières.*

DÉMONSTRATION: Il suffit de le démontrer pour σ. Soit $f_o \in M$. On choisit une base $e_1, ..., e_k$ d'un supplémentaire $\mathfrak{q}$ dans $\mathfrak{g}$ du stabilisateur $\mathfrak{g}(f_o)$ de f. Soit $D = \det([e_i, e_j])_{1 \leq i,j \leq k}$. Soit $f \in M$. On a $D(f) \neq 0$ si et seulement si $\mathfrak{q}$ est supplémentaire de $\mathfrak{g}(f)$ dans $\mathfrak{g}$. Soit $f \in M$ tel que $D(f) \neq 0$. Notons $(c_{ij}(f))_{1 \leq i,j \leq k}$ la matrice inverse de la matrice $(f([e_i, e_j]))_{1 \leq i,j \leq k}$. On vérifie que l'on a

$$\sigma_f = \sum_{i<j} c_{ij}(f) de_i \wedge de_j. \tag{5}$$

Il résulte des formules de Cramer que $D\sigma$ est restriction à M d'une forme différentielle sur $\mathfrak{g}^*$ à coefficients polynomiaux. ∎

On dit qu'un sous-ensemble M de $\mathfrak{g}^*$ est un fermé de Zariski s'il existe une famille finie de polynômes $p_i \in S(\mathfrak{g})$ telle que M soit l'ensemble des zéros de la famille p_i.

LEMME 7. *Soit M une sous-variété différentiable régulièrement plongée dans $\mathfrak{g}^*$. Soit ω une forme différentielle régulière sur M. On suppose que M est un fermé de Zariski de $\mathfrak{g}^*$. Alors il existe une fonction polynomiale $q \in S(\mathfrak{g})$ partout strictement positive sur $\mathfrak{g}^*$ telle que $q\omega$ soit restriction à M d'une forme différentielle sur $\mathfrak{g}^*$ à coefficients polynomiaux.*

DÉMONSTRATION: On emploie les notations p_i et q_j introduites plus haut. On pose $q = \sum_i p_i^2 + \sum_j q_j^2$. La fonction q est strictement positive sur $\mathfrak{g}^*$. D'autre part, on a $q\omega = \sum_j q_j(q_j\omega)$, et donc, si ν_j est une forme différentielle sur $\mathfrak{g}^*$ à coefficients polynomiaux qui prolonge $q_j\omega$, la forme $\nu = \sum_j q_j\nu_j$ prolonge $q\omega$. ∎

REMARQUE. Lorsque les p_i et les q_j n'ont pas de zéros communs dans le complexifié $\mathfrak{g}_{\mathbf{C}}^*$, on peut dans le lemme ci-dessus choisir $q = 1$, c'est-à-dire que ω est restriction à M d'une forme différentielle à coefficients polynomiaux (voir [8] proposition 2.1). En effet, d'après le théorème des zéros de Hilbert, on peut écrire $1 = \sum_i \alpha_i p_i + \sum_j \beta_j q_j$, où les α_i et les β_j sont des polynômes, et ω est la restriction de la forme $\sum_j \beta_j \nu_j$ à M.

PROPOSITION 8. *Soit M une sous-variété de dimension n régulièrement plongée dans $\mathfrak{g}^*$. Soit ω une forme différentielle régulière de degré n sur M. On suppose que M est un fermé de Zariski de $\mathfrak{g}^*$. Alors la mesure positive $|\omega|$ sur M définie par ω est tempérée.*

DÉMONSTRATION: On peut supposer que M est un fermé de Zariski irréductible. D'après le lemme 7, on peut supposer que ω est de la forme $q^{-1}pde_1 \wedge \cdots \wedge de_n$, où $e_1, \ldots, e_n$ sont des vecteurs de $\mathfrak{g}$ tels que la restriction de $de_1 \wedge \cdots \wedge de_n$ à M ne soit pas nulle, p est un polynôme, et q est un polynôme à valeurs strictement positives dans $\mathfrak{g}^*$. Considérons l'application de $\mathfrak{g}^*$ dans $\mathbf{R}^n$ définie par $e_1, \ldots, e_n$. Il existe un entier k tel que l'image réciproque de tout point $l \in \mathbf{R}^n$ dans M ait au plus k points. Il existe un entier N' tel que l'on ait $q(f) \geq (1 + \|f\|)^{-N'}$ pour tout $f \in \mathfrak{g}^*$ (voir [16] exemple A.2.7), et donc un entier N'' tel que l'on ait $q^{-1}p(f) \leq (1+\|f\|)^{N''}$ pour tout $f \in \mathfrak{g}^*$. Soit $N = N'' + n + 1$. Pour un choix convenable de la norme $\|f\|$, on a

$$\int_M ((1 + \|f\|)^{-N} d|\omega|(f) \leq k \int_{\mathbf{R}^n} (1 + (e_1^2 + \cdots + e_n^2)^{1/2})^{-n-1} de_1 \wedge \cdots \wedge de_n < \infty.$$

∎

La proposition 8 s'applique dans la situation suivante. On considère un groupe de Lie G qui est un ouvert (pour la topologie ordinaire) du groupe $\mathbf{G}(\mathbf{R})$ des points réels d'un groupe algébrique affine $\mathbf{G}$ défini sur $\mathbf{R}$. On suppose que $M_{\mathbf{C}}$ est une orbite fermée de $\mathbf{G}(\mathbf{C})$ dans $\mathfrak{g}_{\mathbf{C}}^*$. Alors les orbites de G dans $\mathfrak{g}^* \cap M_{\mathbf{C}}$ sont tempérées. Notons que Charbonnel [9] a démontré que si $\mathfrak{g}$ est unimodulaire, il existe un ouvert de Zariski non vide de $\mathfrak{g}_{\mathbf{C}}^*$ réunion de $\mathbf{G}(\mathbf{C})$-orbites fermées.

*****

D'après la *méthode des orbites* de Kirillov, une orbite coadjointe de G est l'analogue "classique" d'une représentation unitaire irréductible de G, et une orbite tempérée est l'analogue d'une représentation unitaire irréductible fortement traçable au sens de Howe [17], c'est-à-dire, une représentation unitaire irréductible T telle qu'il existe une base $X_1, \ldots, X_l$ de $\mathfrak{g}$, et un entier N tel que la clôture de l'opérateur $T((1 + X_1^2 + \cdots + X_l^2)^N)$ (défini sur l'espace des vecteurs différentiables de T) soit l'inverse d'un opérateur traçable. Dans ce cas, elle est traçable et admet un caractère χ qui est une fonction généralisée sur G. L'espace $\mathfrak{g}^*$ est l'espace cotangent à G au point 1. Le cône fermé $\mathrm{WF}_1(\chi)$ de $\mathfrak{g}^* \backslash 0$ est un invariant intéressant de la représentation T qui peut être adopté comme définition du *front d'onde* de T (voir [17]).

Soient H un sous-groupe fermé de G et T une représentation unitaire irréductible fortement traçable de G. Sur le modèle de la proposition 5 on peut raisonablement espérer, et c'est en tout cas souvent vrai, que lorsque l'on a $\mathrm{WF}_1(\chi) \cap \mathfrak{h}^\perp = \emptyset$ la restriction $T|_H$ de T à H est une représentation de H traçable dont le caractère est égal à $\chi|_H$. Dans ce cas, la représentation $T|_H$ est somme directe de représentations unitaires irréductibles traçables de H intervenant chacune avec multiplicité finie. Dans la situation de l'exemple 4, l'étude de $\theta|_{\mathfrak{h}}$ est donc l'analogue classique de l'étude de la restriction de T à H.

2. Cohomologie équivariante

Nous rappelons quelques notions de cohomologie équivariante: formes différentielles équivariantes, forme symplectique équivariante associée à une action hamiltonnienne, classes de Thom et d'Euler équivariantes, formule de localisation.

2.1 Formes différentielles équivariantes. Soit M une variété C^∞ munie d'une action d'un groupe de Lie G. Soit $\mathfrak{g}$ l'algèbre de Lie de G. Pour $X \in \mathfrak{g}$, rappelons la définition (2) du champ de vecteurs X_M sur M. Soit $\mathcal{A}(M) = \bigoplus_i \mathcal{A}^i(M)$ l' algèbre des formes différentielles à valeurs complexes sur M. On note $\mathcal{A}_c(M)$ la sous-algèbre des formes différentielles à support compact. On note $d : \mathcal{A}^*(M) \to \mathcal{A}^{*+1}(M)$ la différentielle extérieure. Si ξ est un champ de vecteurs sur M, on note $\iota(\xi) : \mathcal{A}^*(M) \to \mathcal{A}^{*-1}(M)$ la contraction par le champ de vecteurs ξ. Les applications d et $\iota(\xi)$ sont des dérivations de carré nul de $\mathcal{A}(M)$.

Remarque. Dans cet article, les algèbres considérées sont naturellement graduées sur $\mathbb{Z}/2\mathbb{Z}$, et nous respectons la règle des signes dans la définition des produits tensoriels, des dérivations, etc...

Pour $\alpha \in \mathcal{A}(M)$ on note $\alpha = \sum_i \alpha_{[i]}$ sa décomposition en une somme d'éléments homogènes $\alpha_{[i]} \in \mathcal{A}^i(M)$. Si M est une variété orientée et α une forme différentielle à support compact (ou intégrable) on note $\int_M \alpha = \int_M \alpha_{[\dim M]}$ l'intégrale de son terme de degré maximal. Si $\alpha = d\beta$ et si β est à support compact, on a $\int_M \alpha = 0$.

Soit $\pi : M \to B$ une fibration de variétés dont la fibre typique est orientée et de dimension n. Soit $\alpha \in \mathcal{A}^i(M)$ une forme différentielle sur M de degré i telle que π soit propre sur le support de α. On note $\int_{M/B} \alpha \in \mathcal{A}^{i-n}(B)$ (et on dit que c'est l'intégrale de α sur la fibre) la forme différentielle sur B telle que

$$\int_B \left(\beta \wedge \int_{M/B} \alpha \right) = \int_M \pi^* \beta \wedge \alpha,$$

pour toute forme différentielle à support compact β sur la base B. L'application $\int_{M/B}$ commute à la différentielle extérieure. Si ξ est un champ de vecteurs sur M projetable en un champ de vecteurs $\pi_*(\xi)$ sur la base, alors

$$\iota(\pi_*\xi)\int_{M/B}\alpha=\int_{M/B}\iota(\xi)\alpha.$$

Si ξ est un champ de vecteurs sur M, on note d_ξ la dérivation $d_\xi = d-\iota(\xi)$ de $\mathcal{A}(M)$. Si X est dans $\mathfrak{g}$, on posera $d_X = d_{X_M}$. Nous étudierons des formes différentielles $\alpha \in \mathcal{A}(M)$ telles que $d_X\alpha = 0$. Cela siginifie que les composantes homogènes de α vérifient la série de relations

$$(6) \qquad 0=\iota(X_M)\alpha_{[1]},\ d\alpha_{[0]}=\iota(X_M)\alpha_{[2]},\dots,\ d\alpha_{[\dim M]}=0.$$

La relation de Lie s'écrit $d_X^2 = -\mathcal{L}(X_M)$. L'endomorphisme d_X^2 est donc nul sur le sous-espace $\mathcal{A}(M)^X$ de $\mathcal{A}(M)$ annulé par $\mathcal{L}(X_M)$. On notera $H_X(M)$ l'algèbre $\ker d_X/d_X(\mathcal{A}(M)^X)$. Lorsque $X = 0$, $H_X(M)$ est la cohomologie de de Rham $H^*(M,\mathbf{C})$ usuelle. En général, $H_X(M)$ est graduée sur $\mathbf{Z}/2\mathbf{Z}$, mais pas sur $\mathbf{Z}$.

On notera $H_{X,c}(M)$ l'algèbre $\{\ker d_X \subset \mathcal{A}_c(M)\}/d_X(\mathcal{A}_c(M)^X)$.

En fait, il est utile de considérer des familles $\underline{\alpha} : X \mapsto \underline{\alpha}(X)$ de telles formes, dépendant du paramètre $X \in \mathfrak{g}$ de manière appropriée.

On introduit l' algèbre $S(\mathfrak{g}^*)\otimes\mathcal{A}(M)$ des fonctions polynômiales sur $\mathfrak{g}$ à valeurs dans $\mathcal{A}(M)$. Le groupe G agit sur $S(\mathfrak{g}^*)\otimes\mathcal{A}(M)$ par $(g\cdot\underline{\alpha})(X) = g\cdot(\underline{\alpha}(g^{-1}\cdot X))$ pour $g\in G$, $\underline{\alpha}\in S(\mathfrak{g}^*)\otimes\mathcal{A}(M)$, $X\in\mathfrak{g}$. Soit $\mathcal{A}_\mathfrak{g}(M) = (S(\mathfrak{g}^*)\otimes\mathcal{A}(M))^G$ la sous-algèbre des éléments G-invariants (nous ne faisons pas figurer G dans la notation car nous nous intéressons surtout au cas où G est connexe). Les éléments de $\mathcal{A}_\mathfrak{g}(M)$ seront appellés formes différentielles G-équivariantes sur M puisque $\underline{\alpha}\in\mathcal{A}_\mathfrak{g}(M)$ satisfait la relation $\underline{\alpha}(g\cdot X) = g\cdot\underline{\alpha}(X)$.

En plus de la graduation par le degré des formes différentielles, on introduit une seconde graduation dans l'algèbre $S(\mathfrak{g}^*)\otimes\mathcal{A}(M)$ par

$$(7) \qquad \deg(P\otimes\alpha) = 2\deg P + \deg\alpha$$

pour $P\in S(\mathfrak{g}^*)$ et $\alpha\in\mathcal{A}(M)$. Nous dirons que le degré défini par (7) est le degré total.

On définit un opérateur $d_\mathfrak{g}$ sur $S(\mathfrak{g}^*)\otimes\mathcal{A}(M)$ par $(d_\mathfrak{g}\underline{\alpha})(X) = d_X(\underline{\alpha}(X))$. Si E^i est une base de $\mathfrak{g}$ et f_i la base duale, on voit que l'on a $\iota(X_M)\underline{\alpha}(X) = \sum_i f_i(X)\iota(E^i_M)\underline{\alpha}(X)$. Considérons la dérivation de carré nul

$$\iota = \sum_i f_i\iota(E^i_M)$$

de l'algèbre $S(\mathfrak{g}^*) \otimes \mathcal{A}(M)$. L'opérateur $d_{\mathfrak{g}}$ s'écrit donc comme différence de deux dérivations de carré nul de l'algèbre $S(\mathfrak{g}^*) \otimes \mathcal{A}(M)$:

$$d_{\mathfrak{g}} = d - \iota. \tag{8}$$

L'opérateur $d_{\mathfrak{g}}$ augmente de 1 le degré total sur $S(\mathfrak{g}^*) \otimes \mathcal{A}(M)$ et préserve $\mathcal{A}_{\mathfrak{g}}(M)$. La relation de Cartan implique $d_{\mathfrak{g}}^2 \underline{\alpha} = 0$ pour tout $\underline{\alpha} \in \mathcal{A}_{\mathfrak{g}}(M)$, donc $(\mathcal{A}_{\mathfrak{g}}(M), d_{\mathfrak{g}})$ est un complexe. Les formes équivariantes telles que $d_{\mathfrak{g}} \underline{\alpha} = 0$ sont appelées formes fermées équivariantes. Celles de la forme $d_{\mathfrak{g}} \underline{\beta}$ sont appelées exactes.

On note $H_{\mathfrak{g}}^*(M)$ la cohomologie du complexe $(\mathcal{A}_{\mathfrak{g}}(M), d_{\mathfrak{g}})$. C'est une algèbre graduée sur $\mathbb{Z}$. Les groupes $H_{\mathfrak{g}}^i(M)$ sont appellés *groupes de cohomologie équivariante*.

REMARQUE. Lorsque G est compact et connexe, ces groupes sont isomorphes aux groupes $H_G^*(M, \mathbf{C})$ définis en topologie algébrique (voir [7]).

Si $p \in S(\mathfrak{g}_{\mathbf{C}}^*)^G$ est un polynôme invariant homogène de degré k sur $\mathfrak{g}$ et si $\underline{\alpha}$ est une forme équivariante fermée de degré total j, l'élément $p\underline{\alpha}$ est une forme fermée de degré total $2k + j$ dans $\mathcal{A}_{\mathfrak{g}}(M)$. Donc $H_{\mathfrak{g}}^*(M)$ est une algèbre graduée sur $S(\mathfrak{g}_{\mathbf{C}}^*)^G$.

On considère l'idéal gradué $\mathcal{A}_{\mathfrak{g},c}(M) = (S(\mathfrak{g}^*) \otimes \mathcal{A}_c(M))^G$ de $\mathcal{A}_{\mathfrak{g}}(M)$ des formes différentielles équivariantes à support compact. Nous notons $H_{\mathfrak{g},c}^*(M)$ l'algèbre de cohomologie correspondante. C'est une algèbre graduée sur $H_{\mathfrak{g}}^*(M)$, en général sans élément unité.

Lorsque G est réduit à l'identité, le complexe de cohomologie équivariante est le complexe de de Rham $(\mathcal{A}(M), d)$ usuel. Dans le cas général, l'évaluation en un point $X \in \mathfrak{g} : \underline{\alpha} \mapsto \underline{\alpha}(X)$ est un homomorphisme d'algèbres différentielles de $\mathcal{A}_{\mathfrak{g}}(M)$ dans $\mathcal{A}(M)^X$ et induit une application $H_{\mathfrak{g}}^*(M) \to H_X(M)$.

Si $\phi : N \to M$ est une application G-équivariante de G-variétés, l'image réciproque ϕ^* induit une application de $\mathcal{A}_{\mathfrak{g}}(M)$ dans $\mathcal{A}_{\mathfrak{g}}(N)$ qui commute à la différentielle $d_{\mathfrak{g}}$. Si M est un point $\bullet$, alors $\mathcal{A}_{\mathfrak{g}}(\bullet) = S(\mathfrak{g}_{\mathbf{C}}^*)^G$ est l'algèbre des polynômes G-invariants sur $\mathfrak{g}$.

Si $\pi : M \to B$ est une fibration de G-variétés dont les fibres sont orientées de manière G-invariante, l'intégration sur la fibre fournit une application notée $\int_{M/B}$ de $\mathcal{A}_{\mathfrak{g},c}(M)$ dans $\mathcal{A}_{\mathfrak{g},c}(B)$ définie par

$$\Big(\int_{M/B} \underline{\alpha}\Big)(X) = \int_{M/B} (\underline{\alpha}(X))$$

commutant à la différentielle $d_{\mathfrak{g}}$. Elle induit un homomorphisme de $S(\mathfrak{g}_{\mathbf{C}}^*)^G$-modules de $H_{\mathfrak{g},c}^*(M)$ dans $H_{\mathfrak{g},c}^*(B)$. En particulier, si M est une

G-variété orientée, l'intégration $\underline{\alpha} \to \int_M \underline{\alpha}(X)$ sur M est une application $\int_M$ de $\mathcal{A}_{\mathfrak{g},c}(M)$ dans $S(\mathfrak{g}_{\mathbf{C}}^*)^G$ et induit une application de $H^*_{\mathfrak{g},c}(M)$ dans $S(\mathfrak{g}_{\mathbf{C}}^*)^G$.

Soit U un ouvert G-invariant de $\mathfrak{g}$. Soit $C^\infty(U, \mathcal{A}(M))$ l'algèbre des formes $\underline{\alpha}(X)$ sur M dépendant de manière C^∞ de $X \in U$. Elle contient l'algèbre $S(\mathfrak{g}^*) \otimes \mathcal{A}(M)$, et la formule (8) permet de prolonger $d_\mathfrak{g}$ en une dérivation de $C^\infty(U, \mathcal{A}(M))$. Nous noterons $\mathcal{A}_U^\infty(M)$ la sous-algèbre de $C^\infty(U, \mathcal{A}(M))$ formée des éléments G-invariants. Comme plus haut, l'opérateur $d_\mathfrak{g}$ est de carré nul dans $\mathcal{A}_U^\infty(M)$.

On peut encore étendre cette construction. Soit $C^{-\infty}(U, \mathcal{A}(M))$ l'espace des fonctions généralisées sur U à valeurs formes différentielles sur M. Par définition, si $\underline{\alpha}$ est un élément de $C^{-\infty}(U, \mathcal{A}(M))$ et si Φ est une densité C^∞ à support compact sur U, $\int_\mathfrak{g} \underline{\alpha}(X) d\Phi(X)$ est une forme différentielle sur M. Dans un système $x = (x_1, x_2, \ldots, x_n)$ de coordonées locales sur M, un élément $\underline{\alpha}$ de $C^{-\infty}(U, \mathcal{A}(M))$ s'écrit $\underline{\alpha} = \sum_I F_I(x, X) dx_I$ où dx_I est la forme différentielle $dx_{i_1} \wedge dx_{i_2} \wedge \cdots \wedge dx_{i_p}$ indexée par un multi-indice I et $F_I(x, X)$ est une fonction généralisée sur U dépendant de x de manière indéfiniment différentiable. L'espace $C^{-\infty}(U, \mathcal{A}(M))$ contient l'espace $C^\infty(U, \mathcal{A}(M))$. La formule (8) définit un prolongement de l'opérateur $d_\mathfrak{g}$ à l'espace $C^{-\infty}(U, \mathcal{A}(M))$. Le groupe G opère dans $C^{-\infty}(U, \mathcal{A}(M))$. On note $\mathcal{A}_U^{-\infty}(M)$ l'espace des éléments G-invariants. L'opérateur $d_\mathfrak{g}$ est de carré nul dans $\mathcal{A}_U^{-\infty}(M)$. Soit $\pi : M \to B$ une fibration propre de G-variétés dont les fibres sont orientées de manière G-invariante. Comme ci-dessus, l'intégration sur la fibre $\int_{M/B}$ fournit une application notée $\int_{M/B}$ de $\mathcal{A}_U^{-\infty}(M)$ dans $\mathcal{A}_U^{-\infty}(B)$ commutant à la différentielle $d_\mathfrak{g}$. En particulier, si M est une G-variété compacte orientée, l'intégration $\underline{\alpha} \to \int_M \underline{\alpha}(X)$ sur M est une application $\int_M$ de $\mathcal{A}_U^{-\infty}(M)$ dans $C^{-\infty}(U)^G$. On démontre facilement le

Lemme 9. *Soit M une G-variété compacte orientée. Si $\underline{\alpha} = d_\mathfrak{g}\underline{\beta}$, avec $\underline{\beta} \in \mathcal{A}_U^{-\infty}(M)$, alors on a $\int_M \underline{\alpha} = 0$.*

Soit $\underline{\alpha} \in C^\infty(U, \mathcal{A}(M))$, et supposons $\underline{\alpha}$ à valeurs dans les formes paires. Si p est un polynôme en une variable (à coefficients réels ou complexes), on peut définir $p(\underline{\alpha})$ (puisque $C^\infty(U, \mathcal{A}(M))$ est une algèbre associative) et l'on a $d_\mathfrak{g}(p(\underline{\alpha})) = p'(\underline{\alpha})\, d_\mathfrak{g}(\underline{\alpha})$, (car $\underline{\alpha}$ est pair). Supposons de plus $\underline{\alpha}$ à valeurs réelles. Nous allons de même définir $\phi(\underline{\alpha})$ pour $\phi \in C^\infty(\mathbf{R})$. On écrit $\underline{\alpha}_+ = \sum_{i>0} \underline{\alpha}_{[i]}$ de sorte que $\underline{\alpha} = \underline{\alpha}_{[0]} + \underline{\alpha}_+$. La forme $\underline{\alpha}_+$ est nilpotente, et $\underline{\alpha}_{[0]}$ est une fonction C^∞ dans $U \times M$. On définit $\phi(\underline{\alpha}) \in C^\infty(U, \mathcal{A}(M))$ par le développement de Taylor (qui est une somme finie)

$$\phi(\underline{\alpha}) = \sum_k \frac{\phi^{(k)}(\underline{\alpha}_{[0]})}{k!}(\underline{\alpha}_+)^k. \tag{9}$$

On vérifie que $\phi(\underline{\alpha})$ appartient à $\mathcal{A}_U^\infty(M)$ si $\underline{\alpha} \in \mathcal{A}_U^\infty(M)$. Plus généralement, on peut définir $\phi(\underline{\alpha})$ si ϕ est une fonction différentiable dans un ouvert de $\mathbf{R}$ contenant $\underline{\alpha}_{[0]}(U \times M)$. On a facilement

$$d_{\mathfrak{g}}(\phi(\underline{\alpha})) = \phi'(\underline{\alpha})\, d_{\mathfrak{g}}(\underline{\alpha}),$$

de sorte que $\phi(\underline{\alpha})$ est une forme équivariante fermée si $\underline{\alpha}$ en est une.

Nous donnons maintenant des exemples importants de formes équivariantes fermées.

2.2 Action hamiltonnienne. Soit M une variété symplectique de dimension $n = 2d$ et soit σ sa 2-forme symplectique. Soit G un groupe de Lie agissant de manière hamiltonienne sur M. Cela signifie que l'on s'est donné une application linéaire $X \mapsto \mu(X)$ de $\mathfrak{g}$ dans l'espace $C^\infty(M)$ des fonctions C^∞ sur M telle que $d(\mu(X)) = \iota(X_M)\sigma$ et $g \cdot \mu(X) = \mu(g \cdot X)$ pour tout $g \in G$ et tout $X \in \mathfrak{g}$. La fonction $\mu(X)$ est appelée la fonction moment. On vérifie immédiatement que la formule

$$X \mapsto \sigma_{\mathfrak{g}}(X) = \mu(X) + \sigma$$

définit une forme fermée équivariante $\sigma_{\mathfrak{g}}$ homogène de degré total 2. Nous dirons que $\sigma_{\mathfrak{g}}$ est la *forme symplectique équivariante* associée à l'action de hamiltonnienne de G dans M. On note

$$\beta_M = (2\pi)^{-d}\frac{\sigma^d}{d!}$$

la forme de Liouville. La forme $e^{i\sigma_{\mathfrak{g}}}$ est une forme différentielle équivariante fermée et on a

$$(e^{i\sigma_{\mathfrak{g}}(X)})_{[n]} = (2i\pi)^d e^{i\mu(X)}\beta_M.$$

La forme de Liouville définit une orientation de M. Nous considérons la fonction généralisée F_M sur $\mathfrak{g}$ définie (si possible) par la formule

$$F_M(X) = \int_M e^{i\mu(X)}\beta_M = (2i\pi)^{-d}\int_M e^{i\sigma_{\mathfrak{g}}(X)}, \tag{10}$$

c'est-à-dire la fonction généralisée F_M sur $\mathfrak{g}$ telle que pour toute densité $\Phi \in \mathcal{M}_c^\infty(\mathfrak{g})$ on ait

$$(11) \qquad \int_{\mathfrak{g}} F_M(X)d\Phi(X) = \int_M \{\int_{\mathfrak{g}} e^{i\mu(X)(m)} d\Phi(X)\} d\beta_M(m).$$

Pour donner un sens à la formule (10), on peut considérer l'application moment $\tilde{\mu}$ de M dans $\mathfrak{g}^*$ telle que $\mu(X)(m) = \tilde{\mu}(m)(X)$ pour tout $m \in M$ et tout $X \in \mathfrak{g}$. Soit $\tilde{\mu}_*(\beta_M)$ la mesure positive sur $\mathfrak{g}^*$ image de β_M. *Supposons la mesure* $\tilde{\mu}_*(\beta_M)$ *tempérée.* Alors on peut écrire le second membre de (11) comme l'intégrale absolument convergente

$$\int_{\mathfrak{g}^*} \{\int_{\mathfrak{g}} e^{if(X)} d\Phi(X)\} d\tilde{\mu}_*(\beta_M)(f).$$

La fonction généralisée F_M est bien définie: c'est la transformée de Fourier de la mesure $\tilde{\mu}_*(\beta_M)$.

Un exemple d'action hamiltonienne est celui de l'action coadjointe. Soit G un groupe de Lie d'algèbre de Lie $\mathfrak{g}$. Soit $\mathfrak{g}^*$ l'espace vectoriel dual de l'espace vectoriel $\mathfrak{g}$. Soit M une orbite de la représentation coadjointe. Les champs de vecteurs X_M, $X \in \mathfrak{g}$, engendrent l'espace tangent en chaque point f de M et on a $(X_M)_f = -Xf$, avec $-(Xf)(Y) = f([X,Y])$. Soit $\mu(X)$ la restriction à M de la fonction $f \to f(X)$ sur $\mathfrak{g}^*$. Rappelons que la forme $\sigma(X_M, Y_M)_f = -f([X,Y])$ définit sur M une forme symplectique G-invariante. L'action coadjointe de G sur M est hamiltonnienne et le moment de X est $\mu(X)$. Lorsque M est tempérée et munie de l'orientation définie par β_M, la fonction généralisée F_M est la transformée de Fourier de l'orbite M comme définie par la formule (3).

2.3 Forme de Thom équivariante. Soit V un espace vectoriel réel orienté de dimension n. Soit X un endomorphisme elliptique de V, c'est-à-dire semi-simple à valeurs propres imaginaires pures. L'orientation permet de définir de manière canonique une racine carrée de $\det_V X$ que nous noterons $\det_V^{1/2}(X)$. En voici la description. Nous pouvons supposer X inversible. Dans ce cas, n est pair. Posons $n = 2d$. On choisit une base orientée e_i de V telle que les vecteurs $e_{2j-1} - ie_{2j}$, pour $j = 1, \cdots, d$, soient propres de valeur propre $i\lambda_j$ avec $\lambda_j \in \mathbf{R}$. On pose

$$(12) \qquad \det_V^{1/2}(X) = \lambda_1 \cdots \lambda_d.$$

Par exemple, si $n = 2$ et si X est représenté par la matrice $\begin{pmatrix} 0 & -\lambda \\ \lambda & 0 \end{pmatrix}$, on a $\det_V^{1/2}(X) = \lambda$.

Soit V un espace vectoriel euclidien orienté de dimension n muni de l'action naturelle du groupe $\mathrm{SO}(V)$. Soit G un sous-groupe fermé de $\mathrm{SO}(V)$, et soit $\mathfrak{g} \subset \mathfrak{so}(V)$ l'algèbre de Lie de G. La proposition ci-dessous est la version équivariante du "lemme de Poincaré à support compact".

PROPOSITION 10. *Il existe dans $H^n_{\mathfrak{g},c}(V)$ un élément et un seul, noté u_V, tel que l'on ait $\int_V u_V(X) = 1$ pour tout $X \in \mathfrak{g}$. De plus, $H^*_{\mathfrak{g},c}(V)$ est un $S(\mathfrak{g}^*_{\mathbf{C}})^G$-module libre de base u_V.*

DÉMONSTRATION: Le lemme de Poincaré à supports compacts affirme que toute forme différentielle fermée $\gamma \in \mathcal{A}^k_c(V)$ est de la forme $d\beta$ avec $\beta \in \mathcal{A}^{k-1}_c(V)$ si $k < n$, ou si $k = n$ et $\int_V \gamma = 0$. On en déduit facilement que tout élément $\underline{\gamma} \in S(\mathfrak{g}^*) \otimes \mathcal{A}^k_c(V)$ annulé par d est de la forme $d\underline{\beta}$ avec $\underline{\beta} \in S(\mathfrak{g}^*) \otimes \mathcal{A}^{k-1}_c(V)$ si $k < n$, ou si $k = n$ et $\int_V \underline{\gamma}(X) = 0$ pour tout $X \in \mathfrak{g}$. De plus, si $\underline{\gamma}$ est G-équivariante, on peut supposer, en remplaçant au besoin $\underline{\beta}$ par sa moyenne sous l'action du groupe compact G, que $\underline{\beta}$ est G-équivariante.

Montrons qu'il existe une classe équivariante fermée $\underline{\alpha}$ à support compact de degré total n et d'intégrale identiquement égale à 1. On choisit une forme différentielle G-invariante $\alpha_{[n]}$ sur V à support compact de degré n et d'intégrale 1. Posons $\underline{\alpha}_{[n]} = 1 \otimes \alpha_{[n]}$. Nous allons définir par récurrence descendante des formes équivariantes $\underline{\alpha}_{[n-2]}, \underline{\alpha}_{[n-4]}, \cdots \in (S(\mathfrak{g}^*) \otimes \mathcal{A}_c(M))^G$ de degré $n-2, n-4, \ldots$ et de degré total n, de telle sorte que la forme équivariante $\underline{\alpha} = \underline{\alpha}_{[n]} + \underline{\alpha}_{[n-2]} + \underline{\alpha}_{[n-4]} + \ldots$ soit fermée. Considérons $\iota(\underline{\alpha}_{[n]})$. On a $d(\iota(\underline{\alpha}_{[n]})) = -\iota(d(\underline{\alpha}_{[n]}))$ (car la dérivation $\iota d + d\iota$ est nulle sur l'espace des formes équivariantes) et donc $d(\iota(\underline{\alpha}_{[n]})) = 0$. On peut donc trouver une forme équivariante à support compact $\underline{\alpha}_{[n-2]}$ de degré $n-2$ et degré total n tel que $d\underline{\alpha}_{[n-2]} = \iota\underline{\alpha}_{[n]}$. De la même manière on a $d\iota\underline{\alpha}_{[n-2]} = -\iota d\underline{\alpha}_{[n-2]} = -\iota\iota\underline{\alpha}_{[n]} = 0$, et donc on peut trouver $\underline{\alpha}_{[n-4]}$ tel que $d\underline{\alpha}_{[n-4]} = \iota\underline{\alpha}_{[n-2]}$, etc... Les relations (6) sont vérifiées et $\underline{\alpha}$ est fermée.

Soit $\underline{\beta} = \underline{\beta}_{[k]} + \underline{\beta}_{[k-2]} + \cdots$ une forme équivariante fermée à support compact. On suppose que $k < n$ ou que $k = n$ et $\int_V \underline{\beta}_{[n]}(X) = 0$ pour tout $X \in \mathfrak{g}$. Comme plus haut on construit une forme équivariante à support compact $\underline{\gamma} = \underline{\gamma}_{[k-1]} + \underline{\gamma}_{[k-3]} + \cdots$ telle que $d_{\mathfrak{g}}\underline{\gamma} = \underline{\beta}$.

Soit $\underline{\beta}$ une forme équivariante fermée à support compact. Soit $P = \int_V \underline{\beta}$. C'est un élément de $S(\mathfrak{g}^*_{\mathbf{C}})$ et, d'après ce qui précède, $\underline{\beta}$ représente la même classe que $P\underline{\alpha}$. D'autre part, si P est un polynôme invariant non nul, la classe de $P\underline{\alpha}$ est non nulle puisque $\int_V P\underline{\alpha} = P$. ∎

Nous laissons au lecteur le soin de définir des classes de cohomologie équivariantes sur V à décroissance rapide (au sens de Schwartz), et d'énoncer et démontrer une proposition analogue à la proposition 10.

De la même manière que la proposition 10 on démontre la

PROPOSITION 11. *Soit $X \in \mathfrak{g}$. Soit $\alpha \in \mathcal{A}_c(V)$ un élément tel que $d_X\alpha = 0$. Alors il existe $\beta \in \mathcal{A}_c(V)^X$ tel que $\alpha = (\int_V \alpha)u_V(X) + d_X\beta$. En d'autres termes, la classe de $u_V(X)$ est une base de l'espace vectoriel $H_{X,c}(V)$, et l'intégrale $\alpha \mapsto \int_V \alpha$ est la base duale.*

La classe $u_V \in H^n_{\mathfrak{g},c}(V)$ s'appelle la classe de Thom équivariante. Nous donnons ci-dessous une construction explicite d'un représentant de u_V, dans le style de celle de [**18**].

Soit e_i une base orthonormale orientée de V. Soit ΛV l'algèbre extérieure de V. Si $I = \{j_1, j_2, \ldots, j_i | j_1 < j_2 < \cdots < j_i\}$ est un sous ensemble de $\{1, 2, \ldots, n\}$, nous écrivons $e_I = e_{j_1} \wedge \cdots \wedge e_{j_i}$. Les éléments e_I forment une base orthonormée de ΛV. On note I' l'ensemble complémentaire de I dans $\{1, 2, \ldots, n\}$. On note $T : \Lambda V \to \mathbf{R}$ la composante de plus haut degré d'un élément de ΛV, c'est-à-dire

$$T(\sum_I a_I e_I) = a_{\{12\cdots n\}}.$$

La forme T est appelée intégrale de Berezin. Si $e \in V$, on note $\iota_\Lambda(e) : \Lambda V \to \Lambda V$ la contraction par e. C'est une dérivation de l'algèbre ΛV et T vérifie

$$T(\iota_\Lambda(e)\xi) = 0 \tag{13}$$

pour tout $e \in V$ et $\xi \in \Lambda V$.

Si $A \in \Lambda^2(V)$, on note $\exp_\Lambda A$ l'exponentielle de A dans l'algèbre ΛV et on définit le Pfaffien de A par $\mathrm{Pf}(A) = T(\exp_\Lambda A)$.

Soit $\tau : \mathfrak{g} \to \Lambda^2 V$ l'application définie par

$$\tau(X) = \tfrac{1}{2}\sum_{i<j}(Xe_i, e_j)e_i \wedge e_j.$$

Le *pfaffien* de $X \in \mathfrak{so}(V)$ est défini par

$$\mathrm{Pf}(X) = \mathrm{Pf}(2\tau(X)).$$

On vérifie que l'on a $\mathrm{Pf}(X) = \det_V^{1/2}(X)$ (et donc la restriction de la fonction $\det_V^{1/2}$ à $\mathfrak{g}$ est polynomiale). Si V est de dimension impaire le polynôme $X \to \mathrm{Pf}(X)$ est identiquement nul. Si V est de dimension paire n le polynôme $\mathrm{Pf}(X)$ est de degré $d = n/2$.

Nous notons x_i les fonctions coordonnées sur V, et ∂_i la dérivation dans la direction du vecteur e_i.

Considérons l'algèbre $\mathcal{A}(V) \otimes \Lambda V = C^\infty(V) \otimes \Lambda V^* \otimes \Lambda V$ des formes différentielles sur V à valeurs dans ΛV, et l'algèbre $C^\infty(\mathfrak{g}, \mathcal{A}(V) \otimes \Lambda V)$ des fonctions C^∞ sur $\mathfrak{g}$ à valeurs dans $\mathcal{A}(V) \otimes \Lambda V$. Un élément de $\mathcal{A}(V) \otimes \Lambda V$ s'écrit en coordonnées sous la forme $\sum_{IJ} f_{IJ}(x) dx_J e_I$. Nous notons ϵ la fonction identique de V dans V, considérée comme un élément de de $\mathcal{A}(V) \otimes \Lambda V$. Elle s'écrit $\epsilon = \sum_i x_i e_i$. On note encore d l'opérateur $d \otimes I$ de $\mathcal{A}(V) \otimes \Lambda V$. On a $d\epsilon = \sum_i dx_i e_i$. Définissons l'élément f_V de $C^\infty(\mathfrak{g}, \mathcal{A}(V) \otimes \Lambda V)$ par

$$f_V(X) = -\|x\|^2 + d\epsilon + \tau(X), \tag{14}$$

pour $X \in \mathfrak{g}$. En coordonnées

$$f_V(X) = -\sum_i x_i^2 + \sum_i dx_i e_i + \tfrac{1}{2} \sum_{i<j} (Xe_i, e_j) e_i \wedge e_j .$$

Si ξ est un champ de vecteurs sur V on note encore $\iota(\xi)$ l'opérateur $\iota(\xi) \otimes I$ sur $\mathcal{A}(V) \otimes \Lambda V$. C'est une dérivation de $\mathcal{A}(V) \otimes \Lambda V$. Si $e \in V$, on note $\iota_\Lambda(e)$ la dérivation de $\mathcal{A}(V) \otimes \Lambda V$ définie par $\iota_\Lambda(e)(\alpha \otimes \xi) = (-1)^k \alpha \otimes \iota_\Lambda(e)\xi$ si $\alpha \in \mathcal{A}^k(V)$ et $\xi \in \Lambda V$.

On note $\mathcal{R} = \sum_i x_i \partial_i$ le champ d'Euler. On note $\iota_\Lambda(\mathcal{R})$ l'opérateur sur $\mathcal{A}(V) \otimes \Lambda V$ défini par

$$\iota_\Lambda(\mathcal{R}) = \sum_i x_i \iota_\Lambda(e_i).$$

On note δ la dérivation de $C^\infty(\mathfrak{g}, \mathcal{A}(V) \otimes \Lambda V)$ telle que $(\delta(\underline{\alpha}))(X) = (d - \iota(X_V) - 2\iota_\Lambda(\mathcal{R}))\underline{\alpha}(X)$.

Si $X \in \mathfrak{g}$, le champ de vecteurs X_V correspondant à l'action infinitésimale de $\mathfrak{g}$ sur V est $X_V = -\sum_{i<j} (Xe_i, e_j)(x_i \partial_j - x_j \partial_i)$. On en déduit que pour tout $X \in \mathfrak{g}$ on a

$$\delta(f_V) = 0. \tag{15}$$

Comme plus haut, la formule de Taylor (9) permet de définir l'élément $\phi(\underline{\alpha})$ de $C^\infty(\mathfrak{g}, \mathcal{A}(V) \otimes \Lambda V)$ pour $\phi \in C^\infty(\mathbf{R})$ et $\underline{\alpha} \in C^\infty(\mathfrak{g}, \mathcal{A}(V) \otimes \Lambda V)$ réel et pair, et l'on a $\delta(\phi(\underline{\alpha})) = \phi'(\underline{\alpha})\,\delta(\underline{\alpha})$. Si $\delta(\underline{\alpha}) = 0$, on obtient la relation $d_\mathfrak{g} \phi(\underline{\alpha}) = 2\iota_\Lambda(\mathcal{R})\phi(\underline{\alpha})$. On étend par linearité l'intégrale de Berezin en une application T de $C^\infty(\mathfrak{g}, \mathcal{A}(V) \otimes \Lambda V)$ dans $C^\infty(\mathfrak{g}, \mathcal{A}(V))$. Il résulte de (13) que la forme $T(\phi(\underline{\alpha}))$ est une forme annulée par $d_\mathfrak{g}$. On a donc démontré le

LEMME 12. *Soit $\underline{\alpha} \in C^\infty(\mathfrak{g}, \mathcal{A}(V) \otimes \Lambda V)$ un élément réel, pair, G-invariant et annulé par δ. La forme $T(\phi(\underline{\alpha}))$ sur V est fermée équivariante pour l'action de $G = SO(V)$.*

Le lemme s'applique à la forme $u_{\phi,V}$ définie par la formule $u_{\phi,V} = T(\phi(f_V))$. Explicitons $u_{\phi,V}$ en utilisant (9). Il vient

$$u_{\phi,V}(X) = \sum_I \phi^{(n-|I|/2)}(-\|x\|^2) P_I(X/2) dx_{I'}, \tag{16}$$

où, pour tout sous-ensemble d'indices I de cardinal $|I|$ pair, P_I est un polynôme homogène sur $\mathfrak{g}$ de degré $(|I|/2)$ qui coïncide au signe près avec le Pfaffien de la sous-matrice $X_I = \{(Xe_i, e_j)_{ij \in I}\}$. Donc $u_{\phi,V}$ est un élément de $S(\mathfrak{g}^*) \otimes \mathcal{A}(V)$, homogène de degré n pour le degré total. On a

$$u_{\phi,V}(X)_{[n]} = (-1)^{n(n-1)/2} \phi^{(n)}(-\|x\|^2)\, dx_1 dx_2 \cdots dx_n,$$

et, si $n = 2d$ est pair,

$$u_{\phi,V}(X)_{[0]} = \phi^{(d)}(-\|x\|^2)\, \mathrm{Pf}(X/2). \tag{17}$$

Il est clair que $u_{\phi,V}$ est à support compact si ϕ une fonction différentiable sur $\mathbf{R}$ nulle au voisinage de $-\infty$ (resp. à décroissance rapide si ϕ est à décroissance rapide au voisinage de $-\infty$).

Nous avons établi la proposition suivante.

PROPOSITION 13. *Soit G un sous-groupe fermé de $SO(V)$ agissant sur l'espace vectoriel euclidien orienté V. Soit une fonction ψ différentiable sur $\mathbf{R}$ nulle au voisinage de $-\infty$ (resp. à décroissance rapide au voisinage de $-\infty$), et vérifiant*

$$(-1)^{n(n-1)/2} \int_V \psi^{(n)}(-\|x\|^2) dx = 1.$$

La forme équivariante $u_{\psi,V}$ représente la classe de Thom équivariante de V.

Il est souvent commode de choisir la fonction

$$\psi(t) = (-1)^{n(n-1)/2} \pi^{-n/2} e^t. \tag{18}$$

Notons $i : 0 \to V$ l' injection de l'origine dans V.

COROLLAIRE 14. *Si $n = 2d$ est pair on a, pour tout $X \in \mathfrak{g}$*

$$i^*(u_V(X)) = (-\pi)^{-d}\operatorname{Pf}(X/2) = (-2\pi)^{-d}\det_V^{1/2}(X).$$

DÉMONSTRATION: D'après la proposition 10, il suffit de le vérifier sur un représentant de la forme $u_{\psi,V}$. Compte tenu de (17) et de la relation $(-1)^{n(n-1)/2} = (-1)^d$, il s'agit de vérifier la relation $\psi^{(d)}(0) = (-\pi)^{-d}$. On peut le faire en calculant l'intégrale $\int_V \psi^{(n)}(-\|x\|^2)dx$ en coordonnées radiales, ou bien en le vérifiant pour une fonction ψ particulière. On notera que c'est particulièrement facile pour la fonction (18). ∎

COROLLAIRE 15. *Soit $X \in \mathfrak{g}$. Soit $\alpha \in \mathcal{A}_c(V)$ une forme telle que $d_X\alpha = 0$. Si $n = 2d$ est pair on a*

$$(-2\pi)^{-d}\det_V^{1/2}(X)\int_V \alpha = \alpha_{[0]}(0).$$

DÉMONSTRATION: La formule résulte de la proposition 11 et du corollaire 14. ∎

REMARQUE Considérons comme plus haut une fonction différentiable ϕ sur $\mathbf{R}$, nulle ou à décroissance rapide au voisinage de $-\infty$. Soit $t \in \mathbf{R}^\times$. Considérons l'action de la dilatation $\delta(t)(x) = tx$ sur V. Comme les dilatations commutent à l'action de G sur V, la forme $u_{\phi,V}(t)$ définie par $u_{\phi,V}(t,X) = \delta(t)^*u_{\phi,V}(X)$ est toujours une forme équivariante fermée sur V. D'après (16) on a

$$u_{\phi,V}(t,X) = \sum_I \phi^{(n-|I|/2)}(-t^2\|x\|^2)P_I(X/2)t^{|I'|}dx_{I'}. \tag{19}$$

Comme l'intégrale sur V de $u_{\phi,V}(t,X)$ ne dépend pas de t, il résulte de la proposition 10 que, quels que soient t_1 et t_2, la forme $u_{\phi,V}(t_2,X) - u_{\phi,V}(t_1,X)$ s'écrit $d_{\mathfrak{g}}\underline{\beta}$, où $\underline{\beta}$ est une forme équivariante de degré total $n-1$ à support compact (ou à décroissance rapide). Bien que cela ne soit pas indispensable pour la suite, il est amusant de donner une formule explicite pour $\underline{\beta}$. Celle-ci se déduit immédiatement de la formule (20) ci-dessous.

Comme en (14) on introduit l'élément $f_V(t,X) = -t^2\|x\|^2 + td\epsilon + \tau(X)$ de $C^\infty(\mathfrak{g}, \mathcal{A}(V)\otimes\Lambda V)$. On a

$$\frac{d}{dt}u_{\phi,V}(t) = d_{\mathfrak{g}}(T(\epsilon\phi'(f_V(t)))). \tag{20}$$

DÉMONSTRATION: Considérons la dérivation $\delta_t = d - \iota(X_V) - 2t\iota_\Lambda(\mathcal{R})$ de $C^\infty(\mathfrak{g}, \mathcal{A}(V) \otimes \Lambda V)$. On a

$$\delta_t(f_V(t)) = 0 \tag{21}$$

et $\frac{d}{dt} f_V(t, X) = -2t\|x\|^2 + d\epsilon = \delta_t \epsilon$. En calculant comme plus haut, on obtient $u_{\phi,V}(t) = T(\phi(f_V(t)))$ et donc

$$\frac{d}{dt} u_{\phi,V}(t) = T(\frac{d}{dt}\phi(f_V(t))) = T((\delta_t\epsilon)\phi'(f_V(t))).$$

D'après (21), on a $\delta_t\phi'(f_V(t)) = 0$, et donc $\frac{d}{dt}u_{\phi,V}(t) = T(\delta_t(\epsilon\phi'(f_V(t)))$. On conclut grâce à la formule (13). ∎

Considérons un fibré vectoriel $p : E \mapsto M$ de base M et dont la fibre est un espace euclidien orienté de dimension n. On suppose que G est un groupe de Lie compact qui opère dans E et dans M en préservant toutes ces structures. La proposition suivante généralise la proposition 10 et peut se démontrer de manière analogue.

PROPOSITION 16. a) *Il existe une forme différentielle équivariante fermée u_E sur E, de degré total n, telle que p soit propre sur le support de u_E et telle que l'on ait $\int_{E/M} u_E = 1$. La classe de u_E est unique dans le sens suivant: une forme vérifiant les mêmes propriétés ne diffère de u_E que par l'addition d'une forme $d_\mathfrak{g}\underline{\beta}$, où $\underline{\beta}$ est une forme équivariante de degré total $n-1$ telle que p soit propre sur le support de $\underline{\beta}$.*

b) L'application $\underline{\beta} \mapsto p^*(\underline{\beta})u_E$ est un isomorphisme de $H^*_\mathfrak{g}(M)$-modules de $H^*_{\mathfrak{g},c}(M)$ sur $H^*_{\mathfrak{g},c}(E)$.

La classe de u_E s'appelle la *classe de Thom équivariante* du fibré E, et l'assertion b est connue sous le nom d'isomorphisme de Thom.

Voici l'analogue de la proposition 11, et encore une fois, la démonstration est similaire à celle de la proposition 10.

PROPOSITION 17. *Soit $X \in \mathfrak{g}$. Soit $\alpha \in \mathcal{A}_c(E)$ tel que $d_X\alpha = 0$. Alors il existe $\beta \in \mathcal{A}_c(E)^X$ tel que $\alpha = p^*(\int_{E/M}\alpha)u_E(X) + d_X\beta$.*

Soit i_M l'injection de M dans E donnée par la section nulle. La classe de la forme $i_M^*(u_E)$ est un élément de $H^n_\mathfrak{g}(M)$ qui s'appelle *la classe d'Euler équivariante* du fibré E. On la notera Eul_E.

Par exemple, si M est un point, E est juste un espace vectoriel orienté V de dimension paire $2d$ et, d'après le corollaire 14, Eul_V est égal au polynôme $(-2\pi)^{-d}\det_V^{1/2}(X)$.

Plus généralement, soient $m \in M$ et $X \in \mathfrak{g}$ tels que l'on ait $X_M(m) = 0$. Soit $G(m)$ le stabilisateur du point m dans G. Notons E_m l'espace vectoriel euclidien orienté $p^{-1}(m)$. On obtient un homomorphisme de $G(m)$ dans $\mathrm{SO}(E_m)$, et donc un homomorphisme, noté J_m, de l'algèbre de Lie $\mathfrak{g}(m)$ de $G(m)$ dans $\mathfrak{so}(E_m)$. En particulier, le nombre $\det_{E_m}^{1/2}(J_m(X))$ est défini.

LEMME 18. *Supposons la fibre de E paire de dimension $2d$. Si $X_M(m) = 0$, on a*

$$\mathrm{Eul}_E(X)_{[0]}(m) = (-2\pi)^{-d} \det{}_{E_m}^{1/2}(J_m(X)).$$

DÉMONSTRATION: Le lemme résulte de la remarque précédente lorsqu'on considère la restriction au fibré de base $\{m\}$ et de fibre E_m. ∎

L'*homomorphisme de Chern-Weil équivariant* fournit une expression analogue permettant de calculer Eul_E en fonction de la forme de courbure d'une connection G-invariante pour le fibré E ("Théorème de Gauss-Bonnet généralisé équivariant"). Nous renvoyons le lecteur intéressé au ch. 7 du livre **[3]**.

La proposition 17 a un corollaire analogue au corollaire 15.

COROLLAIRE 19. *Soit $X \in \mathfrak{g}$. Soit $\alpha \in \mathcal{A}_c(M)$ un élément tel que $d_X\alpha = 0$. On suppose que $\mathrm{Eul}_E(X)$ est inversible dans $H_X(M)$. On a*

$$\int_E \alpha = \int_M i_M^*(\alpha)\,\mathrm{Eul}_E(X)^{-1}.$$

DÉMONSTRATION: La proposition 17 entraine que dans l'algèbre $H_X(M)$ on a $i_M^*(\alpha) = (\int_{E/M} \alpha)\,\mathrm{Eul}_E(X)$, et donc

$$\int_E \alpha = \int_M (\int_{E/M} \alpha) = \int_M i_M^*(\alpha)\,\mathrm{Eul}_E(X)^{-1}.$$

∎

2.4 Formule de localisation. Dans ce paragraphe, G un groupe de Lie *compact* agissant sur une variété différentiable orientée M. Nous supposons que G est connexe, et nous fixons un élément S dans le centre de $\mathfrak{g}$. On note $M_0(S)$ l'ensemble des zéros du champ de vecteurs S_M. Chaque composante connexe de $M_0(S)$ est une sous-variété fermée de M, invariante par G. Le complexe des formes équivariantes possède des propriétés d'exactitude qui vont permettre éventuellement le calcul des intégrales des formes différentielles equivariantes fermées à support compact. Nous rappelons le fait principal.

PROPOSITION 20. *Soit $S \in \mathfrak{g}$ et soit $\alpha \in \mathcal{A}(M)$ une forme différentielle telle que $d_S\alpha = 0$. Alors la forme $\alpha_{[\dim M]}$ est exacte en dehors de $M_0(S)$.*

DÉMONSTRATION: Nous aurons besoin du

LEMME 21. *Soit $\theta \in \mathcal{A}(M)$ une forme sur M telle que $\mathcal{L}(S_M)\theta = 0$ et telle que $d_S\theta$ soit inversible en dehors de $M_0(S)$ (c'est-à-dire telle que l'on ait $-\theta_{[1]}(S_M)(m) \neq 0$ si $S_M(m)$ est non nul). Alors sur $M\backslash M_0(S)$ on a*

$$\alpha_{[\dim M]} = d\left((\theta(d_S\theta)^{-1}\alpha(S))_{[\dim M-1]}\right).$$

DÉMONSTRATION: Comme $d_S\alpha = 0$ et $d_S^2\theta = 0$, on a

$$\alpha = d_S(\theta(d_S\theta)^{-1}\alpha),$$

et le lemme exprime cette égalite pour le degré maximum. ∎

Pour montrer la proposition nous construisons une telle forme θ. Puisque le groupe G est compact il existe une métrique G-invariante g sur M et la 1-forme

$$\theta(\xi) = g(S_M, \xi) \tag{22}$$

satisfait aux conditions du lemme. ∎

REMARQUE. Notons qu'il est essentiel de supposer G compact. Par exemple considérons sur le tore $M = \mathbf{R}/\mathbf{Z} \times \mathbf{R}/\mathbf{Z}$ le champ de vecteurs $\xi = (1 + \frac{1}{2}\sin x)\partial_y$. Il ne s'annule jamais. Considérons la forme

$$\alpha = \tfrac{1}{2}(7\cos x + \sin 2x) - (1 - 4\sin x)dx \wedge dy.$$

On vérifie que $d_\xi\alpha = 0$. Mais comme $\int_M \alpha = -(2\pi)^2$, la forme $\alpha_{[2]}$ n'est pas exacte.

Soit $\alpha \in \mathcal{A}_c(M)$ une forme différentielle à support compact sur M telle que $d_S\alpha = 0$. La proposition 20 suggère que l'intégrale $\int_M \alpha$ ne dépend que de la restriction de α à $M_0(S)$, et en effet, la formule de localisation de Berline-Vergne [4] (dont nous allons redonner une démonstration) permet de calculer $\int_M \alpha$ en fonction de la forme $i^*_{M_0(S)}\alpha$.

Nous avons besoin de quelques notations. Soit $X \in \mathfrak{g}$. Soit $m \in M$ un zéro de X_M. Si ξ est un champ de vecteurs sur M défini au voisinage de m, la valeur du crochet de Lie $[X_M, \xi]$ au point m ne dépend que de $\xi(m)$. Notons $J(X)_m$ l'endomorphisme $\xi(m) \mapsto [X_M, \xi](m)$ de l'espace tangent $T_m(M)$. Par exemple, si G est le groupe $SO(V)$ opérant dans un espace euclidien orienté V de dimension n, et si X est dans $\mathfrak{so}(V)$, on a $X_V(v) = -Xv$, et $J(X)_0$ est égal à X.

Nous commençons par le cas particulier où les zéros de S_M sont des points isolés. Soit $m \in M_0(S)$. Le groupe G laisse m fixe et opère dans $T_m(M)$ par l'application tangente. Le point m admet un voisinage ouvert G-invariant W_m qui est G-isomorphe à $T_m(M)$. On en déduit que $J(S)_m$ est un endomorphisme elliptique inversible de l'espace vectoriel orienté $T_m(M)$. Si $M_0(S)$ est non vide, la dimension n de M est paire et le nombre $\det^{1/2}_{T_m(M)}(J(S)_m)$ est non nul.

PROPOSITION 22. *On suppose que les zéros de S_M sont des points isolés. Notons $n = 2d$ la dimension de M. Soit α une forme différentielle à support compact vérifiant $d_S\alpha = 0$. On a*

$$\int_M \alpha = (-2\pi)^d \sum_{m\in M_0(S)} \frac{\alpha_{[0]}(m)}{\det^{1/2}_{T_m(M)}(J(S)_m)}.$$

DÉMONSTRATION: Soit ϕ une fonction différentiable sur **R**, telle que $\phi(0) = 1$. Soit θ la forme différentielle (22) de degré 1 sur M. Soit ψ la fonction différentiable sur **R** telle que l'on ait $\psi(t) = (1 - \phi(t))/t$ pour $t \neq 0$. Posons $\beta = \theta\psi(d_S\theta)$. On a $(1 - \phi(d_S\theta))\alpha = d_S(\beta\alpha)$, et donc on peut remplacer la forme α par la forme $\phi(d_S\theta)\alpha$. Choisissons ϕ à support dans un intervalle $(-a, a)$, où a est un réel strictement positif. Il résulte de la formule (9) que $\phi(d_S\theta)$ est nulle dans l'ouvert de M formé des points m tels que $g(S_M, S_M)(m) > a$. En choisissant a assez petit, on voit que l'on est ramené au cas où α est à support dans la réunion des ouverts W_m introduits ci-dessus. On est donc ramené au cas d'un sous-groupe fermé du groupe $SO(T_m(M))$. Dans ce cas, le calcul a déjà été fait dans le corollaire 15. ∎

Le cas particulier de l'intégrale (10) est dû à Duistermaat et Heckman [**12**]:

COROLLAIRE 23. *On suppose que M est une variété symplectique compacte, et que l'on s'est donné une action hamiltonnienne de G sur M. On suppose que les zéros de S_M sont des points isolés. Notons $n = 2d$ la dimension de M. Rappelons la définition (10) de F_M. On a*

$$F_M(S) = i^d \sum_{m\in M_0(S)} \frac{e^{i\mu(S)(m)}}{\det^{1/2}_{T_m(M)}(J(S)_m)}.$$

Nous ne supposons plus que $M_0(S)$ est réunion de points isolés. Soit α une forme à support compact. Suivant Bismut [**6**], nous allons calculer $\int_M \alpha$ en fonction de la forme $i^*_{M_0(S)}\alpha$ non seulement lorsque $d_S\alpha = 0$, mais plus généralement lorsque $d_X\alpha = 0$ pour un élément X de $\mathfrak{g}$

suffisamment voisin de S en un sens que nous allons préciser. Lorsque $X \mapsto \underline{\alpha}(X)$ est une forme différentielle équivariante fermée à support compact, ceci permet le calcul de $\int_M \underline{\alpha}(X)$ pour X voisin de S par une intégrale sur $M_0(S)$.

Comme nous ne supposons pas M compact, il faut faire un peu attention. Nous fixons un ouvert relativement compact C de M, et nous considérons des formes à support compact contenu dans C.

Nous introduisons encore quelques notations. Si N est une sous-variété localement fermée de M de dimension constante, on note $T_N(M)$ le fibré normal à N dans M: c'est le fibré de base N dont la fibre en un point $m \in N$ est $T_N(M)_m = T_m(M)/T_m(N)$. Soit N une composante connexe de $M_0(S)$. Comme $T_m(N)$ est le noyau de l'opérateur $J(S)_m$, celui-ci induit dans $T_N(M)_m$ un endomorphisme elliptique inversible. On en déduit que $T_N(M)$ admet une orientation G-invariante et on choisit une telle orientation (par exemple l'orientation telle que l'on ait $\det^{1/2}_{T_N(M)_m}(J(S)_m) > 0$ pour tout $m \in N$). On choisit sur N l'orientation telle que l'orientation de M en tout point $m \in N$ soit produit de l'orientation de N et de celle de $T_N(M)$. On notera $\mathrm{Eul}_{M/N}$ la classe d'Euler équivariante du fibré $T_N(M)$.

LEMME 24. *Il existe un voisinage V de S dans $\mathfrak{g}$ tel que, pour tout $X \in V$ et toute composante connexe N de $M_0(S)$, $\mathrm{Eul}_{M/(C\cap N)}(X)$ soit inversible dans $H_X(C \cap N)$.*

DÉMONSTRATION: Il n'y a qu'un nombre fini de composantes N qui rencontrent C. Il suffit de prouver le lemme pour chacune d'elle. Soit N une telle composante. Choisissons une forme équivariante fermée e_N sur N, homogène de degré total $(\dim M - \dim N)/2$, qui représente la classe $\mathrm{Eul}_{M/N}$. Le terme de degré 0 de e_N est une fonction continue $(X, m) \mapsto e_{N[0]}(X, m)$ sur $\mathfrak{g} \times N$. Il résulte du lemme 18 que l'on a

$$e_{N[0]}(S, m) = (-2\pi)^{-d} \det{}^{1/2}_{T_N(M)_m}(J_m(S)). \tag{23}$$

Ceci est non nul quel que soit $m \in N$. On choisit V de sorte que $e_{N[0]}(X, m)$ soit non nul pour tout $X \in V$ et $m \in C \cap N$. La restriction à $C \cap N$ de la forme $e_N(X)$ est inversible dans $\mathcal{A}(C \cap N)$ et représente l'inverse de $\mathrm{Eul}_{M/(C\cap N)}(X)$. ∎

Conservons les notations de la démonstration qui précède. Soit $\beta \in \mathcal{A}_c(C \cap N)$ une forme annulée par d_X. La forme $e_{N[0]}(X)^{-1}\beta$ est un élément de $\mathcal{A}_c(C \cap N)$. On notera $\mathrm{Eul}_{M/N}(X)^{-1}\beta$ l'élément de $\mathcal{A}_c(N)$ obtenu en prolongeant la forme précédente par 0. Sa classe dans $H_{X,c}(N)$ ne dépend pas des choix faits.

PROPOSITION 25. *([4] [6]) Soient C et V comme dans le lemme 24. Soit $X \in V$. Soit $\alpha \in \mathcal{A}_c(M)$ une forme telle que $d_X\alpha = 0$ et dont le support est contenu dans C. On a*

$$\int_M \alpha = \sum_N \int_N \mathrm{Eul}_{M/N}(X)^{-1} i_N^* \alpha,$$

où la somme est prise sur l'ensemble des composantes connexes de $M_0(S)$.

DÉMONSTRATION: La démonstration est analogue à celle de la proposition 22. Pour chaque composante N on choisit un voisinage ouvert G-invariant W_N de $C \cap N$ dans C qui soit G-isomorphe à $T_{C\cap N}(M)$. Comme dans la démonstration de la proposition 22, on se ramène au cas d'une forme à support compact α sur le fibré $T_{C\cap N}(M)$, annulée par d_X. On applique alors le corollaire 19. ∎

REMARQUE. La même démonstration montre que pour $X \in V$, l'espace vectoriel $H_{X,c}(C)$ est isomorphe à la somme directe des espaces $H_{X,c}(C\cap N)$ (voir [2]).

3. COHOMOLOGIE ÉQUIVARIANTE ET ORBITES COADJOINTES

Nous étudions la cohomologie équivariante, et plus particulièrement la cohomologie équivarianteà coefficients fonctions généralisées, d'un espace homogène de groupe compact. Nous étudions la restriction à l'algèbre de Lie d'un sous-groupe compact maximal d'un groupe de Lie semi-simple connexe de centre fini G des transformées de Fourier des orbites de la représentation coadjointe.

3.1 Cohomologie équivariante d'un espace homogène. Soient K un groupe compact et H un sous-groupe fermé de K. Posons $M = K/H$ et $e = H \in M$. Nous notons $\mathfrak{r}$ un supplémentaire H-invariant de $\mathfrak{h}$ dans $\mathfrak{k}$. Pour $\underline{\alpha} \in \mathcal{A}_{\mathfrak{k}}(M)$, on définit un polynôme $\tilde{E}(\underline{\alpha}) \in S(\mathfrak{h}_{\mathbf{C}}^*)^H$ par $\tilde{E}(\underline{\alpha})(X) = (\underline{\alpha}(X)_{[0]})_e$ pour $X \in \mathfrak{h}$. Si $X \in \mathfrak{h}$, le champ de vecteurs X_M s'annule en e. Par passage au quotient, $\tilde{E}$ définit donc une application E de $H_{\mathfrak{k}}^*(M)$ dans $S(\mathfrak{h}_{\mathbf{C}}^*)^H$.

PROPOSITION 26. *L'application E est un isomorphisme d'algèbres de $H_{\mathfrak{k}}^*(M)$ sur $S(\mathfrak{h}_{\mathbf{C}}^*)^H$.*

DÉMONSTRATION: On identifie $\mathfrak{r}$ et $\mathfrak{k}/\mathfrak{h} = T_e(M)$. Soit $\underline{\alpha} \in \mathcal{A}_{\mathfrak{k}}(M)$. La forme $\underline{\alpha}$ est entièrement déterminée par sa valeur en e, $\underline{\alpha}(X)(e) \in \Lambda T_e^* M$. On note v_e: $\mathcal{A}_{\mathfrak{k}}(M) \to (S(\mathfrak{k}^*) \otimes \Lambda \mathfrak{r}^*)^H \otimes \mathbf{C}$ l'isomorphisme d'espaces vectoriels ainsi obtenu. Notons $\iota_{\mathfrak{r}}$ la différentielle de Koszul partielle sur l'espace $S(\mathfrak{k}^*) \otimes \Lambda \mathfrak{r}^*$. Si E^a est une base de $\mathfrak{r}$ de base duale f_a, on a

$\iota_{\mathfrak{r}} = \sum_a f_a \otimes \iota(E^a)$. La différentielle $\iota_{\mathfrak{r}}$ augmente de 1 le degré polynomial et préserve l'espace $(S(\mathfrak{k}^*) \otimes \Lambda\mathfrak{r}^*)^H$. On sait que la cohomologie du complexe de Koszul $S(\mathfrak{r}*) \otimes \Lambda\mathfrak{r}^*$ est égale à $\mathbf{R}$. Comme H est compact, il en résulte que la cohomologie de $(S(\mathfrak{k}^*) \otimes \Lambda\mathfrak{r}^*)^H$ muni de la différentielle $\iota_{\mathfrak{r}}$ est isomorphe à $S(\mathfrak{h}^*)^H$. Considérons l'espace $S(\mathfrak{k}^*)$ comme un module sur $\mathfrak{k}$ grace à l'action adjointe. Le complexe de cohomologie relative $(S(\mathfrak{k}^*) \otimes \Lambda\mathfrak{r}^*)^H$ est muni d'une différentielle $d_{\mathfrak{r}}$ qui préserve le degré polynomial et on voit facilement que l'application v_e induit un isomorphisme du complexe $\mathcal{A}_{\mathfrak{k}}(M)$ muni de la différentielle $d_{\mathfrak{k}}$ sur l'espace vectoriel $(S(\mathfrak{k}^*) \otimes \Lambda\mathfrak{r}^*)^H \otimes \mathbf{C}$ muni de la différentielle $d = d_{\mathfrak{r}} - \iota_{\mathfrak{r}}$. On filtre le complexe $((S(\mathfrak{k}^*) \otimes \Lambda\mathfrak{r}^*)^H, d)$ par les sous-complexes F_k formés des éléments dont le degré total (défini ici sans doubler le degré polynomial) est $\geq k$. Comme le gradué associé est le complexe $((S(\mathfrak{k}^*) \otimes \Lambda\mathfrak{r}^*)^H, -\iota_{\mathfrak{r}})$ dont nous avons calculé la cohomologie, la proposition s'en déduit par un argument standard, comme dans la démonstration de la proposition 10. ∎

On trouvera dans Arabia [1] une autre démonstration de l'isomorphisme de la proposition 26.

Remarque 1. Plus généralement, soit N une variété dans laquelle le groupe H opère. On considère la variété $M = K \times_H N$. L'opération de restriction à N induit un isomorphisme de $H^*_{\mathfrak{k}}(M)$ sur $H^*_{\mathfrak{h}}(N)$ (voir [2]).

Remarque 2. Lorsque $H = \{1\}$, le complexe $S(\mathfrak{k}^*) \otimes \Lambda\mathfrak{k}^*$ est l'algèbre de Weil, et le fait que sa cohomologie soit de dimension 1 est fondamental dans [7].

Nous utiliserons une généralisation de la proposition 26 dont nous laissons la démonstration au lecteur: l'application de restriction $\tilde{E}$ induit un isomorphisme

$$(24) \qquad E: (\ker d_k \subset \mathcal{A}^\infty_k(K/H))/d_k(\mathcal{A}^\infty_k(K/H)) \mapsto C^\infty(\mathfrak{h})^H.$$

Nous noterons W l'application inverse. Il est possible de décrire W par une formule à la Chern-Weil en utilisant la forme de courbure de la connection K-invariante sur K/H associée à la décomposition $\mathfrak{k} = \mathfrak{h} \oplus \mathfrak{r}$, mais nous ne nous en servirons pas.

Dans la suite, nous supposons K et H connexes et de même rang. On choisit une orientation sur la variété $M = K/H$. L'orientation de $\mathfrak{r}$ détermine un polynôme H-invariant $\Pi_{\mathfrak{k}/\mathfrak{h}}(X) = \det_{\mathfrak{r}}^{1/2}(adX)$ sur $\mathfrak{h}$. Il est non nul à cause de l'égalité des rangs. Soit T un sous-groupe de Cartan de H et soit $\mathfrak{t}$ son algèbre de Lie. Soient $W_{\mathfrak{k}}$ le groupe de Weyl de K et $W_{\mathfrak{h}} \subset W_{\mathfrak{k}}$ celui de H. Soit $\Delta = \Delta(\mathfrak{k}_{\mathbf{C}}, \mathfrak{t}_{\mathbf{C}})$ le système de racines de $\mathfrak{t}_{\mathbf{C}}$ dans $\mathfrak{k}_{\mathbf{C}}$. Si $\alpha \in \Delta$, on note $H_\alpha \in i\mathfrak{t}$ la coracine et $(\mathfrak{k}_{\mathbf{C}})_\alpha$ le sous-espace radiciel correspondants. Soient $\Delta_{\mathfrak{k}/\mathfrak{h}} = \{\alpha \in \Delta; (\mathfrak{k}_{\mathbf{C}})_\alpha \subset \mathfrak{r}_{\mathbf{C}}\}$ et

$\Delta^+_{\mathfrak{k}/\mathfrak{h}}$ un ordre sur $\Delta_{\mathfrak{k}/\mathfrak{h}}$. On pose $d = |\Delta^+_{\mathfrak{k}/\mathfrak{h}}| = \dim(M)/2$. Nous dirons que l'orientation sur M est définie par $\Delta^+_{\mathfrak{k}/\mathfrak{h}}$ si on a

$$\Pi_{\mathfrak{k}/\mathfrak{h}}(X) = \prod_{\alpha \in \Delta^+_{\mathfrak{k}/\mathfrak{h}}} i\alpha(X)$$

pour tout $X \in \mathfrak{t}$. Nous fixons un tel ordre $\Delta^+_{\mathfrak{k}/\mathfrak{h}}$ et nous supposons que l'orientation sur M est définie par $\Delta^+_{\mathfrak{k}/\mathfrak{h}}$. Pour chaque $\alpha \in \Delta^+_{\mathfrak{k}/\mathfrak{h}}$ on choisit un vecteur non nul $X_\alpha \in \mathfrak{r}_{\mathbf{C}}$ de poids α et on écrit $X_\alpha = e_\alpha + i f_\alpha$. Si $\alpha_1, \alpha_2, \ldots$ est une énumération de $\Delta^+_{\mathfrak{k}/\mathfrak{h}}$, il résulte de la définition (12) que la base $(e_{\alpha_1}, f_{\alpha_1}, e_{\alpha_2}, f_{\alpha_2}, \ldots)$ de $\mathfrak{r}$ est orientée.

LEMME 27. *Soit $\underline{\alpha} \in \mathcal{A}^\infty_{\mathfrak{k}}(M)$ une forme équivariante fermée et soit $p = E(\underline{\alpha}) \in C^\infty(\mathfrak{h})^H$. Soit X un élément régulier de $\mathfrak{t}$. On a*

$$\int_M \underline{\alpha}(X) = \frac{(i2\pi)^d}{|W_{\mathfrak{h}}|} \sum_{w \in W_{\mathfrak{k}}} \frac{p(wX)}{\prod_{\alpha \in \Delta^+_{\mathfrak{k}/\mathfrak{h}}} \alpha(wX)}$$

DÉMONSTRATION: On applique la formule de localisation (proposition 22). Les points de M où le champ de vecteurs X_M s'annule sont les points $w^{-1}e$ où w parcourt $W_{\mathfrak{h}} \backslash W_{\mathfrak{k}}$. La contribution de chaque point fixe est facile à calculer. ∎

COROLLAIRE 28. *Soit $p \in C^\infty(\mathfrak{h})^H$. Il existe une unique fonction $q \in C^\infty(\mathfrak{k})^K$ telle que l'on ait*

$$q(X) = \frac{1}{|W_{\mathfrak{h}}|} \sum_{w \in W_{\mathfrak{k}}} \frac{p(wX)}{\prod_{\alpha \in \Delta^+_{\mathfrak{k}/\mathfrak{h}}} \alpha(wX)}$$

pour tout élément régulier $X \in \mathfrak{t}$.

DÉMONSTRATION: L'unicité résulte de l'égalité des rangs. L'existence résulte du lemme 27 appliqué à la forme équivariante $(i2\pi)^{-d} W(p)$ car on a $q(X) = (i2\pi)^{-d} \int_M W(p)$. ∎

REMARQUE En considérant la fonction $p(X) \prod_{\alpha \in \Delta^+_{\mathfrak{k}}} \alpha(X)$ on voit que le corollaire résulte du cas particulier où $H = T$. Nous donnerons plus bas une formule explicite pour q. Il serait en fait plus naturel d'établir celle-ci en utilisant la construction de Chern-Weil pour $W(p)$, cela éviterait l'emploi peu naturel de la transformation de Fourier dans la démonstration de la proposition 29.

Soit $\mu \in i\mathfrak{t}^*$. Soit f_μ la fonction H-invariante sur $\mathfrak{h}$ telle que pour $X \in \mathfrak{t}$ on ait

$$f_\mu(X) = \frac{1}{|W_\mathfrak{h}|} \sum_{w \in W_\mathfrak{h}} e^{\mu(wX)}.$$

On note $F^{K/H}(\mu)$ la fonction K-invariante sur $\mathfrak{k}$ telle que pour tout élément régulier $X \in \mathfrak{t}$ on ait

$$\begin{aligned} F^{K/H}(\mu, X) &= \frac{1}{|W_\mathfrak{h}|} \sum_{w \in W_\mathfrak{k}} \frac{f_\mu(wX)}{\prod_{\alpha \in \Delta^+_{\mathfrak{k}/\mathfrak{h}}} \alpha(wX)} \\ &= \frac{1}{|W_\mathfrak{h}|} \sum_{w \in W_\mathfrak{k}} \frac{e^{\mu(wX)}}{\prod_{\alpha \in \Delta^+_{\mathfrak{k}/\mathfrak{h}}} \alpha(wX)}. \end{aligned}$$

Les fonctions $F^{K/H}(\mu, X)$ joue un rôle important dans le paragraphe suivant.

Soit $\lambda \in \mathfrak{t}^*$ et supposons que H soit égal au stabilisateur $K(\lambda)$ de λ dans K. Soit $\Lambda = i\lambda$ et soit $\Delta^+(\Lambda) = \{\alpha \in \Delta | (\Lambda, H_\alpha) > 0\}$. C'est un ordre sur $\Delta_{\mathfrak{k}/\mathfrak{h}}$. L'application $k \to k\lambda$ définit un isomorphisme de K/H avec l'orbite $K\lambda$. On vérifie que l'orientation de K/H définie par $\Delta^+(\Lambda)$ est l'orientation définie par la structure symplectique canonique σ de $K\lambda$. Appliquons le lemme 27 à la forme équivariante $e^{i\sigma_\mathfrak{t}}$ du paragraphe 2.2. On a $E(e^{i\sigma_\mathfrak{t}})(H) = e^{i\lambda(H)}$ pour $H \in \mathfrak{h}$. Soit $F_{K\lambda}$ la transformée de Fourier de l'orbite $K\lambda$. Soit X un élément régulier de $\mathfrak{t}$. On obtient la formule d'Harish-Chandra

$$F_{K\lambda}(X) = \sum_{w \in W_\mathfrak{k}/W_\mathfrak{k}(\lambda)} \frac{e^{iw\lambda(X)}}{\prod_{\alpha \in \Delta^+(\Lambda)} w\alpha(X)}, \tag{25}$$

où $W_\mathfrak{k}(\lambda)$ est le sous-groupe de $W_\mathfrak{k}$ qui stabilise λ.

Si H est égal au stabilisateur $K(\Lambda)$ de Λ dans K, on voit que la fonction $F^{K/H}(\Lambda)$ coïncide au signe près avec la transformée de Fourier $F_{K\lambda}$ de l'orbite dans $\mathfrak{k}^*$ du point $\lambda = -i\Lambda$. Le signe est $+$ si et seulement si l'orientation de K/H définie par $\Delta^+_{\mathfrak{k}/\mathfrak{h}}$ est égale à celle définie par $\Delta^+(\Lambda)$. En particulier, pour $H = T$, la fonction $F^{K/T}(\Lambda)$ est soit nulle (si Λ n'est pas régulier), soit égale au signe près à une transformée de Fourier d'orbite coadjointe.

En plus de l'orientation sur $\mathfrak{k}/\mathfrak{t}$ nous choisissons une mesure de Lebesgue dX. Soit E^a, $a = 1, \ldots, 2d$, une base orientée de $\mathfrak{k}/\mathfrak{t}$, x_a les fonctions coordonnées. On suppose que dX est la mesure définie par la forme

$dx_1 \wedge \cdots \wedge dx_{2d}$. Il existe un polynôme $\varpi \in S(\mathfrak{t})$ homogène de degré d tel que, pour tout λ régulier dans $\mathfrak{t}^*$ on ait, en identifiant $\mathfrak{k}/\mathfrak{t}$ et l'espace tangent en λ à l'orbite $K\lambda$,

$$\frac{\sigma(\lambda)^d}{d!} = \varpi(\lambda) dx_1 \wedge \cdots \wedge dx_{2d}.$$

Nous notons dk la mesure sur K/T tangente à dX, ∂_ϖ l'opérateur différentiel sur $\mathfrak{t}$ défini par ϖ et pr la projection T-invariante de $\mathfrak{k}$ sur $\mathfrak{t}$.

PROPOSITION 29. *Soit p une fonction différentiable sur $\mathfrak{t}$. Soit X un élément régulier de $\mathfrak{t}$. On a*

$$\int_{K/T} (\partial_\varpi p)(\mathrm{pr}(kX)) dk = (2\pi)^d \sum_{w \in W_\mathfrak{k}} \frac{p(wX)}{\Pi_{\mathfrak{k}/\mathfrak{t}}(wX)}.$$

DÉMONSTRATION: Soit λ un élément régulier de $\mathfrak{t}^*$. La formule (25) devient

$$\int_{K/T} i^d \varpi(\lambda) e^{i\lambda(kX)} dk = (2\pi)^d \sum_{w \in W_\mathfrak{k}} \frac{e^{i\lambda(wX)}}{\Pi_{\mathfrak{k}/\mathfrak{t}}(wX)}.$$

ce qui est la formule à démontrer lorsque $p(X) = e^{i\lambda(X)}$. Par des arguments laissés au lecteur la proposition s'en déduit. ∎

Soit $P \in S(\mathfrak{t}^*)^{W_\mathfrak{k}}$ un polynôme $W_\mathfrak{k}$-invariant sur $\mathfrak{t}$ et soit $X \in \mathfrak{t}$. En appliquant cette proposition à la fonction $\Pi_{\mathfrak{k}/\mathfrak{t}} P$ on obtient

$$|W_\mathfrak{k}|(2\pi)^d P(X) = \int_{K/T} \partial_\varpi(\Pi_{\mathfrak{k}/\mathfrak{t}} P)(\mathrm{pr}(kX)) dk.$$

En particulier, si $P = 1$, on trouve la formule classique

$$\mathrm{vol}(K/T) = \frac{|W_\mathfrak{k}|(2\pi)^d}{\partial_\varpi(\Pi_{\mathfrak{k}/\mathfrak{t}})}.$$

Notons $d\tilde{k}$ la mesure de Haar sur l'espace K pour laquelle K est de volume 1. On a donc

$$P(X) = \frac{1}{\partial_\varpi(\Pi_{\mathfrak{k}/\mathfrak{t}})} \int_K \partial_\varpi(\Pi_{\mathfrak{k}/\mathfrak{t}} P)(\mathrm{pr}(kX)) d\tilde{k}.$$

Soit ρ la demi-somme des racines positives. On vérifie que l'on a

$$\frac{\partial_\varpi(\Pi_{\mathfrak{k}/\mathfrak{t}} P)}{\partial_\varpi(\Pi_{\mathfrak{k}/\mathfrak{t}})} = \frac{1}{|W_\mathfrak{k}| \prod_{\alpha \in \Delta^+} \rho(H_\alpha)} \Big(\prod_{\alpha \in \Delta^+} \partial_{H_\alpha} \Big) \Big(\Big(\prod_{\alpha \in \Delta^+} \alpha \Big) P \Big).$$

L'intégrale par rapport à $d\tilde{k}$ est juste la projection K-invariante sur l'espace des éléments K-invariants. On trouve ainsi une formule explicite pour le polynôme K-invariant Q sur $\mathfrak{k}$ qui prolonge P, et donc une démonstration "explicite" du théorème de Chevalley.

PROPOSITION 30. *Soit $P \in S(\mathfrak{t}^*)^{W_\mathfrak{k}}$. Considérons l'élément*

$$\tilde{P} = \frac{1}{|W_\mathfrak{k}| \prod_{\alpha\in\Delta^+} \rho(H_\alpha)} \left(\prod_{\alpha\in\Delta^+} \partial_{H_\alpha}\right)\left(\left(\prod_{\alpha\in\Delta^+} \alpha\right)P\right)$$

de $S(\mathfrak{t}^*)$. On considère $S(\mathfrak{t}^*)$ comme un sous-espace de $S(\mathfrak{k}^*)$ grâce à la projection pr. *La projection de $\tilde{P}$ sur le sous-espace $S(\mathfrak{k}^*)^K$ de $S(\mathfrak{k}^*)$ est le polynôme K-invariant Q sur $\mathfrak{k}$ qui prolonge P.*

On peut faire des remarques analogues pour des fonctions $W_\mathfrak{k}$-invariantes sur $\mathfrak{t}$ appartenant à différentes classes, C^∞, analytiques, etc...

Nous considérons dans la fin de ce paragraphe des formes équivariantes à coefficients fonctions généralisées $\underline{\alpha} \in \mathcal{A}_\mathfrak{k}^{-\infty}(K/H)$. Une telle forme est déterminée par sa valeur en e que nous noterons $\underline{\alpha}_e$. C'est une fonction généralisée sur $\mathfrak{k}$ à valeurs dans l'espace vectoriel $\Lambda\mathfrak{r}^* \otimes \mathbf{C}$, et l'espace $\mathcal{A}_\mathfrak{k}^{-\infty}(K/H)$ est isomorphe à $(C^{-\infty}(\mathfrak{k}) \otimes \Lambda\mathfrak{r}^*)^H$. Choisissons une base E^a, $a = 1, \ldots, n = 2d$ de $\mathfrak{r}$ et soit f_a la base duale de $\mathfrak{r}^*$. Si $I = \{a_1 < a_2 < \ldots\}$ est un sous-ensemble d'indices, on pose $f_I = f_{a_1} \wedge f_{a_2} \wedge \ldots$. Chaque élément γ de $C^{-\infty}(\mathfrak{k}) \otimes \Lambda\mathfrak{r}^*$ s'écrit $\sum_I \gamma_I f_I$ avec $\gamma_I \in C^{-\infty}(\mathfrak{k})$.

Soit $\underline{\alpha} \in \mathcal{A}_\mathfrak{k}^{-\infty}(K/H)$. Le terme de degré 0 de $\underline{\alpha}_e$ est une fonction généralisée H-invariante sur $\mathfrak{k}$, mais en géneral on ne peut pas la restreindre à $\mathfrak{h}$ de sorte que les applications analogues à E et $\tilde{E}$ sont pas définies. Nous dirons que $\underline{\alpha}_e$ admet une restriction à $\mathfrak{h}$ si le front d'onde de chacune des composantes $\underline{\alpha}_{eI}$ est transverse à $\mathfrak{h}$. Nous allons voir que dans ce cas le lemme 27, convenablement formulé, reste valable.

On choisit des mesures de Lebesgue compatibles sur $\mathfrak{k}$, $\mathfrak{h}$ et $\mathfrak{r}$. On notera dk la mesure invariante sur K/H tangente à la mesure sur $\mathfrak{r}$. Si Φ est une fonction intégrable sur $\mathfrak{k}$, on a la formule de Weyl

$$\int_\mathfrak{k} \Phi(X)\, dX = \frac{|W_\mathfrak{h}|}{|W_\mathfrak{k}|} \int_{K/H} \left(\int_\mathfrak{h} \Phi(kH) \det{}_\mathfrak{r}(adH) dH\right) dk.$$

Soit $\Phi \in C_c^\infty(\mathfrak{k})$. Soit $\underline{\alpha} \in \mathcal{A}_\mathfrak{k}^\infty(M)$ une forme équivariante fermée et soit $p = E(\underline{\alpha}) \in C^\infty(\mathfrak{h})^H$. On peut écrire le lemme 27 sous la forme

$$\int_M \left(\int_\mathfrak{k} \underline{\alpha}(X)\Phi(X)\right) dX = (-2\pi)^d \int_{K/H} \left(\int_\mathfrak{h} \Pi_{\mathfrak{k}/\mathfrak{h}}(H) p(H) \Phi(kH) dH\right) dk.$$

PROPOSITION 31. *Soit $\underline{\alpha} \in \mathcal{A}_\mathfrak{k}^{-\infty}(M)$ une forme équivariante généralisée fermée. Supposons qu'on puisse restreindre $(\underline{\alpha})_e$ à $\mathfrak{h}$ et soit $p \in C^{-\infty}(\mathfrak{h})^H$ la restriction du terme de degré 0. Soit Φ une fonction C^∞ à support compact sur $\mathfrak{k}$. Alors la formule ci-dessus est encore vraie.*

DÉMONSTRATION: Fixons un produit scalaire H-invariant sur $\mathfrak{r}$. On suppose que la base E^a est orthonormée et que la mesure de Lebesgue choisie sur $\mathfrak{r}$ est la mesure $df_1 df_2 \ldots df_{2d}$. La variété M est donc munie d'une structure riemannienne K-invariante. Posons $\gamma = \underline{\alpha}_e$ et $\gamma_n = \gamma_{\{1,\ldots,n\}}$. On voit facilement que l'on a

$$\int_M \int_{\mathfrak{k}} \underline{\alpha}(X)\Phi(X)dX = \int_{K/H} \int_{\mathfrak{k}} \gamma_n(X)\Phi(kX)dXdk.$$

Ecrivons $\Psi(k, X) = \Phi(kX)$. C'est une fonction test sur la variété $K \times \mathfrak{k}$. Notons encore dk la mesure de Haar sur K telle que $\mathrm{vol}(K) = \mathrm{vol}(K/H)$. On a donc

$$\int_M \int_{\mathfrak{k}} \underline{\alpha}(X)\Phi(X)dX = \int_K \int_{\mathfrak{k}} \gamma_n(X)\Psi(k, X)dXdk.$$

On définit une 1-forme $\underline{\theta}$ équivariante sur M par la formule $\underline{\theta}(X)(\xi) = (X_M, \xi)$ pour $X \in \mathfrak{k}$. On a $(d_{\mathfrak{k}}\underline{\theta})(X) = -(X_M, X_M) + d\underline{\theta}(X)$. Soit $\phi \in C^\infty(\mathbf{R})$ une fonction telle que $\phi(0) = 1$. Comme dans la proposition 22, on voit que les formes $\underline{\alpha}$ et $\phi(d_k\underline{\theta})\underline{\alpha}$ sont dans la même classe. Soit $t > 0$. Nous appliquons ce qui précède à la fonction $\phi(x) = e^{tx}$ (on peut remplacer l'exponentielle par n'importe quelle fonction suffisamment décroissante au voisinage de $-\infty$). On pose $\underline{\alpha}(t, X) = e^{td_k\underline{\theta}(X)}\underline{\alpha}(X)$ et $\gamma(t, X) = \underline{\alpha}(t, X)_e$. Donc $\gamma_n(t, X)dkdX$ est une distribution sur $K \times \mathfrak{k}$ qui dépend du paramètre $t > 0$ et on a

$$\int_M \int_{\mathfrak{k}} \underline{\alpha}(X)\Phi(X)dX = \int_{K\times\mathfrak{k}} \gamma_n(t, X)\Psi(k, X)dkdX$$

pour tout $t > 0$.

Pour terminer la démonstration, nous allons montrer que $\gamma_n(t, X)dkdX$ tend vers la distribution de support $K \times \mathfrak{h}$

$$(-2\pi)^d \int_{K\times\mathfrak{h}} \Pi_{\mathfrak{k}/\mathfrak{h}}(H)p(H)\Psi(k, H)dkdH.$$

Posons $\underline{\theta}_e = \nu$. Pour $X \in \mathfrak{k}$, on note X_0 et X_1 ses composantes dans $\mathfrak{h}$ et $\mathfrak{r}$ respectivement. Alors ν est l'élement de $\mathfrak{k}^* \otimes \mathfrak{r}^*$ qui représente la forme bilinéaire $X, Y \mapsto -(X_1, Y)$. On emploie les notations d, $d_{\mathfrak{r}}$, etc... de la démonstration de la proposition 26. On a $d\nu = d_{\mathfrak{r}}\nu - \iota_{\mathfrak{r}}\nu$. L'élément $-\iota_{\mathfrak{r}}\nu$ est le polynôme $X \mapsto -\|X_1\|^2$, et $d_{\mathfrak{r}}\nu$ est élément de $\mathfrak{k}^* \otimes \Lambda^2\mathfrak{r}^*$ tel que $d_{\mathfrak{r}}\nu(X)(Y, Z) = ([Y, X]_1, Z) - ([Z, X]_1, Y) + (X_1, [Y, Z])$ pour $X \in \mathfrak{k}$, $Y, Z \in \mathfrak{r}$. En particulier, on a $d_{\mathfrak{r}}\nu(X)(Y, Z) = -2(ad_{\mathfrak{r}}X(Y), Z)$ pour $X \in \mathfrak{h}$, $Y, Z \in \mathfrak{r}$. La forme $e^{td\nu(X)} \in S(\mathfrak{k}^*) \otimes \Lambda(\mathfrak{r})^*$ s'écrit donc $e^{td\nu(X)} = e^{-t\|X_1\|^2} \sum_I P_I(X)t^{|I|/2} f_I$, où P_I est un polynôme homogène de degré $|I|/2$. En particulier, si $I = \{1, \ldots, n\}$ et si $X \in \mathfrak{h}$, on a

$P_I(X) = (-2)^d \det_{\mathfrak{r}}^{1/2}(adX) = (-2)^d \Pi_{\mathfrak{k}/\mathfrak{h}}(X)$. La fonction généralisée $\gamma_n(t,X)$ est combinaison linéaire de termes $t^j e^{-t\|X_1\|^2} P_I(X)\gamma_{I'}(X)$ où I est un ensemble d'indices de cardinal $2j$ et I' l'ensemble complémentaire. On voit facilement que si μ est une fonction généralisée sur $\mathfrak{k}$ dont le front d'onde est transverse à $\mathfrak{h}$ on a

$$\lim_{t\to\infty} t^d \int_{K\times\mathfrak{k}} e^{-t\|X_1\|^2}\mu(X)\Psi(k,X)dkdX = \pi^d \int_{K\times\mathfrak{h}} \mu|_{\mathfrak{h}}(X_0)\Psi(k,X_0)dkdX_0$$

pour toute fonction test Ψ sur $K \times \mathfrak{k}$.

Soit $j = 0,\ldots,d$. Comme $t^j = t^d/t^{d-j}$, chacune des intégrales

$$t^j \int_{K\times\mathfrak{k}} e^{-t\|X_1\|^2} P_I(X)\gamma_{I'}\Psi(k,X)dkdX$$

tend vers une limite quand t tend vers ∞ et cette limite est non nulle seulement pour $j = d$. Dans ce cas l'ensemble I' est vide, et comme p est la restriction de γ_0 à $\mathfrak{h}$, on obtient le résultat annoncé. ∎

3.2 Transformées de Fourier d'orbites semi-simples. Soit G un groupe de Lie réel semi-simple connexe de centre fini. Soit $\mathfrak{g}$ son algèbre de Lie et soit $\mathfrak{g} = \mathfrak{k} \oplus \mathfrak{p}$ la décomposition de Cartan de $\mathfrak{g}$. On écrit $\mathfrak{g}^* = \mathfrak{k}^* \oplus \mathfrak{p}^*$. Si $f \in \mathfrak{g}^*$ on écrit $f = f_0 + f_1$ avec $f_0 \in \mathfrak{k}^*$, $f_1 \in \mathfrak{p}^*$. On peut choisir des structures euclidiennes $\|.\|$ K-invariantes sur $\mathfrak{k}^*$ et $\mathfrak{p}^*$ telles que $q(f) = \|f_1\|^2 - \|f_0\|^2$ soit une forme quadratique G-invariante sur $\mathfrak{g}^*$. On note simplement $(.,.)$ la forme bilinéaire associée. Soit K le sous-groupe connexe de G d'algèbre de Lie $\mathfrak{k}$. Choisissons une orientation sur $\mathfrak{p}^*$. Le groupe K opère par l'action adjointe dans l'espace euclidien orienté $\mathfrak{p}^*$. On note $\Pi_{\mathfrak{g}/\mathfrak{h}}(X) = \det_{\mathfrak{p}}^{1/2}(X)$ le polynôme K-invariant sur $\mathfrak{k}$ associé. Il est non nul si et seulement si le rang de $\mathfrak{g}$ est égal au rang de $\mathfrak{k}$. Dans toute cette section, nous supposons être dans ce cas. La dimension de $\mathfrak{p}$ est paire. Elle est notée $2d_{\mathfrak{p}}$.

Soit $\mathfrak{t}$ une sous-algèbre de Cartan de $\mathfrak{k}$. C'est une sous-algèbre de Cartan de $\mathfrak{g}$. Soit $\Delta = \Delta(\mathfrak{g}_{\mathbf{C}}, \mathfrak{t}_{\mathbf{C}})$ le système de racines de $\mathfrak{t}_{\mathbf{C}}$ dans $\mathfrak{g}_{\mathbf{C}}$. Soit H_α l'unique élément de $[(\mathfrak{g}_{\mathbf{C}})_\alpha, (\mathfrak{g}_{\mathbf{C}})_{-\alpha}]$ tel que $\alpha(H_\alpha) = 2$. Soit $\mathfrak{t}_r = \{H \in \mathfrak{t} | \alpha(H) \neq 0 \text{ pour tout } \alpha \in \Delta\}$. Posons $\Delta_{\mathfrak{k}} = \{\alpha \in \Delta | (\mathfrak{g}_{\mathbf{C}})_\alpha \subset \mathfrak{k}_{\mathbf{C}}\}$ et $\Delta_p = \{\alpha \in \Delta | (\mathfrak{g}_{\mathbf{C}})_\alpha \subset \mathfrak{p}_{\mathbf{C}}\}$. Une racine de $\Delta_{\mathfrak{k}}$ est appelée racine compacte et une racine de $\Delta_{\mathfrak{p}}$ est appelée racine non compacte.

Soit $W = W(\mathfrak{g}_{\mathbf{C}}, \mathfrak{t}_{\mathbf{C}})$ le sous-groupe de transformations de $\mathfrak{t}$ engendré par les réflexions s_α pour $\alpha \in \Delta$. Soit $W_{\mathfrak{k}} = W(\mathfrak{k}_{\mathbf{C}}, \mathfrak{t}_{\mathbf{C}})$ le sous-groupe engendré par les réflexions par rapport aux racines compactes. On dira que W est le groupe de Weyl et que $W_{\mathfrak{k}}$ est le groupe de Weyl compact.

Soit M l'orbite d'un élément semi-simple de la représentation coadjointe munie de sa forme symplectique canonique σ. C'est une sous-variété fermée de $\mathfrak{g}^*$ et c'est une composante connexe d'une sous-variété algébrique de $\mathfrak{g}^*$. La mesure de Liouville β_M est donc tempérée et la transformée de Fourier F_M de β_M est une fonction généralisée G-invariante sur $\mathfrak{g}$. Remarquons que sur M la fonction $q(f) = \|f_1\|^2 - \|f_0\|^2$ est constante. On en déduit que le cône asymptote $\mathrm{CA}(M)$ est contenu dans l'ensemble des $f \in \mathfrak{g}^*$ tels que $q(f) = 0$. En particulier il ne rencontre pas $\mathfrak{p}^* = \mathfrak{k}^\perp$ et, d'après ce qu'on a rappelé dans le premier chapitre, le front d'onde de F_M est transverse à $\mathfrak{k}^\perp$. Directement, ou bien en appliquant la proposition 5, on voit que si Φ est une fonction C^∞ à support compact sur $\mathfrak{k}$ l'intégrale

$$\int_M \left(\int_{\mathfrak{k}} e^{if(X)} \Phi(X) dX \right) \beta_M$$

est finie et définit une fonction généralisée sur $\mathfrak{k}$. D'après la proposition 5, c'est la restriction $F_M|_{\mathfrak{k}}$ de F_M à $\mathfrak{k}$. Le premier résultat sur le calcul de $F_M|_{\mathfrak{k}}$ est la

PROPOSITION 32. *Soit M une orbite fermée de la représentation coadjointe. La fonction généralisée $\Pi_{\mathfrak{g}/\mathfrak{k}} F_M|_{\mathfrak{k}}$ sur $\mathfrak{k}$ est analytique. Elle est nulle si $M \cap \mathfrak{k}^* = \emptyset$.*

DÉMONSTRATION: Soit ψ une fonction C^∞ sur $\mathbf{R}$ nulle au voisinage de $-\infty$ et vérifiant la condition de la proposition 13. Comme dans cette proposition, soit $u_{\psi,\mathfrak{p}^*}$ le représentant de la classe de Thom K-équivariante sur $\mathfrak{p}^*$ associé à ψ. Notons u_ψ la forme qui s'en déduit sur M grâce à la projection de M sur $\mathfrak{p}^*$. Choisissons une base de $\mathfrak{p}$ et notons f_{1j}, $j = 1, \ldots, 2d_{\mathfrak{p}}$ les fonctions coordonnées sur $\mathfrak{p}^*$ correspondantes. Avec les notations de la formule (16) on a

$$u_\psi(X) = \sum_I \psi^{(n-|I|/2)}(-\|f_1\|^2) P_I(X/2) df_{1_I},$$

pour tout $X \in \mathfrak{k}$. La forme $u_{\psi,\mathfrak{p}^*}$ est à support compact car la projection de M sur $\mathfrak{p}^*$ est propre. Soit $\sigma_{\mathfrak{k}}$ la forme K-équivariante sur M obtenue en restreignant $\sigma_{\mathfrak{g}}$ à $\mathfrak{k}$. Donc $\sigma_{\mathfrak{k}}(X)(f) = f_1(X) + \sigma(f)$ si $X \in \mathfrak{k}$ et $f \in M$. Posons $\underline{\alpha} = e^{i\sigma_{\mathfrak{k}}}$. La forme $\underline{\alpha} u_\psi$ est équivariante, fermée et à support compact. Sa classe est indépendante de ψ. Son intégrale $B(X) = \int_M \underline{\alpha}(X) u_\psi(X)$ est une fonction analytique sur $\mathfrak{k}$ qui ne dépend pas du choix de ψ. Si M ne rencontre pas $\mathfrak{k}^*$, on peut choisir le support de ψ assez voisin de 0 pour que u_ψ soit nulle et donc, dans ce cas, on a $B(X) = 0$.

Pour établir la proposition il reste à démontrer l'assertion suivante. Soit $n = 2d$ la dimension de M. Soit $\Phi \in C_c^\infty(\mathfrak{k})$. Considérons

$$F_M(\Pi_{\mathfrak{g}/\mathfrak{k}} \Phi dX) = \int_M \left(\int_{\mathfrak{k}} e^{if(X)} \Pi_{\mathfrak{g}/\mathfrak{k}}(X) \Phi(X) dX \right) d\beta_M(f).$$

On a

$$(26) \qquad (-2\pi)^{-d_{\mathfrak{p}}}(2i\pi)^d F_M(\Pi_{\mathfrak{g}/\mathfrak{k}}\Phi dX) = \int_{\mathfrak{k}} B(X)\Phi(X)dX.$$

Soit $t > 0$. Comme en (19), on considère la forme

$$u_t(X) = \sum_{j=0}^{d_{\mathfrak{p}}} t^{n-2j} \sum_{|I|=2j} \psi^{(n-j)}(-t^2\|f_1\|^2)P_I(X/2)df_{1I'}$$

déduite de u_ψ par la transformation $(f_0, f_1) \to (f_0, tf_1)$. Comme la classe de u_t ne dépend pas de t, on a

$$(27) \qquad \int_{\mathfrak{k}} B(X)\Phi(X)dX = \int_M \int_{\mathfrak{k}} \underline{\alpha}(X)u_t(X)\Phi(X)dX$$

pour tout $t > 0$. Rappelons la définition $\underline{\alpha}(X) = e^{if_1(X)}\sum_{j=0}^{d}\frac{(i\sigma)^j}{j!}$. La forme

$$\int_{\mathfrak{k}} \underline{\alpha}(X)u_t(X)\Phi(X)dX$$

est somme finie de termes $t^{n-2j}\psi^{(n-j)}(-t^2\|f_1\|^2)\Psi_I(f_1)\omega_I$, où Ψ_I est la transformée de Fourier de la fonction $P_I(X/2)\Phi(X)$ et ω_I une forme régulière sur M. Chacune des intégrales $\int_M \Psi_I(f_1)\omega_{I[2d]}$ est absolument convergente (proposition 8). Le théorème de convergence dominée assure que l'on a

$$\lim_{t\to 0}\int_M \psi^{(n-j)}(-t^2\|f_1\|^2)\Psi_I(f_1)\omega_{I[2d]} = \int_M \psi^{(n-j)}(0)\Psi_I(f_1)\omega_{I[2d]}.$$

Pour $j = d_{\mathfrak{p}}$, on a $I = \{1, \ldots, 2d_{\mathfrak{p}}\}$ et

$$\psi^{(d_{\mathfrak{p}})}(0)P_I(X/2) = (-2\pi)^{-d_{\mathfrak{p}}}\Pi_{\mathfrak{g}/\mathfrak{k}}(X)$$

(voir la démonstration du corollaire 14). On a $\omega_{I[2d]} = \frac{(i\sigma)^d}{d!} = (2i\pi)^d\beta_M$. On obtient donc (26) en faisant tendre t vers 0 dans (27). ∎

Cette proposition peut être considérée comme l'analogue “classique” d'un résultat de W. Schmid ([20], 8.26). Il est naturel que le rôle joué dans [20] par la représentation spinorielle de K déduite de l'homomorphisme de K dans SO($\mathfrak{p}$) soit joué ici par la classe d'Euler $(-2\pi)^{d_{\mathfrak{p}}}\Pi_{\mathfrak{g}/\mathfrak{k}}$.

Nous noterons simplement $\Pi_{\mathfrak{g}/\mathfrak{k}}(X)F_M(X)$ la valeur de $\Pi_{\mathfrak{g}/\mathfrak{k}}F_M|_{\mathfrak{k}}$ en un point $X \in \mathfrak{k}$. Lorsque $M \cap \mathfrak{k}$ est non vide on complète la proposition 32 en calculant $\Pi_{\mathfrak{g}/\mathfrak{k}}(X)F_M(X)$. Posons $N = M \cap \mathfrak{k}$. C'est une orbite de K dans $\mathfrak{k}^*$ et la restriction de la forme symplectique de M à N est égale à celle de N. Nous noterons σ^0 sa forme symplectique et $\sigma^0_{\mathfrak{k}}$ la forme équivariante fermée associée. On a $\sigma^0_{\mathfrak{k}} = i^*_N\sigma_{\mathfrak{k}}$. Si $f \in N$, on pose $\mathfrak{p}(f) = \mathfrak{g}(f) \cap \mathfrak{p}$. L'espace $\mathfrak{p}/\mathfrak{p}(f)$ est isomorphe au quotient $T_f(M)/T_f(N)$. On le munit de l'orientation qui s'en déduit. Comme $\mathfrak{p}$ est orienté cela fournit sur $\mathfrak{p}(f)$ une orientation naturelle.

Soit $\tau \in H^*_{\mathfrak{k}}(N)$ la classe d'Euler équivariante du fibré orienté E de base N et de fibre $\mathfrak{p}(f)$. Comme la somme directe de E et du fibré normal $T_N(M)$ est le fibré trivial $N \times \mathfrak{p}$, dans l'algèbre $H^*_{\mathfrak{k}}(N)$ on a l'égalité

$$(-2\pi)^{-d_{\mathfrak{p}}}\Pi_{\mathfrak{g}/\mathfrak{k}} = \tau\, \mathrm{Eul}_{M/N}\,. \tag{28}$$

THÉORÈME 33. *Avec les notations ci-dessus, on a*

$$(-2\pi)^{-d_{\mathfrak{p}}}(2i\pi)^d\Pi_{\mathfrak{g}/\mathfrak{k}}(X)F_M(X) = \int_N e^{i\sigma^0_{\mathfrak{k}}(X)}\tau(X)$$

pour tout $X \in \mathfrak{k}$.

DÉMONSTRATION: Il suffit de démontrer l'égalité lorsque X vérifie $\Pi_{\mathfrak{g}/\mathfrak{k}}(X) \neq 0$ ce que nous supposons ci-dessous. On emploie les notations de la démonstration de la proposition précédente. D'après la formule (26), le côté gauche de l'égalité à établir est égal à $\int_M \underline{\alpha}(X)u_\psi(X)$. Si l'on choisit le support de ψ assez voisin de 0, la forme $\underline{\alpha}(X)u_\psi(X)$ est à support dans un voisinage de N que l'on peut supposer K-invariant et isomorphe comme K-variété au fibré normal $T_N(M)$. D'après (28) la classe d'Euler $\mathrm{Eul}_{M/N}(X)$ est inversible dans $H_X(N)$ et on applique le corollaire 19. ∎

Lorsque le centralisateur $G(f)$ d'un élément $f \in M$ est compact —par exemple si M est une orbite régulière— l'espace $\mathfrak{p}(f)$ est nul pour tout $f \in N$. En choisissant l'orientation de $\mathfrak{p}^*$ de telle sorte qu'en un point $f \in N$ l'orientation de M soit produit de celles de N et $\mathfrak{p}^*$, on obtient la formule particulièrement agréable

$$\Pi_{\mathfrak{g}/\mathfrak{k}}(X)F_M(X) = i^{d_{\mathfrak{p}}}F_N(X).$$

Nous allons reformuler le théorème 33. On choisit un point $\lambda \in \mathfrak{t}^* \cap M$ de sorte que $M = G\lambda$. La fonction $F_M(X)$ est analytique dans l'ouvert de $\mathfrak{k}$ où $\Pi_{\mathfrak{g}/\mathfrak{k}}$ est non nul et elle est déterminée dans cet ouvert par sa restriction à l'ouvert $\mathfrak{t}_r$ de $\mathfrak{t}$. Posons $\Lambda = i\lambda$. Soit $\Delta^+(\Lambda) = \{\alpha \in \Delta | (\Lambda, H_\alpha) > 0\}$. On note $\Delta^+_{\mathfrak{k}}(\Lambda) = \Delta^+(\Lambda) \cap \Delta_{\mathfrak{k}}$, $\Delta^+_{\mathfrak{p}}(\Lambda) = \Delta^+(\Lambda) \cap \Delta_{\mathfrak{p}}$. Soit d_λ le cardinal de l'ensemble $\Delta^+_{\mathfrak{p}}(\Lambda)$.

Lorsque λ est un élément régulier de $\mathfrak{t}^*$, la proposition ci-dessous est due à Rossmann [19].

PROPOSITION 34. *Soit $\lambda \in \mathfrak{t}^*$ et soit $M = G\lambda$. Pour tout $X \in \mathfrak{t}_r$ on a*

$$F_M(X) = (-1)^{d_\lambda} \sum_{w \in W_\mathfrak{k}/W_\mathfrak{k}(\lambda)} \frac{e^{iw\lambda(X)}}{\prod_{\alpha \in \Delta^+(\Lambda)} w\alpha(X)}$$

où $W_\mathfrak{k}(\lambda)$ est le sous-groupe de $W_\mathfrak{k}$ qui stabilise λ.

DÉMONSTRATION: Les points de $N \subset \mathfrak{k}^*$ où le champ de vecteurs X_M s'annule sont les points $w\lambda$ où λ parcourt $W_\mathfrak{k}/W_\mathfrak{k}(\lambda)$. On applique la formule de localisation (proposition 22) à l'intégrale $\int_N e^{i\sigma_\mathfrak{k}} \tau$ du théorème 33. Le seul point délicat dans le calcul de la contribution de chaque point fixe est la détermination des signes. Le fait essentiel, facile à vérifier, est le suivant.

Soit $G(\lambda)$ le centralisateur de λ dans G et soit $\mathfrak{g}(\lambda)$ son algèbre de Lie. L'espace $\mathfrak{g}/\mathfrak{g}(\lambda)$ s'identifie à l'espace tangent $T_\lambda(M)$. On le munit de l'orientation qui s'en déduit. On a

$$\det{}^{1/2}_{\mathfrak{g}/\mathfrak{g}(\lambda)}(adX) = (-1)^{d_\lambda} \prod_{\alpha \in \Delta^+(\Lambda)} i\alpha(X).$$

Si on préfère éviter l'emploi de la classe d'Euler τ d'un fibré peut-être non trivial, on peut appliquer directement la formule de localisation à l'intégrale $\int_M \underline{\alpha}(X)u_\psi(X)$ qui apparaît dans la démonstration de la proposition 32 ce qui est possible car la forme $\underline{\alpha}(X)u_\psi(X)$ est à support compact. ∎

Nous allons donner une formule pour la fonction généralisée $F_M|_\mathfrak{k}$. A priori, il n'est même pas clair comment énoncer une telle formule. Par exemple, lorsque $M \cap \mathfrak{k}^*$ est vide, la fonction généralisée $F_M|_\mathfrak{k}$ a son support dans l'ensemble des zéros du polynôme $\Pi_{\mathfrak{g}/\mathfrak{k}}$. Lorsque M est l'orbite d'un élément régulier de $\mathfrak{t}^*$ nous avons exprimé $F_M|_\mathfrak{k}$ comme une *série* de transformées de Fourier d'orbites de la représentation coadjointe de K. C'est l'analogue classique de la "formule de Blattner" donnant la décomposition en représentations irréductibles de K des représentations unitaires irréductibles de carré intégrable de G (voir [11]). Ce résultat a été étendu à toutes les orbites régulières par Sengupta [21].

Nous donnons ici une nouvelle démonstration des résultats de [11] qui s'adapte à toutes les orbites elliptiques, et nous les formulons de deux manières. Cette démonstration, ainsi que la première formulation (théorème 40) sont naturelles dans le cadre de la cohomologie équivariante. La seconde formulation (théorème 41) est due à Cohen et Enriques (mémoire de D.E.A.). Elle consiste à écrire $F_M|_{\mathfrak{k}}$ comme une *intégrale* de transformées de Fourier d'orbites de la représentation coadjointe de K plutôt que comme une série, ce qui est plus naturel tant que l'on ne considère pas de représentations des groupes.

La méthode sera la suivante. Si θ est une 1-forme K-équivariante sur M choisie de manière appropriée, nous introduisons pour tout $s \geq 0$ la fonction généralisée sur $\mathfrak{k}$

$$F_M(s) = (2i\pi)^{-d} \int_M e^{i\sigma_{\mathfrak{k}}} e^{isd_{\mathfrak{k}}\theta}.$$

Comme $e^{ist} = 1$ pour $t = 0$ les formes $e^{i\sigma_{\mathfrak{k}}}$ et $e^{i\sigma_{\mathfrak{k}}} e^{isd_{\mathfrak{k}}\theta}$ sont dans la même classe et il est naturel de penser que le résultat est indépendant de s. La méthode des phases stationnaires va nous permettre le calcul exact de $F_M|_{\mathfrak{k}}$ en étudiant le comportement asymptotique dans la variable s de l'intégrale précédente.

Dans la suite de cet article, M est l'orbite $G\lambda$ d'un élément $\lambda \in \mathfrak{t}^*$. Identifions $\mathfrak{g}^*$ et $\mathfrak{g}$ à l'aide de la forme bilinéaire G-invariante non dégénérée associée à la forme quadratique q et écrivons un point de $m \in \mathfrak{g} = \mathfrak{k} \oplus \mathfrak{p}$ sous la forme $x(m) + y(m)$. On définit une 1-forme K-invariante sur $\mathfrak{g}$ par

$$\theta = -([x, y], dy).$$

Si ξ est un champ de vecteurs sur $\mathfrak{g}$, on a donc $\theta(\xi)(x+y) = -([x, y], \xi(x+y))$. Soit $X \in \mathfrak{k}$. Calculons la forme $(d_{\mathfrak{k}}\theta)(X) = d\theta - \iota(X_{\mathfrak{g}})\theta$ sur $\mathfrak{g}$. Comme $X_{\mathfrak{g}}(x + y) = -[X, x] - [X, y]$, On a $(d_{\mathfrak{k}}\theta)(X) = (X, [[x, y], y]) - ([dx, y], dy) - ([x, dy], dy])$.

Soit $N = K\lambda$. Considérons le fibré normal $E = T_N(M)$ à N dans M. Comme l'espace tangent à M en un point $x \in N$ est somme directe de $\mathfrak{k} \cdot x$ et de $\mathfrak{p} \cdot x$, on identifie E à un sous ensemble de $N \times \mathfrak{p}$,

$$E = \{(x, z) | x \in N, z \in \mathfrak{p} \cdot x\}.$$

On munit E de l'orientation définie par l'orientation canonique de M. On note encore θ la restriction de θ à $E \subset \mathfrak{g}$. Soit $\sigma_{\mathfrak{k}}^0$ la forme symplectique équivariante de N. Nous notons encore $\sigma_{\mathfrak{k}}^0$ la forme équivariante sur E image réciproque par la projection de E sur N.

Nous prouvons tout d'abord

Proposition 35. *On a l'égalité de fonctions généralisées de $X \in \mathfrak{k}$*

$$F_M|_{\mathfrak{k}}(X) = (2i\pi)^{-d} \int_E e^{i\sigma_{\mathfrak{k}}^0(X)} e^{id_{\mathfrak{k}}\theta(X)}.$$

DÉMONSTRATION: Si Φ est une fonction C^∞ à support compact sur $\mathfrak{k}$, considérons

$$\hat{\Phi}(s,x,y) = \int_{\mathfrak{k}} e^{i(X,x)} e^{is(X,[[x,y],y])} \Phi(X) dX.$$

C'est la transformée de Fourier de Φ calculée au point $x + s([[x,y],y])$ de $\mathfrak{k}$. Remarquons que

$$\|x + s([[x,y],y])\|^2 \geq \|x\|^2 + 2s\|[x,y]\|^2.$$

En effet

$$\|x + s([[x,y],y])\|^2 = \|x\|^2 + 2sq([[x,y],y],x) + s^2\|[[x,y],y]\|^2$$

et $q([[x,y],y],x) = -q([x,y],[x,y]) = \|[x,y]\|^2$ car $[x,y] \in \mathfrak{k}$. Quel que soit $k > 0$, il existe une constante $C_k > 0$ telle que l'on ait

$$\hat{\Phi}(s,x,y) \leq C_k(1 + \|x\|^2 + 2s\|[x,y]\|^2)^{-k} \tag{29}$$

pour tout $x \in \mathfrak{k}$, $y \in \mathfrak{p}$ et $s \geq 0$, en particulier $\hat{\Phi}(s,x,y) \leq C_k(1 + \|x\|^2)^{-k}$.

Soit $s \geq 0$. Considérons la forme équivariante fermée $\underline{\alpha}(s)$ sur M définie par

$$\underline{\alpha}(s) = e^{i\sigma_{\mathfrak{k}}} e^{isd_{\mathfrak{k}}\theta}$$

c'est-à-dire

$$\underline{\alpha}(s,X) = e^{i\sigma_{\mathfrak{k}}(X)} e^{is(d_{\mathfrak{k}}\theta)(X)} = e^{i(X,x)} e^{is(X,[[x,y],y])} e^{i\sigma} e^{isd\theta}.$$

On a

$$\frac{d}{ds}\underline{\alpha}(s) = d_{\mathfrak{k}}(\underline{\beta}(s))$$

avec

$$\underline{\beta}(s) = (i\theta) e^{i\sigma_{\mathfrak{k}}} e^{isd_{\mathfrak{k}}\theta}$$

c'est-à-dire

$$\underline{\beta}(s,X) = i\theta e^{i\sigma_{\mathfrak{k}}(X)} e^{is(d_{\mathfrak{k}}\theta)(X)} = e^{i(X,x)} e^{is(X,[[x,y],y])} (i\theta) e^{i\sigma} e^{isd\theta}.$$

Soit $\Phi = \Phi(X)dX$ une densité C^∞ à support compact sur $\mathfrak{k}$. On pose $\alpha(s,\Phi) = \int_{\mathfrak{k}} \underline{\alpha}(s,X)\Phi(X)dX$ et $\beta(s,\Phi) = \int_{\mathfrak{k}} \underline{\beta}(s,X)\Phi(X)dX$. Ce sont des formes différentielles sur M et on a

$$\alpha(s,\Phi) = \hat{\Phi}(s,x,y) e^{i\sigma} e^{isd\theta}$$

$$\beta(s,\Phi) = \hat{\Phi}(s,x,y)(i\theta) e^{i\sigma} e^{isd\theta}.$$

Les formes $\alpha(s,\Phi)$ et $\beta(s,\Phi)$ sont produit d'une fonction à décroissance rapide par une forme régulière sur M et leur intégrale sur M est absolument convergente. La forme linéaire $\Phi \to \int_M \alpha(s,\Phi)$ est une fonction généralisée K-invariante sur $\mathfrak{k}$. On la note $\int_M \underline{\alpha}(s,X)$. On pose $F(s,X) = (2i\pi)^{-d} \int_M \alpha(s,X)$. Pour $s = 0$, $F(s,X) = F_M|_{\mathfrak{k}}(X)$. On a $\frac{d}{ds}\alpha(s,\Phi)_{[2d]} = d(\beta(s,\Phi)_{[2d-1]})$ et la discussion précédente montre que $F(s,X)$ est indépendante de $s \geq 0$.

Lemme 29. *Soit χ une fonction continue à support compact sur M identiquement égale à 1 sur un voisinage de N. Alors*

$$(2i\pi)^{(\dim M)/2} F_M(\Phi) = \lim_{s\to\infty} \int_M \alpha(s,\Phi) = \lim_{s\to\infty} \int_M \chi\alpha(s,\Phi).$$

Démonstration: Soit $\tilde{\chi} = 1 - \chi$. La fonction $\tilde{\chi}$ est continue bornée, et nulle dans un voisinage de N. Il faut montrer que l'on a

$$\lim_{s\to\infty} \int_M \tilde{\chi}\alpha(s,\Phi) = 0.$$

Pour $j = 1,\ldots,d$ posons $\omega_j = e^{i\sigma^0} \frac{(id\theta)^j}{j!}$. C'est une forme régulière sur M. L'intégrale à calculer est somme pour $j = 1,\ldots,d$ des

$$s^j \int_M \tilde{\chi}(x,y)\hat{\Phi}(s,x,y)\omega_j. \tag{30}$$

Nous allons montrer que chacune de ces intégrales est une fonction de s décroissant plus vite que toutes les fonctions s^{-l} au voisinage de ∞, ce qui terminera la démonstration du lemme.

Si $x + y \in M$ et si $[x,y] = 0$, alors $x \in \mathfrak{k}$ et $y \in \mathfrak{p}$ sont semi-simples et commutent. Comme les points de M sont elliptiques, ceci implique $y = 0$ et $x \in N$. Soit $t > 0$. Posons $\psi(t) = \inf_{\|x\|^2=t, x+y\in M} \|[x,y]\|^2$. On a $\psi(t) > 0$ pour tout $t > 0$, et il résulte de **[16]**, appendice A.2, qu'il existe des constantes $a \in \mathbf{R}$ et $b > 0$ telles que $\psi(t)$ soit équivalent à bt^a au voisinage de ∞. Il existe donc une constante $c > 0$ telle que l'on ait $\|[x,y]\|^2 \geq c\|x\|^{2a}$ si $x + y \in \text{supp}(\tilde{\chi})$. Quitte à diminuer a, on peut supposer $a \leq 0$.

Soit $k > 0$ tel que $\int_M (1 + \|x\|^2)^{-k}\omega_j < \infty$. Rappelons qu'un tel k existe d'après la proposition 8 et parce que les fonctions $\|x\|^2$ et $\|x+y\|^2$ sont équivalentes sur M. Soit $l > 0$ et soit $k' = k - la + l$. Si $x + y$ est dans le support de $\tilde{\chi}$ on a

$$\begin{aligned}|\hat{\Phi}(s,x,y)| &\leq C_{k'}(1 + \|x\|^2 + 2s\|([x,y])\|^2)^{-k'}\\ &\leq C_{k'}(1 + \|x\|^2)^{-k+la}(2s\|([x,y])\|^2)^{-l}\\ &\leq D_l s^{-l}(1 + \|x\|^2)^{-k},\end{aligned}$$

où D_l est une constante. En portant cette inégalité dans l'intégrale (30) notre assertion est démontrée. ∎

Nous sommes donc ramenés à paramétrer un voisinage de N dans M. Il existe un difféomorphisme I d'un voisinage de N dans E sur un voisinage de N dans M qui est l'identité sur N et qui induit l'identité sur l'espace tangent $T_b(M) = T_b(E) \subset \mathfrak{g}$ en tout point de $b \in N$. Ecrivons encore $x = x \circ I$, $y = y \circ I$. Donc, pour $b \in N$, $z \in E_b = \mathfrak{p} \cdot b$ on a $I(b,z) = x(b,z) + y(b,z)$ avec $x(b,z) \in \mathfrak{k}$ et $y(b,z) \in \mathfrak{p}$. Posons $t = s^{-1/2}$. On a ramené le calcul de $F_M|_{\mathfrak{k}}(\Phi)$ au calcul de la limite quand $t \to 0$ de $I(t) = \int_E I^*(\alpha(t^{-2}, \Phi)\chi)$. On a

$$I(t) = \int_{E\times\mathfrak{k}} e^{it^{-2}(X,[[x,y],y])} e^{i(X,x)} e^{iI^*\sigma} e^{it^{-2}dI^*\theta} \Phi(X)\chi(x,y)dX.$$

Pour évaluer cette limite, nous faisons le changement de variable $\delta(t)$ défini par $\delta(t)(b,z) = (b,tz)$. On a, en écrivant $x = x(b,tz)$, $y = y(b,tz)$,

(31)

$$I(t) = \int_E \left\{ \left(\int_k e^{it^{-2}(X,[[x,y],y])} e^{i(X,x)} \Phi(X)dX \right) \chi(x,y)\delta(t)^* I^* e^{i\sigma + it^{-2}d\theta} \right\}$$

Le lemme suivant montre que chacun des termes de cette intégrale tend vers une limite.

LEMME 37. *Lorsque t tend vers 0*

a) $x(b,z)$ tend vers b,

b) $y(b,z)$ tend vers 0,

c) $t^{-2}[[x(b,tz), y(b,tz)], y(b,tz)]$ tend vers $[[b,z],z]$,

d) la forme $\delta(s)^* I^*\sigma$ tend vers la forme σ^0 ,

e) la forme $t^{-2}\delta(t)^* I^*\theta$ tend vers θ.

De plus, il existe un voisinage V de N dans E et une constante $c > 0$ tels que l'on ait $\|[x(b,z), y(b,z)]\|^2 \geq c\|z\|^2$ pour tout $(b,z) \in V$.

DÉMONSTRATION: Si $b \in N$ on a $I(b,z) = b + z + O(|z|^2)$ et donc $x = b + O(|z|^2)$, $y = z + O(|z|^2)$. Les premières assertions sont claires.

Choisissons des bases de $\mathfrak{k}$ et $\mathfrak{p}$, et écrivons x_i et y_j les fonctions coordonnées. On a $dx_i = db_i + \sum_l 0(|z|^2)db_l + \sum_k 0(|z|)dz_k$ et $dy_j = dz_j + \sum_l 0(|z|^2)db_l + \sum_k 0(|z|)dz_k$.

On écrit $i^*\sigma$ sous la forme

$$\sum_{I,J} r_{IJ}(x,y)dx_I dy_J = \sum_{I,J} R_{IJ}(b,z)db_I dz_J$$

où les r_{IJ} et les R_{IJ} sont des fonctions différentiables. Donc $\delta(t)^* I^*\sigma = \sum_{I,J} t^{|J|} R_{IJ}(b,tz)db_I dz_J$. A la limite, seuls les termes où J est vide donnent une contribution non nulle et l'on trouve σ^0.

La forme $I^*\theta$ est combinaison linéaire de termes $x_i dy_l dy_j$ et de termes $y_l dy_j dx_i$. On voit que $t^{-1}\delta(t)^* I^* y_j$ tend vers z_j et $t^{-1}\delta(t)^* I^* dy_j$ vers dz_j. L'assertion e) en résulte.

La fonction $\|[x(b,z), y(b,z)]\|^2$ est égale à $\|[b,z]\|^2 + 0(|z|^4)$ uniformément en b. Comme b est semi-simple, $ad(b)$ est inversible dans l'espace $E_b = ad(b)(\mathfrak{p})$. La dernière assertion du lemme en résulte. ∎

Comme $(d_{\mathfrak{k}}\theta)(X) = ([[b,z],z],X) + d\theta$ et $\sigma^0_{\mathfrak{k}}(X) = e^{(b,X)} + \sigma^0$, le terme entre accolades dans (31) tend vers $\int_{\mathfrak{k}} \Phi(X) e^{i\sigma^0_{\mathfrak{k}}(X)} e^{id_{\mathfrak{k}}\theta(X)}$. Pour démontrer la proposition il nous suffit de voir qu'on a le droit de passer à la limite dans l'intégrale (31). La forme $\delta(t)^* I^* e^{i\sigma + it^{-2}d\theta}$ est combinaison linéaire de termes $Q_{IJ}(b,z,t) db_I dz_J$, et les fonctions Q_{IJ} sont bornées indépendamment de b, t et z quand $\chi(x,y) \neq 0$. On suppose que le support de $\chi \circ I$ est contenu dans le voisinage V du lemme ci-dessus. Pour $x = x(b,tz)$, $y = y(b,tz)$ tels que $\chi(x,y) \neq 0$ on a $\|x + t^{-2}[[x,y],y]\|^2 \geq 2t^{-2}\|[x,y]\|^2 \geq 2c\|z\|^2$. Pour tout $k > 0$ il existe une constante $C_k > 0$ telle que l'on ait

$$|\chi(x,y) \int_{\mathfrak{k}} e^{it^{-2}(X,[[x,y],y])} e^{i(X,x)} \Phi(X) dX| \leq C_k (1 + \|z\|^2)^{-k}.$$

Pour k assez grand, chacune des intégrales $\int_E (1 + \|z\|^2)^{-k} db_I dz_J$ converge. On peut donc appliquer le théorème de convergence dominée ce qui termine la démonstration de la proposition. ∎

Considérons la fibration $E \to N$. Formellement nous écrivons

$$\int_E e^{i\sigma^0_{\mathfrak{k}}(X)} e^{id_{\mathfrak{k}}\theta(X)} = \int_N e^{i\sigma^0_{\mathfrak{k}}(X)} \Big(\int_{E/N} e^{id_{\mathfrak{k}}\theta(X)}\Big).$$

Nous allons justifier cette procédure.

PROPOSITION 38. *La formule*

$$\underline{\nu}(X) = \int_{E/N} e^{id_{\mathfrak{k}}\theta(X)}$$

définit une forme équivariante fermée sur N à coefficients fonctions généralisées $\underline{\nu} \in \mathcal{A}^{-\infty}_{\mathfrak{k}}(N)$.

DÉMONSTRATION: Soit $\Phi(X)dX$ une densité C^∞ à support compact sur $\mathfrak{k}$. On considère la forme

$$\int_{\mathfrak{k}} e^{id_{\mathfrak{k}}\theta(X)} \Phi(X) dX$$

sur $\mathfrak{g}$. Elle est égale au produit de $e^{id\theta}$ par la fonction

$$\int_{\mathfrak{k}} e^{i([[x,y],y],X)}\Phi(kX)dX.$$

Si x_i et y_j sont des fonctions coordonnées sur $\mathfrak{k}$ et $\mathfrak{p}$ respectivement, nous avons vu que $d\theta$ est combinaison linéaire de formes $y_i dx_l dy_k$ et $x_i dy_l dy_k$. La forme $e^{id\theta}$ est combinaison linéaire de formes $q_{IJ}(x,y)dy_J dx_I$ où $q_{IJ}(x,y)$ est un polynôme en x et y. La forme $\int_{\mathfrak{k}} e^{id_t\theta(X)}\Phi(X)dX$ est combinaison linéaire de formes

$$(\int_{\mathfrak{k}} e^{i([[x,y],y],X)}q_{IJ}(x,y)\Phi(X)dX)dy_J dx_I.$$

Nous devons montrer que l'intégrale

$$\int_{E_x}(\int_{\mathfrak{k}} e^{i([[x,y],y],X)}q_{IJ}(x,y)\Phi(X)dX)dy_J$$

définit une fonction généralisée de X dépendant de manière différentiable de $x \in N$. Par invariance sous l'action de K, on voit qu'il suffit de vérifier que pour $x = \lambda$ cette formule définit une fonction généralisée de X. Ceci résulte du lemme suivant pour lequel nous introduisons quelques notations.

Rappelons que nous avons défini $\Lambda = i\lambda$ et $\Delta_{\mathfrak{p}}^+(\Lambda)$. Soit $\mathfrak{p}_\lambda = \mathfrak{p}\dot{\lambda} \subset \mathfrak{p}$. On munit $\mathfrak{p}_\lambda$ du produit scalaire $(.,.)$ induit par celui de $\mathfrak{p}$. On choisit une base orthonormée e_α, f_α de $\mathfrak{p}_\lambda$ indexée par $\alpha \in \Delta_{\mathfrak{p}}^+(\Lambda)$ telle que pour $T \in \mathfrak{t}$ on ait

$$[T, e_\alpha] = i(\alpha, T)f_\alpha,$$
$$[T, f_\alpha] = -i(\alpha, T)e_\alpha,$$

c'est-à-dire $e_\alpha + if_\alpha \in (\mathfrak{g}_{\mathbf{C}})_\alpha$. Si $y \in \mathfrak{p}_\lambda$ on écrit $y = \sum_{\alpha \in \Delta_{\mathfrak{p}}^+(\Lambda)} u_\alpha e_\alpha + v_\alpha f_\alpha$.

Lemme 39. *a) La projection $T(y)$ de $[[\lambda, y], y]$ sur $\mathfrak{t}$ est égale à*

$$T(y) = \sum_{\alpha \in \Delta_{\mathfrak{p}}^+(\Lambda)} (\Lambda, \alpha)(u_\alpha^2 + v_\alpha^2)(-i\alpha).$$

b) Il existe une constante $c > 0$ telle que $\|T(y)\| \geq c\|y\|^2$ pour tout $y \in \mathfrak{p}_\lambda$.

c) Notons dy la mesure euclidienne sur $\mathfrak{p}_\lambda$. Soit $y \to q(y)$ une fonction polynomiale en y. L'intégrale $\int_{\mathfrak{p}_\lambda} e^{i([[\lambda,y],y],X)}q(y)dy$ définit une fonction généralisée de X sur $\mathfrak{k}$. Son front d'onde est transverse à $\mathfrak{t}$.

DÉMONSTRATION: a) Il est clair que

$$T(y) = \sum_{\alpha\in\Delta_{\mathfrak{p}}^+(\lambda)} (\Lambda,\alpha)[u_\alpha f_\alpha - v_\alpha e_\alpha, u_\alpha e_\alpha + v_\alpha f_\alpha].$$

Mais

$$q([e_\alpha, f_\alpha], H_\alpha) = q(e_\alpha, [f_\alpha, H_\alpha]) = 2iq(e_\alpha, e_\alpha) = 2i,$$

et donc $[e_\alpha, f_\alpha] = i\alpha$.

b) résulte de a).

c) Considérons l'application $J(y) = [[\lambda, y], y]$ de $\mathfrak{p}_\lambda$ dans $\mathfrak{k}$. Son image est un cône dans $\mathfrak{k}$. La mesure image $J_*(p(y)dy)$ est tempérée à cause de b). La fonction généralisée est la transformée de Fourier de l'image de la mesure $J_*(p(y)dy)$ par l'application $y \mapsto [[\lambda, y], y]$ de $\mathfrak{p}_\lambda$ dans $\mathfrak{k}$ (identifié comme d'habitude à $\mathfrak{k}^*$ grâce à $(.,.)$). Cette mesure est tempérée à cause de b). Le front d'onde est contenu dans le cône $J(\mathfrak{p}_\lambda)$. Il résulte de b) que $\mathfrak{t}^\perp \cap J(\mathfrak{p}_\lambda) = \{0\}$, ce qui termine la démonstration de c). ∎

Pour terminer la démonstration de la proposition, il reste à voir que la forme $\underline{\nu}$ est fermée. Nous laissons ce soin au lecteur. ∎

Nous avons finalement démontré le théorème suivant

THÉORÈME 40. *Soit M une orbite elliptique de la représentation coadjointe de G. Soit $2d$ sa dimension. Soit $N = \mathfrak{t}^* \cap M$. Soit σ la forme symplectique de M et $\sigma_\mathfrak{t}$ la forme K-équivariante fermée associée. Soit $\sigma_\mathfrak{t}^0 = \iota_N^*\sigma_\mathfrak{t}$ la forme symplectique K-équivariante de N. Soit $\underline{\nu} \in \mathcal{A}_\mathfrak{t}^{-\infty}(N)$ la forme K-équivariante fermée à coefficients fonctions généralisées introduite dans la proposition 38. Soit $F_M|_\mathfrak{t}(X) = (2i\pi)^{-d}\int_M e^{i\sigma_\mathfrak{t}(X)}$ la restriction à $\mathfrak{t}$ de la transformée de Fourier de la mesure de Liouville β_M de M. On a*

$$F_M|_\mathfrak{t}(X) = (2i\pi)^{-d}\int_N e^{i\sigma_\mathfrak{t}^0(X)}\underline{\nu}(X).$$

REMARQUE. Si l'on compare cette formule avec celles du corollaire 19 ou du théorème 33, on voit que la forme $\underline{\nu}(X)$ joue le rôle d'inverse au sens distribution de la forme d'Euler $\mathrm{Eul}_E(X)$ du fibré normal E à N dans M. De même, la démonstration du théorème montre que l'on peut espérer prouver une égalité de fonctions généralisées sur $\mathfrak{t}$

$$\int_M \underline{\alpha}(X) = \int_N \iota_N^*(\underline{\alpha})\underline{\nu}(X)$$

pour beaucoup d'autres formes K-équivariantes $\underline{\alpha}$ sur M. Nous espérons revenir sur ces questions dans un autre article.

Nous allons calculer l'intégrale $\int_N e^{i\sigma_{\mathfrak{k}}^0(X)}\underline{\nu}(X)$ grâce à la proposition 31 de la section précédente. Commençons par énoncer le résultat. Soit $H = K(\lambda)$ le stabilisateur de λ dans K de sorte que $N = K\lambda = K/H$. Soit $\Lambda = i\lambda$. On munit K/H de l'orientation définie par $\Delta_{\mathfrak{k}}^+(\Lambda)$. Soient $\alpha_1, \alpha_2, \ldots, \alpha_{d_\lambda}$ les éléments de l'ensemble $\Delta_{\mathfrak{p}}^+(\Lambda)$. Nous exprimons $F_M|_{\mathfrak{t}}$ comme une intégrale des fonctions analytiques K-invariantes $F^{K/H}(\mu)$, $\mu \in i\mathfrak{t}^*$ définies dans la section précédente. On note $\mathbf{R}_+ = \{t \in \mathbf{R} | t \geq 0\}$.

THÉORÈME 41. *Soit $\lambda \in \mathfrak{t}^*$, soit $M = G\lambda$, et soit $H = K(\lambda)$ le stabilisateur de λ dans K. Soit $\Lambda = i\lambda$. Alors*

$$F_M|_{\mathfrak{t}} = (-1)^{d_\lambda} \int_{\mathbf{R}_+^{d_\lambda}} F^{K/H}(\Lambda + \sum_{i=1}^{d_\lambda} t_i\alpha_i)dt_1dt_2\cdots dt_{d_\lambda}.$$

DÉMONSTRATION: Soit $\mu \in i\mathfrak{t}^*$. Le groupe de Weyl $W_{\mathfrak{h}}$ est égal à $W_{\mathfrak{t}}(\lambda)$. Rappelons que $f(\mu)(H)$ est la fonction analytique H-invariante sur $\mathfrak{h}$ telle que

$$f(\mu)(H) = \frac{1}{|W_{\mathfrak{h}}|} \sum_{w \in W_{\mathfrak{h}}} e^{iw\mu(H)}.$$

Il résulte du lemme 39, assertion c, que la valeur $\underline{\nu}_\lambda$ en λ de $\underline{\nu}$ (qui est un élément de $C^{-\infty}(\mathfrak{k}) \otimes \Lambda(\mathfrak{k}/\mathfrak{h})^*$) admet une restriction à $\mathfrak{t}$ et a fortiori à $\mathfrak{h}$. On note $r \in C^{-\infty}(\mathfrak{h})^H$ la restriction du terme de degré zéro $\underline{\nu_{[0]}}_\lambda$. Nous aurons besoin de la

PROPOSITION 42. *On a l'égalité de fonctions généralisées sur $\mathfrak{h}$*

$$(2i\pi)^{-d_\lambda} r = (-1)^{d_\lambda} \int_{\mathbf{R}_+^{d_\lambda}} f(\sum_i t^i\alpha_i)dt_1dt_2\cdots dt_{d_\lambda}.$$

DÉMONSTRATION: Le terme de degré extérieur 0 par rapport à la variable x de $e^{id\theta}$ au point λ est

$$e^{-i([\lambda,dy],dy)}.$$

L'orientation de $\mathfrak{p}_\lambda$ définie par la restriction de la forme symplectique σ de M est donnée par l'élément $\prod_{\alpha\in\Delta_{\mathfrak{p}}^+(\Lambda)} e_\alpha \wedge f_\alpha$ car $\sigma_\lambda(e_\alpha, f_\alpha) = (\Lambda, \alpha)^{-1}$. Le terme de degré $2d_\lambda$ de $e^{-i([\lambda,dy],dy)}$ sur $\mathfrak{p}_\lambda$ s'identifie donc à la densité $(-2i)^{d_\lambda} \prod_{\alpha\in\Delta_{\mathfrak{p}}^+(\Lambda)}(\Lambda, \alpha)dy$ où dy est la densité fournie par la structure euclidienne de $\mathfrak{p}_\lambda$. Posons $\Pi_\Lambda = \prod_{\alpha\in\Delta_{\mathfrak{p}}^+(\Lambda)}(\Lambda, \alpha)$. Soit $\Phi \in C_c^\infty(\mathfrak{h})$. On a donc

$$\int_{\mathfrak{h}} r(H)\Phi(H)dH = (-2i)^{d_\lambda}\Pi_\Lambda \int_{\mathfrak{p}_\lambda} \left(\int_{\mathfrak{h}} e^{i([[\lambda,y],y],H)}\Phi(H)dH\right) dy$$

D'après le lemme 39 $\int_{\mathfrak{t}} e^{i([[\lambda,y],y],T)}\Phi(T)dT$ est aussi une fonction rapidement décroissante de $y \in \mathfrak{p}_\lambda$. Par la formule d'intégration de Hermann Weyl, on a

$$\int_{\mathfrak{h}} e^{i([[\lambda,y],y],H)}\Phi(H)dH$$
$$= \frac{1}{|W_{\mathfrak{h}}|}\int_{H/T}(\int_{\mathfrak{t}} e^{i([[\lambda,h^{-1}y],h^{-1}y],T)}\det{}_{\mathfrak{h}/\mathfrak{t}}(T)\Phi(hT)dT)dh.$$

On obtient donc

$$\int_{\mathfrak{p}_\lambda}\left(\int_{\mathfrak{h}} e^{i([[\lambda,y],y],H)}\Phi(H)dH\right)dy$$
$$= \frac{1}{|W_{\mathfrak{h}}|}\int_{\mathfrak{p}_\lambda}\left(\int_{H/T}(\int_{\mathfrak{t}} e^{i([[\lambda,y],y],T)}\det{}_{\mathfrak{h}/\mathfrak{t}}(T)\Phi(hT)dT)dh\right)dy$$

Utilisons la formule du lemme 39 pour $T(y)$. On a

$$dy = du_{\alpha_1}dv_{\alpha_1}\dots du_{\alpha_{d_\lambda}}dv_{\alpha_{d_\lambda}}.$$

Posons $t_\alpha = (\Lambda,\alpha)(u_\alpha^2+v_\alpha^2)$. En passant en coordonnées radiales dans les plans (u_α,v_α) on trouve

$$\Pi_\Lambda\int_{\mathfrak{p}_\lambda}\left(\int_{\mathfrak{h}} e^{i([[\lambda,y],y],H)}\Phi(H)dH\right)dy$$
$$= \frac{\pi^{d_\lambda}}{|W_{\mathfrak{h}}|}\int_{\mathbb{R}_+^{d_\lambda}}\left(\int_{H/T}(\int_{\mathfrak{t}} e^{(\sum_i t^i\alpha_i,T)}\det{}_{\mathfrak{h}/\mathfrak{t}}(T)\Phi(hT)dT)d\bar{h}\right)dt_1dt_2\cdots dt_{d_\lambda}$$

et donc

$$\Pi_\Lambda\int_{\mathfrak{p}_\lambda}\left(\int_{\mathfrak{h}} e^{i([[\lambda,y],y],H)}\Phi(H)dH\right)dy$$
$$= \pi^{d_\lambda}\int_{\mathbb{R}_+^{d_\lambda}}\left(\int_{\mathfrak{h}} f(\sum_i t^i\alpha_i)(H)\Phi(H)dH\right)dt_1dt_2\cdots dt_{d_\lambda},$$

ce qui est l'assertion à démontrer. ∎

Posons $\underline{\omega} = e^{i\sigma_{\mathfrak{t}}^0}\underline{\nu}$. Comme λ est invariant par H par définition de H, la restriction p de $\underline{\omega}_{[0]_\lambda}$ est égale à $e^{i\lambda(X)}r(X)$. On a donc

$$(2i\pi)^{-d_\lambda}p = (-1)^{d_\lambda}\int_{\mathbb{R}_+^{d_\lambda}} f(\Lambda+\sum_i t^i\alpha_i)dt_1dt_2\cdots dt_{d_\lambda}.$$

Le théorème résulte alors de la définition de $F^{K/H}$ et de la proposition 31. ∎

RÉFÉRENCES

[1] A. Arabia, *Cycles de Schubert et cohomologie K-équivariante de K/T*, Thèse de doctorat, Université Paris 7 (1985).

[2] M. Atiyah et R. Bott, *The moment map and equivariant cohomology*, Topology **23** (1984), 1-28.

[3] N. Berline, E. Getzler et M. Vergne, "Heat kernel and Dirac operator," En préparation, 1989.

[4] N. Berline et M. Vergne, *Classes caractéristiques équivariantes. Formule de localisation en cohomologie équivariante*, C. R. Acad. Sci. Paris **295** (1982), 539-541.

[5] N. Berline et M. Vergne, *Fourier transforms of orbits of the coadjoint representation*, in "Representation theory of reductive groups," Birkhauser, 1983, pp. 53-57.

[6] J.-M. Bismut, *Localization formulas, superconnections, and the index theorem for families*, Comm. Math. Phys. **103** (1986), 127-166.

[7] H. Cartan, *La transgression dans un groupe de Lie et dans un espace fibré principal. In "Colloque de Topologie"*, C. B. R. M., Bruxelles (1950), 57-71.

[8] J.-Y. Charbonnel, *Sur les caractères des groupes de Lie*, J. of Functional Analysis **72** (1987), 94-150.

[9] J.-Y. Charbonnel, *Sur les orbites de la représentation coadjointe*, Compositio Math. **46** (1982), 273-305.

[10] J.-Y. Charbonnel, *Orbites fermées et orbites tempérées*, A paraître aux Annales de l'E.N.S.

[11] M. Duflo, G. Heckman et M. Vergne, *Projection d'orbites, formule de Kirillov et formule de Blattner*, Mem. Soc. Math. Fr. **15** (1984), 65-128.

[12] J. J. Duistermaat et G. Heckman, *On the variation of the cohomology of the reduced phase space*, Inventiones Math. **69** (1982), 259-268.

[13] J. J. Duistermaat et G. Heckman, *On the variation of the cohomology of the reduced phase space. Addendum*, Inventiones Math. **72** (1983), 153-158.

[14] V. A. Ginzburg, *Fast decreasing functions and characters of real algebraic groups*, Funct. Anal. Appl. **16** (1982), 53-54.

[15] L. Hörmander, "The analysis of linear partial differential operators I, Distribution theory and Fourier analysis," Springer Verlag, Berlin, Heidelberg, New York, Tokyo, 1983.

[16] L. Hörmander, "The analysis of linear partial differential operators II, Differential operators with constant coefficients," Springer Verlag, Berlin, Heidelberg, New York, Tokyo, 1983.

[17] R. Howe, *Wave front sets of representations of Lie groups*, in "In Automorphic forms, representation theory and arithmetic, Bombay Colloquium," Springer Verlag, Berlin, Heidelberg, New York, Tokyo, 1981.

[18] V. Mathai et D. Quillen, *Superconnections, Thom classes, and equivariant differential forms*, Topology **25** (1986), 85-110.

[19] W. Rossmann, *Kirillov's character formula for reductive groups*, Inventiones Math. **48** (1978), 207-220.

[20] W. Schmid, *On the characters of discrete series*, Inventiones math. **30** (1975), 47-144.

[21] I. Sengupta, *Projection of orbits and K-multiplicities*, J. of Functional Analysis **84** (1989), 215-225.

[22] V. S. Varadarajan, "Harmonic analysis on real reductive groups," Springer Lecture Notes in Mathematics 576, Springer Verlag, Berlin, Heidelberg, New York, Tokyo, 1977.

M. Duflo
Université Paris 7 et UA 748 du C.N.R.S.
2 place Jussieu
F-75251 Paris Cedex 05
FRANCE

M. Vergne
E.N.S. et UA 762 du C.N.R.S.
45 rue d'Ulm
F-75005 Paris
FRANCE

Représentations Monomiales des Groupes de Lie Résolubles Exponentiels

HIDÉNORI FUJIWARA

0. Soit G un groupe de Lie résoluble exponentiel d'algèbre de Lie $\mathfrak{g}$. Cela signifie que l'application exponentielle est un difféomorphisme de $\mathfrak{g}$ sur G: nous le notons $G = \exp \mathfrak{g}$. C'est aux représentations monomiales de G que nous nous intéressons. Soient H un sous-groupe connexe de G et χ son caractère unitaire. Le but de cette étude est de décrire dans le cadre de la méthode des orbites la désintégration centrale canonique de la représentation induite $\tau = \mathrm{ind}_H^G \chi$. Soit $\mathfrak{h}$ l'algèbre de Lie de H. Alors il existe $f \in \mathfrak{g}^*$, une forme linéaire sur $\mathfrak{g}$, telle que f s'annule sur l'algèbre dérivée $[\mathfrak{h}, \mathfrak{h}]$ de $\mathfrak{h}$ et que χ s'écrive $\chi(\exp X) = e^{if(X)}$ ($i = (-1)^{1/2}$, $X \in \mathfrak{h}$). Dans cette situation, χ se notera χ_f. Après que l'on avait vigoureusement étudié le cas essentiel où $\mathfrak{h}$ était une polarisation en f, vers '72 Grélaud [9] et Quint [16] ont mis fin au cas uù $\mathfrak{h}$ était un idéal de $\mathfrak{g}$.

De plus, Quint a laissé des conjectures suivantes. Soit μ une mesure positive finie sur l'espace affine $\Xi = f + \mathfrak{h}^\perp$ désignant l'annihilateur de $\mathfrak{h}$ dans $\mathfrak{g}^*$, équivalente à la mesure de Lebesgue sur Ξ. On regarde μ comme une mesure sur $\mathfrak{g}^*$ et prend son image ν par l'application de Kirillov–Bernat $\theta : \mathfrak{g}^* \to \hat{G}$, le dual unitaire de G. $\hat{G}$ s'obtient comme l'espace des orbites coadjointes $\mathfrak{g}^*/G$ de G au moyen de la bijection induite $\overline{\theta} = \overline{\theta}_G : \mathfrak{g}^*/G \to \hat{G}$ (cf. [4], [11]). Pour $\pi \in \hat{G}$ on note $\Omega(\pi)$ l'orbite associée. Soit $Z(\Xi)$ l'ensemble des $l \in \Xi$ telles que l'orbite $G \cdot l$ atteint la dimension maximum parmi les orbites rencontrant Ξ.

Conjecture de Quint [16]: (i) Pour toute $l \in Z(\Xi)$, chaque composante connexe de $\Xi \cap G \cdot l$ est une variété différentielle de dimension supérieure ou égale à $1/2 \dim G \cdot l$.

(ii) On a $\tau = \mathrm{ind}_H^G \chi_f = \int_{\hat{G}}^{\oplus} m(\pi) \pi d\nu(\pi)$. Ici, la multiplicité $m(\pi)$ est égale au nombre des composantes connexes de $\Xi \cap \Omega(\pi)$ si chaque composante est une variété de dimension $1/2 \dim \Omega(\pi)$, sinon $m(\pi) = +\infty$.

Depuis ont été faits des travaux fondamentaux, par Benoist [2], [3] pour le cas symétrique et par Corwin–Greenleaf/Grélaud [6], [10] pour le cas nilpotent, établissant les conjectures de Quint à condition d'une petite modification sans importance: dans (i), "toute $l \in Z(\Xi)$" doit être remplacé par "μ-presque toute $l \in \Xi$". On voit encore des résultats dus à Lipsman [12], [13], [14] qui contiennent la désintégration de τ pour les cas complètement résoluble. En plus ils ont montré que la multiplicité $m(\pi)$ était retrouvée comme le nombre de H-orbites inclues dans $\Xi \cap \Omega(\pi)$. Ici nous allons voir dans la suite le:

THÉORÈME. *Modifiées comme ci-dessus, s'établissent les conjectures de Quint. On peut d'ailleurs choisir pour multiplicité $m(\pi)$ le nombre des H-orbites contenues dans $\Xi \cap \Omega(\pi)$.*

Enfin nous laisserons deux questions ayant rapport à notre sujet et en étudierons certains exemples.

1. Soient $G = \exp \mathfrak{g}$ et $f \in \mathfrak{g}^*$. On note $S(f, \mathfrak{g})$ l'ensemble des sous-algèbres $\mathfrak{h}$ de $\mathfrak{g}$ subordonnées à f, c'est-à-dire que $\mathfrak{h}$ est un sous-espace totalement isotrope pour la forme bilinéaire alternée B_f sur $\mathfrak{g}$ définie par $B_f(X,Y) = f([X,Y])$.

THEORÈME 1. *Soient $\mathfrak{k}$ un idéal de $\mathfrak{g}$, $f \in \mathfrak{k}^*$ et $\mathfrak{h} \in S(f, \mathfrak{k})$. On note $\mathfrak{h}^{\perp,\mathfrak{k}}$ l'annihilateur de $\mathfrak{h}$ dans $\mathfrak{k}^*$, μ une mesure positive finie sur $\mathfrak{k}^*$ équivalente à la mesure de Lebesgue sur $f + \mathfrak{h}^{\perp,\mathfrak{k}}$ et ν l'image de μ par l'application canonique de $\mathfrak{k}^*$ sur l'espace des G-orbites $\mathfrak{k}^*/G$. Pour ν-presque toutes les orbites $\Omega \in \mathfrak{k}^*/G$:*

(i) *Chaque composante connexe C de $(f + \mathfrak{h}^{\perp,\mathfrak{k}}) \cap \Omega$ est une variété.*
(ii) *L'espace tangent de C au point $l \in C$ est égal à $\mathfrak{g} \cdot 1 \cap \mathfrak{h}^{\perp,\mathfrak{k}}$.*
(iii) *Si l'on désintègre μ par rapport à ν, $\mu = \int \mu_\Omega d\nu(\Omega)$, la mesure de fibre μ_Ω restreinte à une carte $(U; x_1, \dots, x_m)$ de C est équivalente à $dx_1 \dots dx_m$.*

DÉMONSTRATION: La démonstration se fait par récurrence sur $\dim \mathfrak{k}$. Soit $\mathfrak{k}_0$ un idéal de $\mathfrak{g}$ contenu dans $\mathfrak{k}$ tel que $\mathfrak{k}/\mathfrak{k}_0$ soit irréductible en tant que $\mathfrak{g}$-module. On identifie $\mathfrak{k}_0^*$ à un sous-espace de $\mathfrak{k}^*$. Si $\dim \mathfrak{k}/\mathfrak{k}_0 = 1$, on a $\mathfrak{k}^* = \mathbf{R}\xi \oplus \mathfrak{k}_0^*$, $\xi|\mathfrak{k}_0 = 0$, $X \cdot \xi = \lambda(X)\xi$ $(X \in \mathfrak{g})$ avec $\lambda \in \mathfrak{g}^*$ et que $\dim \mathfrak{k}/\mathfrak{k}_0 = 2$, on a $\mathfrak{k}^* = \mathbf{R}\xi_1 \oplus \mathbf{R}\xi_2 \oplus \mathfrak{k}_0^*$, $\xi_1|\mathfrak{k}_0 = \xi_2|\mathfrak{k}_0 = 0$, $X \cdot \xi_1 = \lambda(X)(\xi_1 - \alpha\xi_2)$, $X \cdot \xi_2 = \lambda(X)(\xi_2 + \alpha\xi_1)$ $(X \in \mathfrak{g})$ avec $\alpha \in \mathbf{R}\backslash\{0\}$ et $\lambda \in \mathfrak{g}^*$. L'application de restriction $\mathfrak{k}^* \ni l \to l|\mathfrak{k}_0 \in \mathfrak{k}_0^*$ n'est autre que la projection $p : \mathfrak{k}^* \to \mathfrak{k}_0^*$ longeant cette décomposition et envoie des G-orbites dans $\mathfrak{k}^*$ à celles dans $\mathfrak{k}_0^*$.

Pour $l \in \mathfrak{k}^*$, posons $l_0 = l|\mathfrak{k}_0 \in \mathfrak{k}_0^*$ et regardons l'ensemble $p^{-1}(l_0) \cap G \cdot l = G(l_0) \cdot l$, où $G(l_0)$ dénote le stabilisateur, qui est connexe, de $l_0 \in \mathfrak{k}_0^*$ dans G. On convient de noter $\mathfrak{g}(l_0)$ l'algèbre de Lie de $G(l_0)$.

Si $\mathfrak{g}(l_0)$ est contenue dans le noyau de λ, alors $G(l_0)$ laisse invariante ξ (ou ξ_1, ξ_2). Lorsque $\dim \mathfrak{k}/\mathfrak{k}_0 = 1$, en écrivant $g \cdot l = l + a(g)\xi$ $(g \in G(l_0))$ avec $a(g) \in \mathbf{R}$, la fonction $a : G(l_0) \to \mathbf{R}$ vérifie $a(g_1 g_2) = a(g_1) + a(g_2)$. $G(l_0)$ étant connexe, l'image de l'homomorphisme a se réduit à

$\{0\}$ ou coïncide avec $\mathbf{R}$ tout entier. Lorsque $\dim \mathfrak{k}/\mathfrak{k}_0 = 2$, en posant de même $g \cdot l = l + a_1(g)\xi_1 + a_2(g)\xi_2$ $(g \in G(l_0))$, on se procure deux homomorphismes a_1, $a_2 : G(l_0) \to \mathbf{R}$, ce qui entraîne trois possibilités qui ne dépendent que de l_0, non de l; $p^{-1}(l_0) \cap G \cdot l$ peut être le plan $p^{-1}(l_0)$, une ligne droite ou le point $\{1\}$.

Supposons maintenant que λ ne s'annule pas sur $\mathfrak{G}(l_0)$. Soit $X \in \mathfrak{g}(l_0)$ tel que $\lambda(X) = 1$. Si $\dim \mathfrak{k}/\mathfrak{k}_0 = 1$, la relation $\exp(tX) \cdot l = l + a(t)\xi$ $(t \in \mathbf{R})$ exige $a(t)(e^s - 1) = a(s)(e^t - 1)$ pour tous s, $t \in \mathbf{R}$. D'où $a(t) = c(e^t - 1)$ avec une certaine constante $c \in \mathbf{R}$. Par suite $p^{-1}(l_0) \cap G \cdot l$ peut être le point $\{1\}$, une demi-droite ou la droite $p^{-1}(l_0)$. Si $\dim \mathfrak{k}/\mathfrak{k}_0 = 2$, on pose $\exp(tX) \cdot l = l + a_1(t)\xi_1 + a_2(t)\xi_2$ $(t \in \mathbf{R})$ et calcule $\exp(tX)\exp(\pi X/\alpha) \cdot l = \exp(\pi X/\alpha)\exp(tX) \cdot l$ pour arriver à:

$$a_1(t) + a_2(\pi/\alpha)e^t \sin(\alpha t) + a_1(\pi/\alpha)e^t \cos(\alpha t) = a_1(\pi/\alpha) - e^{\pi/\alpha} a_1(t),$$
$$a_2(t) + a_2(\pi/\alpha)e^t \cos(\alpha t) - a_1(\pi/\alpha)e^t \sin(\alpha t) = a_2(\pi/\alpha) - e^{\pi/\alpha} a_2(t).$$

Il s'ensuit que, pour tout $t \in \mathbf{R}$,

$$a_1(t) = c_1 - e^t(c_1 \cos(\alpha t) + c_2 \sin(\alpha t)),$$
$$a_2(t) = c_2 - e^t(c_2 \cos(\alpha t) - c_1 \sin(\alpha t))$$

avec certaines constantes c_1, $c_2 \in \mathbf{R}$. Cela posé, on en voit que $p^{-1}(l_0) \cap G \cdot l$ peut être le point $\{1\}$, une spirale ou le plan $p^{-1}(l_0)$.

Supposons d'abord $\mathfrak{h} \subset \mathfrak{k}_0$. On note Θ l'ensemble des $l \in \Xi = f + \mathfrak{h}^{\perp,\mathfrak{k}}$ dont G-orbite $G \cdot l \subset \mathfrak{k}^*$ ait la dimension maximum parmi des G-orbites rencontrant Ξ. En remplaçant G par le sous-groupe connexe distingué $G' = \exp \mathfrak{g}'$, $\mathfrak{g}'$ étant le noyau de λ, on introduit de même façon Θ' dans Ξ. Puis on prend $\Xi_0 = f_0 + \mathfrak{h}^{\perp,\mathfrak{k}_0}$, $f_0 = f|\mathfrak{k}_0$, au lieu de Ξ pour fabriquer Θ_0, Θ'_0 dans Ξ_0. Posons maintenant $E = \Theta \cap \Theta' \cap p^{-1}(\Theta_0) \cap p^{-1}(\Theta'_0)$. Il se voit alors que $\Xi \cap G \cdot E \subset E$ et que le complément de E dans Ξ est μ-négligeable, ce qui nous permet de restreindre nos attentions à E. Là, pour toutes les $l \in E$, il se réalise dans les situations examinées ci-dessus la même éventualité de $G(l_0) \cdot l$. En appliquant l'hypothèse de récurrence à Ξ_0, on en déduit les assertions (i) $\sim$ (iii).

Supposons désormais $\mathfrak{h} \not\subset \mathfrak{k}_0$ et posons $\mathfrak{h}_0 = \mathfrak{h} \cap \mathfrak{k}_0$. Soient $\dim \mathfrak{k}/\mathfrak{k}_0 = 1$ et $\mathfrak{h} = \mathbf{R}X \oplus \mathfrak{h}_0$, $\mathfrak{k} = \mathbf{R}X \oplus \mathfrak{k}_0$. On identifie ici $\mathfrak{k}_0^*$ à l'hyperplan $\{l \in \mathfrak{k}^*;\ l(X) = f(X)\}$ et applique l'hypothèse de récurrence à $\Xi_0 = f_0 + \mathfrak{h}_0^{\perp,\mathfrak{k}_0}$. Comme précédemment, nous nous servons de G_0 et Ξ_0 pour construire un ouvert de Zariski E de $f + \mathfrak{h}_0^{\perp,\mathfrak{k}}$. Pourvu que $G(l_0) \cdot l$, $l \in E$, soit une droite, il nous ne reste rien à montrer, non plus dans le cas où $G(l_0) \cdot l$ serait une demi-droite pour $l \in E$ mais un seul point pour $l \in \Xi$. Si encore $G(l_0) \cdot l$, $l \in E$, est une demi-droite et que $\Xi \cap E \neq \emptyset$, il suffit de remarquer que chaque composante connexe de $\Xi \cap G \cdot l$ centenue dans celle de $\Xi_0 \cap G \cdot l_0$, une variété par hypothèse, est une sous-variété ouverte de la dernière.

Examinons le cas essentiel où $G(l_0) \cdot l$ est un seul point pour $l \in E$ quelconque. Fixons $l \in \Xi$ telle que des variétés soient les composantes connexes de $M = \Xi_0 \cap G \cdot l_0 \subset \Xi$, l se notant l_0 en tant qu'un élément de $\mathfrak{k}_0^*$. On définit une fonction ϕ sur M par $\phi(\eta) = g^{-1} \cdot \eta(X)$, l'action de G prise dans $\mathfrak{k}^*$, si $\eta = g \cdot l_0$ dans $\mathfrak{k}_0^*$, i.e. $\eta_0 = g \cdot l_0$. Rappelons que $\Xi \cap G \cdot l = \phi^{-1}(l(X))$. Or, ϕ étant analytique, l'ensemble N des points critiques de ϕ dans une composante connexe de M est négligeable par rapport à la mesure de fibre induite sur M à l'issue de la désintégration de μ, à savoir celle équivalente à la mesure de Lebesgue sur des cartes locales, sinon ϕ serait constante sur cette composante connexe.

D'après le théorème de Sard (cf. [**17**]), l'ensemble W des valeurs critiques de ϕ est négligeable pour la mesure de Lebesgue et l'on sait d'ailleurs que chaque composante connexe de $\Xi \cap G \cdot l = \phi^{-1}(l(X))$ est une sous-variété si $l(X) \in W$. Cela dit, on montre (i) et (iii) car au voisinage d'un point régulier de ϕ on peut prendre ϕ comme une des coordonnées locales.

Quant à (ii), la fonction ϕ considérée ci-dessus se définit sur la variéte $p^{-1}(G \cdot l_0) \subset \mathfrak{k}^*$ et y classe des orbites. En la restreignant à Ξ, on trouve que l'espace tangent de $\Xi \cap G \cdot l = \Xi \cap \phi^{-1}(l(X))$ au point 1, supposé régulier, est égal à $\mathfrak{h}^{\perp,\mathfrak{k}} \cap \mathfrak{g} \cdot l$.

Pour tout ce qui précède, on pourrait même évoquer ce que l'on a, pour $Y \in \mathfrak{g}$ quelconque,

$$\begin{aligned} Y \cdot \phi(\eta) &= Y \cdot (g \cdot l_0(g \cdot X)) = Y \cdot \eta_0(g \cdot X) + \eta([Y, g \cdot X]) \\ &= -\eta_0([Y, (g \cdot X)_0]) + \eta([Y, g \cdot X]) = c\eta([Y, X]) \quad (0 \neq c \in \mathbf{R}), \end{aligned}$$

$(g \cdot X)_0$ désignant la $\mathfrak{k}_0$-composante de $g \cdot X \in \mathfrak{k} = \mathbf{R}X \oplus \mathfrak{k}_0$.

Ces raisonnements s'appliquent bien à d'autres cas. Supposons par exemple que $\dim \mathfrak{k}/\mathfrak{k}_0 = 2$. En appliquant l'hypothèse de récurrence au couple $(f_0, \mathfrak{h}_0)$, on fixe $l \in \Xi$ en position générale comme dans le cas précédent. Si $G(l_0) \cdot l$ coïncide avec le plan $p^{-1}(l_0)$, il ne reste plus rien à montrer. Soit $\dim \mathfrak{h}/\mathfrak{h}_0 = 1$. On voit par hypothèse que $M = \Xi \cap p^{-1}(G \cdot l_0)$ est une variété. Lorsque $G(l_0) \cdot l$ est une droite, soit $l + \mathbf{R}\xi_1^*$. Pour tout $\eta \in M$, $G \cdot \eta$ rencontrant la droite $l + \xi_1^* + \mathbf{R}\xi_2^*$ en un seul point, noté $l + \xi_1^* + \phi(\eta)\xi_2^*$, on fabrique une fonction $\phi : M \to \mathbf{R}$. Lorsque $G(l_0) \cdot l$ est une spirale, il existe dans $p^{-1}(l_0)$ le point de jaillissement, ce qui veut signifier le point unique l_1 tel que $G(l_0) \cdot l_1 = \{l_1\}$. On décrit dans le plan $p^{-1}(l_0)$ un cercle S au centre l_1 et de rayon positif arbitrairement déterminé. Alors pour tout $\eta \in M$, $G \cdot \eta$ intersecte S en un unique point $\phi(\eta)$, ce qui nous donne une application $\phi : M \to S$.

Lorsque $G(l_0) \cdot l = \{l\}$, notre application $\phi : M \to p^{-1}(l_0)$ associe à $\eta \in M$ le point d'intersection $p^{-1}(l_0) \cap G \cdot l$.

Enfin, soit $\dim \mathfrak{h}/\mathfrak{h}_0 = 2$. En identifiant $f_0 + \mathfrak{h}_0^{\perp, \mathfrak{k}_0} \subset \mathfrak{k}_0^*$ avec $\Xi = f + \mathfrak{h}^{\perp, \mathfrak{k}} \subset \mathfrak{k}^*$, on fixe $l \in \Xi$ de manière que $M = \Xi \cap G \cdot l_0$ soit une variété par l'hypothèse de récurrence. Puis on construit exactement comme ci-dessus une application ϕ suivant les cas qui se produisent.

En tout cas il nous suffit d'appliquer à telle ϕ le théorème de Sard cité plus haut.

c.q.f.d.

On met $\mathfrak{g}$ lui-même au lieu de $\mathfrak{k}$ dans le théorème et en garde les notations.

Corollaire. *Soient $f \in \mathfrak{g}^*, \mathfrak{h} \in S(f, \mathfrak{G})$ et $\Xi = f + \mathfrak{h}^\perp$.*

(i) *Pour ν-presque toutes les orbites coadjointes $\Omega \in \mathfrak{g}^*/G$, le support de la mesure μ_Ω est égal à $\Xi \cap \Omega$ tout entier.*

(ii) *Pour ν-presque toutes les orbites coadjointes $\Omega \in \mathfrak{g}^*/G$, chaque composante connexe C de $\Xi \cap \Omega$ est une variété ayant la dimension supérieure ou égale à $1/2 \dim \Omega$.*

(iii) *On a $\dim C = 1/2 \dim \Omega$ si et seulement si $H \cdot l = C$ pour toute $l \in C$. S'il en est ainsi, $\mathfrak{h} + \mathfrak{g}(l)$ est un sous-espace Lagrangien pour B_l.*

Démonstration: L'énoncé (i) est clair. Soit $l \in C$. Voici un simple calcul:

$$\begin{aligned} \dim C &= \dim(\mathfrak{g} \cdot l \cap \mathfrak{h}^\perp) = \dim \mathfrak{h}'/\mathfrak{g}(l) \\ &= \dim \mathfrak{g} - \dim \mathfrak{h} + \dim \mathfrak{g}(l) \cap \mathfrak{h} - \dim \mathfrak{g}(l) \\ &= \dim \mathfrak{g} - \dim(\mathfrak{h} + \mathfrak{g}(l)) \geq \dim \mathfrak{g} - 1/2(\dim \mathfrak{g} + \dim \mathfrak{g}(l)) \\ &= 1/2 \dim \Omega, \end{aligned}$$

$\mathfrak{h}'$ se notant l'espace orthogonal à $\mathfrak{h}$ par rapport à la forme bilinéaire B_l. Là, l'égalité s'obtient si et suelement si $\mathfrak{h} + \mathfrak{g}(l)$ est un sous-espace Lagrangien pour B_l, ce qui entraîne

$$\begin{aligned} \dim H \cdot l &= \dim \mathfrak{h}/(\mathfrak{g}(l) \cap \mathfrak{h}) = \dim \mathfrak{h} - \dim \mathfrak{g}(l) \cap \mathfrak{h} \\ &= \dim(\mathfrak{h} + \mathfrak{g}(l)) - \dim \mathfrak{g}(l) = 1/2 \dim \Omega. \end{aligned}$$

D'où viennent les assertions (ii) et (iii).

c.q.f.d.

2. Pour donner la désintégration des représentations monomiales de G, notre méthode sera différente de celle de Lipsman [**14**]. Dans ce qui suit, on confondra parfois les classes d'équivalence des représentations unitaires avec leurs représentations et liera deux représentations équivalentes par le symbole $\simeq$ ou même par le signe d'égalité. Avant d'énoncer le théorème, on se prépare un lemme concernant des groupes à petite dimension.

LEMME. *(i) Soit* $G = \exp \mathfrak{g}_2, \mathfrak{g}_2 = \langle X, Y\rangle_{\mathbf{R}} = \mathbf{R}X + \mathbf{R}Y : [X,Y] = Y$. *Soient* $f \in \mathfrak{g}_2^*, \mathfrak{h} = \mathbf{R}X$ *et* $H = \exp \mathfrak{h}$. *Alors* $\mathrm{ind}_H^G \chi_f \simeq \mathrm{ind}_H^G \chi_{Y^*} \oplus \mathrm{ind}_{H'}^G \chi_{-Y^*}$ avec $H' = \exp(\mathbf{R}Y)$.

(ii) Soit $G = \exp \mathfrak{g}_3(\alpha), \mathfrak{g}_3(\alpha) = \langle T, Y_1, Y_2\rangle_{\mathbf{R}} : [T, Y_1] = Y_1 - \alpha Y_2, [T, Y_2] = Y_2 + \alpha Y_1$ $(0 \neq \alpha \in \mathbf{R})$. *Soient* $f \in \mathfrak{g}_3(\alpha)^*, \mathfrak{h} = \mathbf{R}T$ *et* $H = \exp \mathfrak{h}$. *Alors*

$$\mathrm{ind}_H^G \chi_f \simeq \int_{[0,2\pi]}^{\oplus} \mathrm{ind}_{H'}^G \chi_{\hat\theta} d\theta$$

avec $H' = \exp(\mathbf{R}Y_1 + \mathbf{R}Y_2), \hat\theta = (\cos\theta)Y_1^* + (\sin\theta)Y_2^* \in \mathfrak{g}_3(\alpha)^*$.

(iii) Soient $G = \exp \mathfrak{g}_4, \mathfrak{g}_4 = \langle T, X, Y, Z\rangle_{\mathbf{R}} : [X,Y] = Z, [T,X] = -X, [T,Y] = Y$. *Soient* $f = \alpha T^* + \beta Z^* \in \mathfrak{g}_4^*$ $(\beta \neq 0), \mathfrak{h} = \langle T, X, Z\rangle_{\mathbf{R}}$ *et* $H = \exp \mathfrak{h}$. *Alors* $\mathrm{ind}_H^G \chi_f \simeq \mathrm{ind}_{H'}^G \chi_f$ avec $H' = \exp \mathfrak{h}'$, $\mathfrak{h}' = \langle T, Y, Z\rangle_{\mathbf{R}}$.

(iv) Soit $G = \exp \mathfrak{g}_6, \mathfrak{g}_6 = \langle T, X_1, X_2, Y_1, Y_2, Z\rangle_{\mathbf{R}} : [X_i, Y_j] = \delta_{ij} Z$ $(1 \leq i,j \leq 2), [T, X_1] = -X_1 - \alpha X_2, [T, X_2] = -X_2 + \alpha X_1, [T, Y_1] = Y_1 - \alpha Y_2, [T, Y_2] = Y_2 + \alpha Y_1$ $(0 \neq \alpha \in \mathbf{R})$. *Soient* $f = \beta T^* + \gamma Z^* \in \mathfrak{g}_6^*$ $(\gamma \neq 0)$, $\mathfrak{h} = \langle T, X_1, X_2, Z\rangle_{\mathbf{R}}$ *et* $H = \exp \mathfrak{h}$. *Alors* $\mathrm{ind}_H^G \chi_f \simeq \mathrm{ind}_{H'}^G \chi_f$ avec $H' = \exp \mathfrak{h}'$, $\mathfrak{h}' = \langle T, Y_1, Y_2, Z\rangle_{\mathbf{R}}$.

DÉMONSTRATION: Ces faits sont bien connus (cf. [**4**], [**19**]), peut-être sauf (ii) que l'on va voir ici pour plus de sûreté. Il suffit pour cela que l'on continue un peu plus loin des observations faites à la page 135 de [**4**]. On y emprunte certaines notations.

Au moyen de la bijection $\xi Y_1 + \eta Y_2 \mapsto \xi + i\eta$, $i = (-1)^{-1/2}$, on identifie $\mathfrak{a} = \mathbf{R}Y_1 \oplus \mathbf{R}Y_2$ au plan C des nombres complexes. Soit $\lambda = f(T)$. La représentation monomiale $\tau = \mathrm{ind}_H^G \chi_f$ se réalise dans $L^2(\mathbf{C}) = L^2(\mathbf{R}^2)$ par la formule suivante: pour $\xi_0 \in \mathbf{R}$, z_0, $z \in \mathbf{C}$ et $\phi \in L^2(\mathbf{C})$,

$$\tau(\exp(\xi_0 T)\exp z_0)\phi(z) = e^{\xi_0(i\lambda - 1)}\phi(ze^{-\xi_0(1-i\alpha)} - z_0).$$

Soint $\mathcal{F}$ l'isomorphisme de Fourier de $L^2(\mathbf{C})$, $(\ |\ )$ le produit scalaire de $\mathbf{R}^2 = \mathbf{C}$ et dz la mesure de Lebesgue de $\mathbf{R}^2$. Un calcul direct vérifie: pour $\hat g = \exp(\xi_0 T)\exp z_0 \in G$, $\psi \in L^2(\mathbf{C})$ et $v \in \mathbf{C}$,

$$\mathcal{F}(\tau(\hat g)\psi)(v) = e^{\xi_0(i\lambda+1) + i(v|e^{\xi_0(1-i\alpha)}z_0)}\mathcal{F}\psi(e^{\xi_0(1+i\alpha)}v).$$

Or, à l'aide de la base duale $\{Y_1^*, Y_2^*\}$ de $\mathfrak{a}^*$ et de la bijection $\xi Y_1^* + \eta Y_2^* \mapsto \xi + i\eta$, $\mathfrak{a}^*$ s'identifie à $\mathbf{C}$. Alors $\xi' + i\eta' = \exp(tT)\cdot(\xi + i\eta)$, $t \in \mathbf{R}$, se calcule par

$$\xi' = e^{-t}(\xi\cos(\alpha t) + \eta\sin(\alpha t)),\ \eta' = e^{-t}(\xi\sin(-\alpha t) + \eta\cos(\alpha t)).$$

On en constante que l'orbite coadjointe Ω_θ passant par le point $e^{i\theta}$ s'obtient comme $e^{-t+i(\theta-\alpha t)}$, t décrivant $\mathbf{R}$, et que e^{-2t} est le déterminant fonctionnelle de la transformation des variables $(\xi', \eta') \mapsto (t, \theta)$.

Posons $\tau_1 = \mathcal{F} \circ \tau \circ \mathcal{F}^{-1}$. Comme, pour $\hat{g} = \exp(\xi_0 T)\exp z_0 \in G$,

$$\begin{aligned}&\tau_1(\hat{g})\phi(e^{-t+i(\theta-\alpha t)})\\ &= e^{\xi_0(1+i\lambda)+i(e^{-t+i(\theta-\alpha t)}|e^{\xi_0(\theta-\alpha t)}z_0)}\phi(e^{(-t+\xi_0)(1+i\alpha)+i\theta}),\end{aligned}$$

τ_1 laisse stable $L^2(\Omega_\theta)$. Cela nous donne une représentation $\hat{\tau}$ de G, qui agit par la formule

$$\hat{\tau}(g)\Psi(t) = e^{\xi_0(1+i\lambda)+i(e^{i\theta}|e^{(\xi_0-t)(1-i\alpha)}z_0)}\Psi(t-\xi_0)$$

dans l'espace des fonctions Ψ vérifiant

$$\int_{\mathbf{R}} e^{-2t} \mid \Psi(t) \mid^2 dt < +\infty.$$

Toutefois $\hat{\tau}$ n'est autre que $\pi_\theta = \mathrm{ind}_{H'}^G\, \chi_{\hat{\theta}}$. En effet,

$$\pi_\theta(\hat{g})\phi(x) = e^{i(e^{i\theta}|e^{(\xi_0-x)(1-i\alpha)}z_0)}\phi(x-\xi_0)\ \ (x \in \mathbf{R})$$

pour $\hat{g} \in G$ comme avant et $\phi \in L^2(\mathbf{R})$. Finallement on retrouve $\hat{\tau}$ en transférant π_θ par une application qui à $\phi \in L^2(\mathbf{R})$ associe $\Phi \in L^2(\mathbf{R}; e^{-2x}dx)$, $\Phi(x) = e^{(1+i\lambda)x}\phi(x)$, ce qui achève la démonstration.

c.q.f.d.

Revenant au cas général, nous reprenons les notations précédentes: Soient $G = \exp \mathfrak{g}$ un groupe de Lie résoluble exponentiel, $f \in \mathfrak{g}^*$, $\mathfrak{h} \in S(f, \mathfrak{g})$, $H = \exp \mathfrak{h}$ et $\tau = \mathrm{ind}_H^G\, \chi_f$. Soient encore μ une mesure positive finie sur $\Xi = f + \mathfrak{h}^\perp \subset \mathfrak{g}^*$ équivalente à la mesure de Lebesgue et ν son image par l'application de Kirillov-Bernat.

THÉORÈME 2. *La désintégration de τ s'écrit*

$$\tau \simeq \int_{\hat{G}}^{\oplus} m(\pi)\pi d\nu(\pi)$$

avec la fonction de multiplicités donnée à la façon suivante: $m(\pi)$ *est le nombre des composantes connexes de* $\Xi \cap \Omega(\pi)$ *si chaque composante est une variété de dimension égale à* $1/2 \dim \Omega(\pi)$. *Lorsque cette condition n'est pas remplie,* $m(\pi)$ *est égale à* $+\infty$. *En tout cas* $m(\pi)$ *s'obtient comme le nombre des* H*-orbites contenues dans* $\Xi \cap \Omega(\pi)$.

DÉMONSTRATION: Cela se fait par récurrence sur $\dim G$. En premier lieu, suppossons $\mathfrak{g} \neq \mathfrak{h} + [\mathfrak{g}, \mathfrak{g}]$. Il existe alors un idéal $\mathfrak{g}_0$ contenant $\mathfrak{h}$ et de codimension un dans $\mathfrak{g}$. Posons $G_0 = \exp \mathfrak{g}_0$ auquel s'applique l'hypothèse de récurrence. Par rapport à la restriction $p : \mathfrak{g}^* \to \mathfrak{g}_0^*$, une orbite coadjointe Ω de G est ou bien saturée, ou bien non-saturée et il se passe μ-presque partout la même éventualité de cette alternative (cf. [9], [**16**]). Si Ω est saturée , $p(\Omega)$ se compose d'une famille à un paramètre $\{\omega_t\}_{t\in\mathbf{R}}$ de G_0-orbites, $\dim \Omega = \dim \omega_t + 2$ et $\overline{\theta}_G(\Omega) = \mathrm{ind}_{G_0}^G \overline{\theta}_{G_0}(\omega_t)$ pour tout $t \in \mathbf{R}$. Par contre, si Ω est non-saturée, $\omega = p(\Omega)$ est une G_0-orbite, p donne un difféomorphisme de Ω sur ω, $p^{-1}(\omega)$ se compose d'une famille à un paramètre $\{\Omega_t\}_{t\in\mathbf{R}}$ de G-orbites et que

$$\mathrm{ind}_{G_0}^G \overline{\theta}_{G_0}(\omega) = \int_{\mathbf{R}}^{\oplus} \overline{\theta}_G(\Omega_t)dt.$$

En ce qui concerne la mesure, notre assertion s'est déjà établie (cf. ibid.). Quant aux multiplicités, on raisonne comme suit. Dans le cas où presque toutes les orbites sont non-saturées, il suffit d'appliquer l'hypothèse de récurrence, compte tenu de ce que l'on a vu ci-dessus. Si l'on n'a presque partout que des orbites saturées, posons $f_0 = p(f) \in \mathfrak{g}_0^*$ et $\Xi_0 = f_0 + \mathfrak{h}^{\perp,\mathfrak{g}_0}$. Soit

$$\tau_0 = \int_{\hat{G}_0}^{\oplus} m_0(\pi_0)\pi_0 d\nu_0(\pi_0)$$

la désintégration centrale canonique de $\tau_0 = \mathrm{ind}_H^{G_0} \chi_{f_0}$. Le sous-groupe G_0 étant distingué, G agit dans $\hat{G}_0$ de la façon usuelle et à un ensemble ν-négligeable près $\hat{G}$ s'identifie à l'espace quotient $\hat{G}_0/G$. Or la désintégration de ν_0 relative à ν, qui est l'image de ν_0 par la projection $q : \hat{G}_0 \to \hat{G}_0/G$ mène à

$$\begin{aligned}\tau = \mathrm{ind}_{G_0}^G \tau_0 &= \int_{\hat{G}_0}^{\oplus} m_0(\pi_0)(\mathrm{ind}_{G_0}^G \pi_0)d\nu_0(\pi_0) \\ &= \int_{\hat{G}_0}^{\oplus} m_0(\pi_0) \int_{\hat{G}_0}^{\oplus} m_0(\pi_0)(\mathrm{ind}_{G_0}^G \pi_0)d\nu_0^{\pi}(\pi_0),\end{aligned}$$

où ν_0^π est portée par $q^{-1}(\pi)$. D'après ce que l'on vient de voir, la représentation

$$\int_{\hat{G}_0}^{\oplus} m_0(\pi_0)(\mathrm{ind}_{G_0}^G \pi_0) d\nu_0^\pi(\pi_0)$$

est factorielle et

$$m(\pi)\pi = \int_{\hat{G}_0}^{\oplus} m_0(\pi_0)(\mathrm{ind}_{G_0}^G \pi_0) d\nu_0^\pi(\pi_0)$$

pour ν-presque toute π. Il en résulte que $m(\pi)$ est finie pour telle π si et seulement si ν_0^π est portée des points π_0 en nombre fini tels que $m_0(\pi_0)$ soit finie. Celle-ci est donc donnée par le nombre de composantes connexes de $\Xi_0 \cap \Omega_0(\pi_0)$, $\Omega_0(\pi_0)$ désignant la G_0-orbite $\overline{\theta}_{G_0}^{-1}(\pi_0)$.

En combinant cela avec (i) du corollaire, on déduit que $q^{-1}(\pi)$ est un ensemble fini et que

$$m(\pi) = \sum_{\pi_0 \in q^{-1}(\pi)} m_0(\pi_0).$$

Pour chaque composante connexe C de $\Xi \cap \Omega(\pi)$, $p(C)$ coïncide donc avec une composante connexe de $\Xi_0 \cap \Omega_0(\pi_0)$ pour une certaine $\pi_0 \in q^{-1}(\pi)$. Cette composante-ci étant une variété de dimension $1/2 \dim \Omega_0(\pi_0)$, C l'est aussi de dimension attendue. Enfin il vient que $m(\pi)$ fournit le nombre des composantes connexes de $\Xi \cap \Omega(\pi)$.

Supposons désormais $\mathfrak{g} = \mathfrak{h} + [\mathfrak{g}, \mathfrak{g}]$. Remarquons qu'il suffit de démontrer le théorème pour $\mathfrak{h}$ contenant le centre $\mathfrak{z}$ de $\mathfrak{g}$. D'ailleurs, si $\mathfrak{h}$ contient un idéal $\mathfrak{a}$ de $\mathfrak{g}$ sur lequel f s'annule, on n'a qu'à appliquer l'hypothèse de récurrence au groupe quotient $G/\exp \mathfrak{a}$. Tout sera supposé dans ce qui suit. Il s'ensuit que $\dim \mathfrak{z} \leq 1$. Soit $\mathfrak{a}$ un idéal non central minimal de $\mathfrak{g}$. On en trouve que $\mathfrak{a}$ est abélien et que $\dim \mathfrak{a} \leq 3$. L'action adjointe de $\mathfrak{g}$ sur $\mathfrak{a}/\mathfrak{z} \cap \mathfrak{a}$ nous fournit d'une racine $\lambda \in \mathfrak{g}^*$ et il existe toujours $T \in \mathfrak{h}$, on en prendra un, vérifiant $\lambda(T) = 1$. Le noyau de λ, noté $\mathfrak{g}_0$, est un idéal de $\mathfrak{g}$ et l'on a $\mathfrak{g} = \mathbf{R}T \oplus \mathfrak{g}_0$. Maintenant allons examiner différents cas. Soit $A = \exp \mathfrak{a}$.

(I) Supposons que $\mathfrak{a} \cap \mathfrak{z} = \{0\}$. On en trouve que $\mathfrak{H} \cap \mathfrak{a} = \{0\}$. Soient $\mathfrak{h}' = \mathfrak{h} \cap \mathfrak{g}_0 + \mathfrak{a}$ et $H' = \exp \mathfrak{h}'$. Raisonnant au niveau du sous-groupe $K = \exp \mathfrak{k}$, $\mathfrak{k} = \mathfrak{h} + \mathfrak{a}$, on voit aussitôt que la situation est tout à fait pareille à celle de (i) ou (ii) du lemme conformément à la dimension de $\mathfrak{a}$.

Si $\dim \mathfrak{a} = 1$, soient $\mathfrak{a} = \mathbf{R}Y$, $f_\pm \in \Xi$ telles que $f_\pm(Y) = \pm 1$. Il est évident que $\mathfrak{h}' \in S(f_\pm, \mathfrak{g})$. Puisque $\mathfrak{g} \neq \mathfrak{h}' + [\mathfrak{g}, \mathfrak{g}]$, le théorème s'est déjà établi pour les couples $(f_\pm, \mathfrak{h}')$. Donc

$$\begin{aligned}\tau &\simeq \mathrm{ind}_{H'}^G \chi_{f_+} \oplus \mathrm{ind}_{H'}^G \chi_{f_-} \\ &= \int_{\hat{G}}^{\oplus} m_+(\pi)\pi d\nu_+(\pi) \oplus \int_{\hat{G}}^{\oplus} m_-(\pi)\pi d\nu_-(\pi)\end{aligned} \tag{1}$$

avec les notations sous-entendues. Posons $P_\pm = \{l \in \mathfrak{g}^*; l(Y) \gtrless 0\}$, $\Xi'_\pm = f_\pm + \mathfrak{h}'^\perp$, $\Xi^0_\pm = f_\pm + \mathfrak{k}^\perp$ et $\Xi_\pm = \Xi \cap P_\pm$. Alors $\Xi_\pm = \exp(\mathbf{R}T)\cdot\Xi^0_\pm$, $\Xi'_\pm = A\cdot\Xi^0_\pm$ et $P_\pm$ sont G-invariants, se qui fait de (1) le résultat cherché.

Si $\dim \mathfrak{a} = 2$, soient Y_1, $Y_2 \in \mathfrak{a}$ tels que les trois éléments T, Y_1, Y_2 satisfassent aux relations de crochet données dans (ii) du lemme. Soit $\hat{\theta} \in \Xi$ tel que $\hat{\theta} \mid \mathfrak{a} = (\cos\theta)Y_1^* + (\sin\theta)Y_2^*$. Trivialement $\mathfrak{h}' \in S(\hat{\theta}, \mathfrak{g})$. Comme précédemment,

$$\tau \simeq \int_{[0,2\pi]}^{\oplus} \mathrm{ind}_{H'}^G \chi_{\hat{\theta}} d\theta = \int_{[0,2\pi]}^{\oplus} d\theta \int_{\hat{G}}^{\oplus} m_\theta(\pi)\pi d\nu_\theta(\pi). \tag{2}$$

Posons $\Xi'_\theta = \hat{\theta} + \mathfrak{h}'^\perp$ et $\Xi^0_\theta = \hat{\theta} + \mathfrak{k}^\perp$. Prenons maintenant $\Omega \in \mathfrak{g}^*/G$ vérifiant $l \mid \mathfrak{a} \neq 0$ $(l \in \Omega)$ et donc $\dim(\mathfrak{g}(l) \cap \mathfrak{a}) = 1$. Pour une telle Ω il existe un unique $\theta \in [0, 2\pi)$ tel que $\Xi \cap \Omega = \exp(\mathbf{R}T) \cdot (\Xi^0_\theta \cap \Omega)$, $\Xi'_\theta \cap \Omega = A \cdot (\Xi^0_\theta \cap \Omega)$. Compte tenu de ces observations, la formule (2) s'interprète comme le résultat attendu.

(II) Supposons que $\mathfrak{a} \cap \mathfrak{z} = \mathfrak{z} \neq \{0\}$. Soit d'abord $\mathfrak{a} \subset \mathfrak{h}$. On introduit le propre sous-algèbre $\mathfrak{G}_1 = \mathfrak{a}^f$ qui contient $\mathfrak{h}$. D'après l'hypothèse de récurrence appliquée au sous-groupe $G_1 = \exp \mathfrak{g}_1$, il vient

$$\tau_1 = \mathrm{ind}_H^{G_1} \chi_f \simeq \int_{\hat{G}_1}^{\oplus} m_1(\rho)\rho d\nu_1(\rho).$$

Ensuite

$$\tau = \mathrm{ind}_{G_1}^G \tau_1 \simeq \int_{\hat{G}_1}^{\oplus} m_1(\rho)(\mathrm{ind}_{G_1}^G \rho) d\nu_1(\rho). \tag{3}$$

Ici $\rho|A$ étant un multiple de $\chi_f|A$, ces $\mathrm{ind}_{G_1}^G \rho$ sont toutes irréductibles et inéquivalentes l'une à l'autre. D'autre part $A \cdot l = l + (\mathfrak{a}^l)^\perp$ pour n'importe quelle $l \in \mathfrak{g}^*$. Au bout de compte, c'est la formule (3) que l'on voulait obtenir.

On se place maintenant dans le cas où $\mathfrak{h} \cap \mathfrak{a} = \mathfrak{z}$. Supposons en premier lieu que $\mathfrak{h}_0 = \mathfrak{h} \cap \mathfrak{G}_0 \subset \mathfrak{g}_1$. Car $[\mathfrak{h}_0, \mathfrak{a}] = \{0\}$, $\mathfrak{h}' \in S(f_\pm, \mathfrak{g})$ ou $\mathfrak{h}' \in S(\hat{\theta}, \mathfrak{g})$, ce qui dépend de $\dim \mathfrak{a}$, justement comme dans les cas qui précèdent. Les arguments faits plus haut nous amènent à la désintégration (1) si $\dim \mathfrak{a} = 2$, ou à (2) si $\dim \mathfrak{a} = 3$. Mais cette fois une complication se produit. Soient $\dim \mathfrak{a} = 2$ et $Y \in \mathfrak{a}$ tel que $[T,Y] = Y$. A la formule (1) les deux intégrales

$$\int_{\hat{G}}^{\oplus} m_+(\pi)\pi d\nu_+(\pi) \qquad \text{et} \qquad \int_{\hat{G}}^{\oplus} m_-(\pi)\pi d\nu_-(\pi)$$

peuvent avoir des mêmes composantes irréductibles, car les deux ensembles $P_\pm$, on garde les notations introduites là-bas, ne sont plus G-invariants. En passant au quotient par l'action de G, on montre l'assertion concernant la mesure.

Quant aux multiplicités,

$$m(\pi) = m_+(\pi) + m_-(\pi)$$

d'après (i) du corollaire. Soit $\Omega \in \mathfrak{g}^*/G$ telle que chaque composante connexe C' de $\Xi'_\pm \cap \Omega$ est une H'-orbite et que celle de $\Xi \cap \Omega$ est une variété. On en déduit que $\exp(\mathbf{R}T)\cdot(\Xi^0_\pm \cap C') = H \cdot l$, $l \in \Xi^0_\pm \cap C'$, est la composante connexe de $\Xi \cap \Omega$ contenant $\Xi^0_\pm \cap C'$, ce qui veut dire qu'un tel $H \cdot l$ dans $\Xi_+ \cap \Omega$ et un autre $H \cdot l'$ dans $\Xi_- \cap \Omega$ ne se trouvent pas dans une même composante connexe de $\Xi \cap \Omega$. En effet, cette dernière doit être par hypothèse une variété, sa dimension étant bien nécessairement $1/2 \dim \Omega$. D'après (iii) du corollaire, on a le résultat ν-presque partout. Il en découle finallement que, pour une telle Ω, les composantes connexes de $\Xi \cap \Omega$ contenues dans l'hyperplan $\{l \in \mathfrak{g}^*;\ l(Y) = 0\}$ ne comptent pas à cause de (i) du corollaire, ce qui achève la démonstration dans ce cas.

Soint $\dim \mathfrak{a} = 3$ et $Y_1, Y_2 \in \mathfrak{a}$ tels que les éléments T, Y_1, Y_2 vérifient les relations de crochet dans (ii) du lemme. On sait déjà que

$$\tau \simeq \int_{[0,2\pi)}^{\oplus} d\theta \int_{\hat{G}}^{\oplus} m_\theta(\pi)\pi d\nu_\theta(\pi).$$

Mais des éléments de Ξ'_θ et ceux de Ξ'_{θ_2}, $\theta_1 \neq \theta_2$, se mêlent par l'action de G. Même s'il en est ainsi, l'énoncé sur la mesure se vérifie par passage au quotient suivants la relation d'équivalence.

A propos des multiplicités, $\Theta(\pi)$ se notant l'ensemble de $\theta \in [0, 2\pi)$ tels que $\Xi'_\theta \cap \Omega(\pi) \neq \varnothing$, le fait (i) du corollaire implique que

$$m(\pi) = \sum_{\theta \in \Theta(\pi)} m_\theta(\pi)$$

si $\Theta(\pi)$ est un ensemble dénombrable, sinon $m(\pi) = +\infty$. Examinons la première éventualité un peu plus loin. On la suppose et soit C une composante connexe de $\Xi \cap \Omega(\pi)$. Pour toute $l \in C$, l resstreinte à $V = \mathbf{R}Y_1 \oplus \mathbf{R}Y_2$ appartient à

$$\bigcup_{\theta \in \Theta(\pi)} \exp(\mathbf{R}T) \cdot (\theta | V),$$

sinon $l|V = 0$. Supposons en plus que chaque composante connexe C' de $\Xi'_\theta \cap \Omega(\pi)$, $\theta \in \Theta(\pi)$, est une H'-orbite, i.e. $C' = H' \cdot l'$ avec $l' \in C'$ vérifiant $l'(T) = f(T)$, et que celle de $\Xi \cap \Omega(\pi)$ est une variété. Exactement comme dans le cas précédent, on constate que $H \cdot l'$ est la composante connexe de $\Xi \cap \Omega(\pi)$ contenant l' et que celles inclues dans le sous-espace $\{l \in \mathfrak{g}^*; \; l|V = 0\}$ sont négligeables. De tout ce qu'on vient de voir on gagne la formule de multiplicités.

Allons envisager le dernier cas où $\mathfrak{h}_0 \not\subset \mathfrak{g}_1$. Notre situation sera essentiellement analogue à celle décrite en (iii) ou (iv) du lemme. Si $\dim \mathfrak{a} = 2$, on prend $Z \in \mathfrak{z}$, $Y \in \mathfrak{a} \setminus \mathfrak{z}$ et $X \in \mathfrak{h}_0$ de sorte que les éléments T, X, Y, Z satisfassant aux relations de crochet similaires à celles définies en (iii) du lemme. On y remplace seulement $[T, X] = -X$ par $[T, X] \equiv -X \bmod (\mathfrak{g}_0 \cap \mathfrak{g}_1)$. De même, si $\dim \mathfrak{a} = 3$, on choisit $Z \in \mathfrak{z}$, Y_1, $Y_2 \in \mathfrak{a} \setminus \mathfrak{z}$ et X_1, $X_2 \in \mathfrak{h}_0$ de manière que ces éléments plus T vérifient les relations analogues à celles dans (iv) du lemme. On y exige les égalités $[T, X_1] = -X_1 - \alpha X_2$, $[T, X_2] = -X_2 + \alpha X_1$ modulo le sous-algèbre $\mathfrak{g}_0 \cap \mathfrak{g}_1$.

Soient $\mathfrak{h}' = \mathfrak{h} \cap \mathfrak{g}_1 + \mathfrak{a}$ et $H' = \exp \mathfrak{h}'$. Evidemment $\mathfrak{h}' \in S(f, \mathfrak{g})$. En raisonnant au niveau du sous-groupe $K = \exp \mathfrak{k}$, $\mathfrak{k} = \mathfrak{h} + \mathfrak{a}$, on prouve (cf. [**4**])

$$\tau \simeq \mathrm{ind}_{H'}^G \chi_f$$

justement comme dans le lemme. On est ainsi ramené au cas déjà examiné et par conséquent, l'énoncé du théorème étant applicable à $\mathfrak{h}' \in S(f, \mathfrak{g})$,

$$\tau \simeq \mathrm{ind}_{H'}^G \chi_f = \int_{\hat{G}}^{\oplus} m'(\pi) \pi \, d\nu'(\pi).$$

Pour terminer la démonstration du théorème il ne nous reste plus qu'à remarquer une chose. Etant donnée $\Omega \in \mathfrak{g}^*/G$ quelconque, chaque composante connexe C de $\Xi \cap \Omega$ s'écrit sous la forme $C = \exp U \cdot C_0$. Là, $U = \mathbf{R}X$ ou $U = \mathbf{R}X_1 \oplus \mathbf{R}X_2$ et C_0 désigne une composante connexe de $\Xi_0 \cap \Omega$ avec $\Xi_0 = f + \mathfrak{k}^\perp$. De même pour des composantes connexes C' de $\Xi' \cap \Omega$, $\Xi' = f + \mathfrak{h}'^\perp$; $C' = \exp V \cdot C_0$.

c.q.f.d.

3. Concernant notre représentation monomiale $\tau = \mathrm{ind}_H^G \chi_f$, nous laissons deux questions qui remontent à Penney [**15**]. Pour une représentation unitaire ρ de G, on notera $\mathcal{H}_\rho$ son espace de Hilbert, $\mathcal{H}_\rho^{+\infty}$ l'espace des vecteurs C^∞ muni de la topologie habituelle, et $\mathcal{H}_\rho^{-\infty}$ son antidual. Etant donnés un sous-groupe fermé K de G et son caractère c, nous posons

$$(\mathcal{H}_\rho^{-\infty})^{K,c} = \{a \in \mathcal{H}_\rho^{-\infty}; \rho(k)a = c(k)a,\ k \in K\}.$$

Reprenons notre représentation monomiale

$$\tau \simeq \mathrm{ind}_H^G \chi_f = \int_{\hat{G}}^{\oplus} m(\pi)\pi d\nu(\pi).$$

En désignant par e l'élément neutre de G, par Δ_G la fonction module de G, nous voyons que la mesure de Dirac

$$\delta_\tau : \mathcal{H}_\tau^{+\infty} \ni \psi \mapsto \overline{\psi(e)} \in \mathbf{C}$$

définit un élément de $(\mathfrak{H}_\tau^{-\infty})^{H,\chi_f \Delta_{H,G}^{1/2}}$, où $\Delta_{H,G} = \Delta_H / \Delta_G$. Alors, suivant la désintégration de τ, δ_τ s'écrit

$$\delta_\tau = \int_{\hat{G}}^{\oplus} \left(\sum_{k=1}^{m(\pi)} a_\pi^k \right) d\nu(\pi)$$

avec $a_\pi^k \in (\mathcal{H}_\pi^{-\infty})^{H,\chi_f \Delta_{H,G}^{1/2}}$ (cf. [5], [15]). Ce qui veut dire que, pour toute $\phi \in C_c^\infty(G)$,

$$\phi_H^f(e) = \int_{\hat{G}} \sum_{k=1}^{m(\pi)} \langle \pi(\phi) a_\pi^k, a_\pi^k \rangle d\nu(\pi),$$

où $\phi_H^f(g) = \int_H \phi(gh)\chi_f(h)\Delta_{H,G}^{-1/2}(h)dh$ avec une mesure de Haar dh sur H.

Cela poseé, voici nos questions.

(1) Réciprocité: Peut-on choisir $\dim(\mathcal{H}_\pi^{-\infty})^{H,\chi_f \Delta_{H,G}^{1/2}}$ pour la multiplicité $m(\pi)$?
(2) Formule de Plancherel concrète: Explicitez les a_π^k intervenant dans la désintégration de δ_τ.

Depuis que les travaux de Benoist [2], [3] nous ont incités à étudier ces questions, nous en avons envisagé certains cas (cf. [7], [8]). Dans toute le suite de ces notes nous allons y ajouter quelques exemples.

EXEMPLE 1. $G = G_3(\alpha) = \exp \mathfrak{g}_3(\alpha)$, $\mathfrak{g}_3(\alpha) = \langle T, Y_1, Y_2\rangle_{\mathbf{R}} : [T, Y_1] = Y_1 - \alpha Y_2$, $[T, Y_2] = \alpha Y_1 + Y_2$. Soient $f = Y_1^* \in \mathfrak{g}_3(\alpha)^*$ et $\mathfrak{h} = \mathbf{R}Y_1$. Ici α peut être supposé négatif. On a l'expression paramétrée de l'orbite passant $l = (1, \lambda) \in \mathfrak{a}^* = \mathbf{R}Y_1^* \oplus \mathbf{R}Y_2^*$; y utilisant (x, y)-coordonneés

$$x(t) = e^t(\cos(\alpha t) - \lambda \sin(\alpha t)), \tag{1}$$
$$y(t) = e^t(\sin(\alpha t) + \lambda \cos(\alpha t)). \tag{2}$$

Si la ligne directe $x = 1$ est tangente à l'orbite de l, on voit $(dx/dt)_{t=0} = 1 - \alpha\lambda = 0$, ce qui donne $\lambda = 1/\alpha$, noté λ_0. Soit t^* le premier nombre positif t vérifiant $e^t(\cos(\alpha t) - (1/\alpha)\sin(\alpha t)) = 1$, et la y-coordonnée du point d'intersection se notant λ_1, nous prenons $l = (1, \lambda) \in \mathfrak{a}^*$, $\lambda_0 \leq \lambda < \lambda_1$, comme représentants des orbites qui rencontrent l'espace affine $\Xi = f + \mathfrak{h}^\perp$. Ces orbites sont toutes saturées pour $\mathbf{R}T^*$.

Nous utilisons la paramétrisation de l'orbite, et cherchons, en posant $e^t(\cos(\alpha t) - \lambda \sin(\alpha t)) = 1$, des points d'intersection avec Ξ. D'où $e^t(1 + \lambda^2)^{1/2} \cos(\alpha t + \theta) = 1$ avec θ tel que $\sin\theta = \lambda(1 + \lambda^2)^{-1/2}, \cos\theta = (1+\lambda^2)^{-1/2}$, ci qui entraîne $e^t(\sin(\alpha t) + \lambda\cos(\alpha t)) = (1+\lambda^2)^{1/2} e^t \sin(\alpha t + \theta) = \tan(\alpha t + \theta)$. On en trouve les points $l_n = (1, \tan(\alpha t_n + \theta))$, $l_0 = l$. Choisissons arbitrairement des $g_n \in G$ tels que $g_n : l \to l_n$, et fabriquons l'application

$$a_n : \phi \mapsto \oint_{H/H\cap g_n B g_n^{-1}} \overline{\phi(hg_n)\chi_f(h)} \Delta_{H,G}^{-1/2}(h) d\dot{\nu}(h), \quad \phi \in \mathcal{H}_\pi, \tag{$*$}$$

ce dernier n'est autre que $\overline{\phi(g_n)}$, ici l'on a noté $B = \exp \mathfrak{b}$, associé à la polarisation $\mathfrak{b} = \mathbf{R}Y_1 \oplus \mathbf{R}Y_2$, $\pi = \pi_l = \mathrm{ind}_B^G \chi_l$ dont l'espace se notant $\mathcal{H}_\pi$. On voit facilement que $a_n \in (\mathcal{H}_\pi^{-\infty})^{H, \chi_f \Delta_{H,G}^{1/2}}$. Par suite, pour $\psi \in C_c^\infty(G)$, un calcul mène à

$$(\pi(\psi)a_n)(g) = \Delta_G(g_n)^{-1} \int_B \psi(gbg_n^{-1})\chi_l(b)db \ \ (g \in G).$$

Ensuite

$$\langle \pi(\psi)a_n, a_n\rangle = \int_{\mathbf{R}} \psi_H^f(\exp sY_2)\exp(is\tan(\alpha t_n + \theta))ds.$$

Les formules (1) et (2) entraînent, $y(t)$ s'écrivant simplement y,

$$(1 - \alpha y)(dt/d\lambda) = e^t \sin(\alpha t), \tag{3}$$
$$(dy/d\lambda) - (y + \alpha)(dt/d\lambda) = e^t \cos(\alpha t). \tag{4}$$

D'autre part, d'après $e^{2t} = (1+y^2)/(1+\lambda^2)$,

$$[(1+\lambda^2)(dt/d\lambda)+\lambda](1+y^2)/(1+\lambda^2) = y(dy/d\lambda). \tag{5}$$

Les égalités (3) et (4) impliquent

$$\lambda(dy/d\lambda)+(1-\alpha y-\lambda y-\alpha\lambda)(dt/d\lambda) = y,$$
$$(dy/d\lambda)+(\lambda\alpha y-\lambda-y-\alpha)(dt/d\lambda) = 1,$$
$$\text{donc } (\lambda-y)(dy/d\lambda)+(1-\alpha\lambda)(1+y^2)(dt/d\lambda) = 0.$$

Par substitution de (5),

$$(\lambda-y)(dy/d\lambda)+(1-\alpha\lambda)[y(dy/d\lambda)-\lambda(1+y^2)/(1+\lambda^2)] = 0,$$

c'est-à-dire

$$\lambda(1-\alpha y)(dy/d\lambda) = \lambda(1+\alpha\lambda)(1+y^2)/(1+\lambda^2).$$

En conséquence, pour λ non nul,

$$(dy/d\lambda) = (1-\alpha\lambda)(1+y^2)/(1-\alpha y)(1+\lambda^2).$$

Notre formule à montrer revient à la suivante, $y_n = y_n(\lambda)$ désignant $\tan(\alpha t_n+\theta)$,

$$\psi_H^f(e) = \int_{\lambda_0}^{\lambda_1}\xi(\lambda)d\lambda\left(\sum_{n=0}^{\infty}\kappa(n)\int_{\mathbf{R}}\psi_H^f(\exp sY_2)\exp(is\tan(\alpha t_n+\theta))\right)ds$$
$$= (2\pi)^{-1/2}\int_{\lambda_0}^{\lambda_1}\xi(\lambda)\sum_{n=0}^{\infty}\kappa(n)(\psi_H^f)^\wedge(y_n)d\lambda$$

avec certaines fonctions mesurables $\xi(\lambda)$ et $\kappa(n)\geq 0$, car on multiplie au besoin a_n par un scalaire convenable, et $(\psi_H^f)^\wedge$ signifiant la transformée de Fourier inverse de $\psi_H^f\circ\exp$.

En effet, soint $\xi(\lambda) = (1-\alpha\lambda)/2\pi(1+\lambda^2)$ et $\kappa(n) = (1+y_n^2)/|1-\alpha y_n|$, ce qui veut dire que l'on prend $[(1+y_n^2)/|1-\alpha y_n|]^{1/2}a_n$. Alors,

$$(2\pi)^{-1/2}\int_{\lambda_0}^{\lambda_1}\left(\sum_{n=0}^{\infty}(\psi_H^f)^\wedge(y_n)\kappa(n)\right)\xi(\lambda)d\lambda$$
$$=(2\pi)^{-1/2}\sum_{n=0}^{\infty}\int_{\lambda_0}^{\lambda_1}(\psi_H^f)^\wedge(y_n)(1-\alpha\lambda)(1+y_n^2)/|1-\alpha y_n|(1+\lambda^2)d\lambda$$
$$=(2\pi)^{-1/2}\int_{\mathbf{R}}(\psi_H^f)^\wedge(s)ds = (\psi_H^f\circ\exp)(0) = \psi_H^f(e).$$

EXEMPLE 2. Soit $G = G_3(\alpha) = \exp \mathfrak{g}_3(\alpha)$ comme dans l'exemple 1. Etant données cette fois $\mathfrak{h} = \mathbf{R}T$ et $f \in \mathfrak{g}_3(\alpha)^*$ arbitraire. Alors $f + \mathfrak{h}^\perp = f(T)T^* + \mathbf{R}Y_1^* + \mathbf{R}Y_2^*$ et l'on y trouve que les orbites générales ont leur représentant $\hat{\theta} = (\cos\theta)Y_1^* + (\sin\theta)Y_2^*$ à laquelle s'associe la représentation irréductible $\pi_\theta = \operatorname{ind}_B^G \chi_{\hat{\theta}}$ de G, où B est le sous-groupe analytique correspondant à la polarisation $\mathfrak{b} = \mathbf{R}Y_1 \oplus \mathbf{R}Y_2$.

Dans ce cas, notre formule habituelle $(*)$ pour obtenir des vecteurs généralisés H-semi-invariants nous offre

$$a_\theta : \phi \mapsto \int_H \overline{\phi(h)\chi_f(h)}\Delta_{H,G}^{-1/2}(h)dh.$$

Des raisonnements analogues à ceux fait pour le cas $ax + b$ montrent que notre a_θ possède les propriétés requises. Il est aisé de voir, pour $\psi \in C_c^\infty(G)$,

$$(\pi_\theta(\psi)a_\theta)(g) = \int_B \psi_H(gb)\chi_{\hat{\theta}}(b)db,$$
$$\langle \pi_\theta(\psi)a_\theta, a_\theta\rangle = \int_{\mathbf{R}} e^{2t}dt \int_B \psi_H^f(b)\chi_l(b)db,$$

avec la notation $l = h\cdot\hat{\theta} = e^t\cos(\alpha t+\theta)Y_1^* + e^t\sin(\alpha t+\theta)Y_2^*$, $h = \exp tT$.

Ceci posé,

$$(2\pi)^{-2}\int_0^{2\pi} \langle \pi_\theta(\psi)a_\theta, a_\theta\rangle d\theta$$
$$= (2\pi)^{-2}\int_0^{2\pi} d\theta \int_{\mathbf{R}} e^{2t}dt \int_{\mathbf{R}^2} \psi_H^f(\exp(b_1Y_1 + b_2Y_2))\cdot$$
$$\exp(ie^t(b_1\cos(\alpha t+\theta) + b_2\sin(\alpha t+\theta)))db_1db_2.$$

Appliquons le changement de variables

$$x = e^t\cos(\alpha t+\theta),$$
$$y = e^t\sin(\alpha t+\theta),$$

dont le Jacobien est

$$\partial(x,y)/\partial(t,\theta) = e^{2t};\ dxdy = e^{2t}dtd\theta.$$

On en déduit que

$$(2\pi)^{-2}\int_0^{2\pi} \langle \pi_\theta(\psi)a_\theta, a_\theta\rangle d\theta = (2\pi)^{-2}\int_{\mathbf{R}^2} dxdy \int_{\mathbf{R}^2} (\psi_H^f)^*(b_1,b_2)\cdot$$
$$\exp(i(xb_1 + yb_2))db_1db_2 = (\psi_H^f)^*(0,0) = \psi_H^f(e),$$

où $(\psi_H^f)^*(b_1,b_2) = \psi_H^f \circ \exp(b_1Y_1 + b_2Y_2)$.

EXEMPLE 3. $G = \exp \mathfrak{g}_4$, $\mathfrak{g}_4 = \langle T,X,Y,Z\rangle_{\mathbf{R}}$; $[T,X] = X$, $[T,Y] = -Y$, $[X,Y] = Z$ (oscillateur complètement résoluble). Soient $f(\alpha,\beta) = \alpha T^* + \beta Z^* \in \mathfrak{g}_4^*$ et $O(\alpha,\beta) = G \cdot f(\alpha,\beta)$ pour $\beta \neq 0$. On se donne $f = f(\alpha_0,\beta_0)$.

(i) Soit premièrement $\mathfrak{h} = \mathbf{R}T + \mathbf{R}X$. Alors

$$\mathrm{ind}_H^G\, \chi_f = \int_{\mathbf{R}} \theta(O(\alpha_0,\beta))d\beta.$$

Au moyen de $l = f(\alpha_0,\beta) \in (f + \mathfrak{h}^\perp) \cap O(\alpha_0,\beta)$ et d'une polarisation $\mathfrak{b} = \mathbf{R}T + \mathbf{R}X + \mathbf{R}Z$ en l, on construit $\pi(\alpha_0,\beta) = \mathrm{ind}_B^G\, \chi_l \simeq \overline{\theta}(O(\alpha_0,\beta))$. Dans cette situation, la façon usuelle propose $a(\beta)$ par $\langle a(\beta),\phi\rangle = \overline{\phi(e)}$, qui satisfait clairement aux conditions requises. Comme $\mathfrak{b} = \mathfrak{h} + \mathbf{R}Z$ et que Z est un élément central,

$$(\mathcal{H}^{-\infty}_{\pi(\alpha_0,\beta)})^{H,\chi_f\Delta_{H,G}^{1/2}} = \mathbf{C}a(\beta).$$

Maitenant pour $\psi \in C_c^\infty(G)$, $\pi(\alpha_0,\beta)(\psi)a(\beta) = \psi_B^l$ i.e.,

$$(\pi(\alpha_0,\beta)(\psi)a(\beta))(g) = \int_B \psi(gb)\chi_l(b)\Delta_{B,G}^{-1/2}(b)db,$$

par suite,

$$\langle \pi(\alpha_0,\beta)(\psi)a(\beta), a(\beta)\rangle = \int_B \psi(b)\chi_l(b)\Delta_{B,G}^{-1/2}(b)db.$$

De tout ce qui précède,

$$\begin{aligned}(2\pi)^{-1} &\int_{\mathbf{R}} \langle \pi(\alpha_0,\beta)a(\beta), a(\beta)\rangle d\beta \\ &= (2\pi)^{-1} \int_{\mathbf{R}} d\beta \int_{\mathbf{R}} \psi_H^f(\exp wZ)\exp(i\beta w)dw = \psi_H^f(e).\end{aligned}$$

(ii) Deuxièmement soit $\mathfrak{h} = \mathbf{R}T + \mathbf{R}Z$. Posons $l = f(\alpha,\beta_0) = \alpha T^* + \beta_0 Z^*$, ce qui nous dit

$$(f(\alpha_0,\beta_0)+\mathfrak{h}^\perp)\cap O(\alpha,\beta_0) = \alpha_0T^* + xX^* + yY^* + \beta_0Z^*;\ xy = \beta_0(\alpha - \alpha_0).$$

Si la valeur

$$\begin{aligned}(\exp(aX)\exp(bY)\cdot l)(T) &= (\exp(bY)\cdot l)(T + aX) \\ &= l(T + aX - bY + abZ) = \alpha + ab\beta_0\end{aligned}$$

est égale à $f(T)$, il vient $\alpha + ab\beta_0 = \alpha_0$, i.e. $ab\beta_0 = \alpha_0 - \alpha$. En modifiant les bases par des scalaires convenables, on peut supposer que $\beta_0 = 1$. L'égalité obtenue ci-dessus devient $ab = \alpha_0 - \alpha$. Par conséquent, pour $f(\alpha, \beta_0)$ telle que $\alpha \neq \alpha_0$, $g_j \cdot f(\alpha, \beta_0) \in f + \mathfrak{h}^\perp$ $(j = 1, 2)$ avec $g_1 = \exp(X)\exp((\alpha_0 - \alpha)Y)$, $g_2 = \exp(-X)\exp((\alpha - \alpha_0)Y)$.

Pareillement au cas (i), on réalise la représentation $\pi(\alpha, \beta_0) = \mathrm{ind}_B^G\, \chi(\alpha, \beta_0) \simeq \overline{\theta}(O(\alpha, \beta_0))$. Pour $\phi \in (\mathcal{H}^{+\infty}_{\pi(\alpha,\beta_0)})$, nous rappelons la formule familière:

$$\begin{aligned}\langle a_1(\alpha), \phi\rangle &= \oint_{H/H\cap gBg} \overline{\phi(hg_1)\chi_f(h)}\Delta_{H,G}^{-1/2}(h)d\hat{\nu}(h)\\ &= \int_{\mathbf{R}} \overline{\phi(\exp(tT)\exp(X))}e^{-it\alpha_0}dt = \int_{\mathbf{R}} \overline{\phi(\exp(e^T X))}e^{it(\alpha-\alpha_0)+t/2}dt\\ &= \int_0^\infty \overline{\phi(\exp(sX))}s^{i(\alpha-\alpha_0)-1/2}ds.\end{aligned}$$

L'intégrand de ce dernier est bien intégrable et il est immédiat que

$$a_1(\alpha) \in (\mathcal{H}^{-\infty}_{\pi(\alpha,\beta_0)})^{H,\chi_f\Delta_{H,G}^{1/2}}.$$

De même la formule

$$\begin{aligned}\langle a_2(\alpha), \phi\rangle &= \oint_{H/H\cap g_2Bg_2^{-1}} \overline{\phi(hg_2)\chi_f(h)}\Delta_{H,G}^{-1/2}(h)d\hat{\nu}(h)\\ &= \int_0^\infty \overline{\phi(\exp(-sX))}s^{i(\alpha-\alpha_0)-1/2}ds\end{aligned}$$

définit un élément non nul

$$a_2(\alpha) \in (\mathcal{H}^{-\infty}_{\pi(\alpha,\beta_0)})^{H,\chi_f\Delta_{H,G}^{1/2}}.$$

Soit a un élément quelconque de celui-ci. La semi-invariance de a par rapport à $h = \exp(tT)$, $t \in \mathbf{R}$, nous donne

$$\langle e^{it(\alpha_0-\alpha)-t/2}a, \phi(\exp(xX))\rangle = \langle a, \phi(\exp(e^t xX))\rangle.$$

On en déduit que

$$\langle a, \phi\rangle = c_1 \int_{\mathbf{R}_+} \overline{\phi(\exp(xX))}x^{i(\alpha-\alpha_0)-1/2}dx \quad (c_1 : \text{constante})$$

si $\operatorname{supp}\phi \subseteq \mathbf{R}_+ = \{s \in \mathbf{R}; s > 0\}$, c'est-à-dire que $a = c_1a_1(\alpha)$ (c_1 : constante) sur $\mathbf{R}_+$. De même, $a = c_2a_2(\alpha)$ (c_2 : constante) sur $\mathbf{R}_-$.

Supposons maintenant supp $a \subseteq B$, et écrivant

$$a = \sum_{j=0}^{m} \lambda_j D_j, \quad \text{où } \langle D_j, \phi \rangle = \overline{(d^j \hat{\phi}/dx^j)(0)}, \ \hat{\phi}(x) = \phi(\exp(xX)).$$

Alors en considérant la semi-invariance de a pour $h = \exp(tT)$, on a

$$\exp(i\alpha_0 t) \sum_{j=0}^{m} \overline{(d^j \hat{\phi})/(dx^j)(0)} = \sum_{j=0}^{m} \lambda_j \overline{(d^j \hat{\phi})/(dx^j)(0)} e^{(j+1/2)t} \exp(i\alpha t).$$

Si l'on y choisit ϕ vérifiant $(d^j \hat{\phi})/(dx^j)(0) = \delta_{jm}$, il s'ensuit que

$$\lambda_m \exp(i\alpha_0 t) = \lambda_m e^{(m+1/2)t} \exp(i\alpha t) \quad (t \in \mathbf{R}).$$

Ceci posé, on conclut que $\lambda_m = 0$, ce qui veut dire que $a = 0$. En somme,

$$(\mathcal{H}^{-\infty}_{\pi(\alpha,\beta_0)})^{H, \chi_f \Delta^{1/2}_{H,G}} = \mathbf{C}a_1 \oplus \mathbf{C}a_2.$$

Passons à la formule de Plancherel concrète pour la représentation monomiale $\mathrm{ind}_H^G \chi_f$, $f = f(\alpha_0, \beta_0)$. Pour alléger les notations, $\pi(\alpha, \beta_0)$, examinée de près pour le moment, sera notée π. Soient $\psi \in C_c^\infty(G)$ et $\phi \in \mathcal{H}_\pi^{+\infty}$ à support compact modulo B. On calcule: $a_j(\alpha)$ $(j = 1, 2)$ se notant simplement a_j,

$$\begin{aligned} \langle \pi(\psi) a_1, \phi \rangle = \oint_{H/H \cap g_1 B g_1^{-1}} d\nu(h) \oint_{G/B} d\nu(g) \int_B \psi(g b g_1^{-1} h^{-1}) \cdot \\ \overline{\phi(g)} \Delta_{B,G}^{-1/2}(h) \chi_l(b) \overline{\chi_f(h)} \Delta_G^{-1}(h) \Delta_{H,G}^{-1/2}(h) db \end{aligned} \tag{6}$$

L'ordre des deux premières intégrales au membre droit s'échange, ce que l'on va voir dans la suite. Tout d'abord $\Delta_G(h) = \Delta_{H,G}(h) = 1$ et qu'à l'expression (6), g se met comme $g = \exp(xX)$, x parcourant un certain intervalle fini J. On note $\Xi(h, g, b)$ l'intégrand dans (6) et écrit $h = \exp(tT)$, $b = \exp(sT)\exp(yY)\exp(wZ)$. Alors

$$\begin{aligned} &\int_B \Xi(h, g, b) db = \overline{\hat{\phi}(x)} \cdot \\ &\int_{\mathbf{R}^3} \psi(\exp(xX)\exp(yY)\exp(wZ)\exp((\alpha - \alpha_0)Y)\exp(-X)\exp(-tT)) \cdot \\ &\qquad e^{-s/2} \exp(i(w + \alpha s - \alpha_0)) ds dy dw. \end{aligned}$$

On y trouve

$$\begin{aligned}&\exp(xX)\exp(sT)\exp(yY)\exp(wZ)\exp((\alpha-\alpha_0)Y)\exp(-X)\exp(-tT)\\&=\exp((w+e^{-s}x(y+\alpha-\alpha_0))Z)\cdot\\&\quad\exp(e^{-s}(y+\alpha-\alpha_0)Y)\exp((x-e^{s})X)\exp((s-t)T).\end{aligned}$$

Compte tenu de cela,

$$\begin{aligned}&\left|\int_B \Xi(h,g,b)db\right|\le|\hat{\phi}(x)|\cdot\\&\quad\int_{\mathbf{R}^3}|\psi(\exp(wZ)\exp(e^{-s}yY)\exp((x-e^{s})X)\exp((s-t)T)|\cdot\\&\quad e^{-s/2}dsdydw\\&=|\hat{\phi}(x)|\int_{\mathbf{R}^3}|\psi(\exp(wZ)\exp(yY)(\exp((x-e^{s+t})X)\exp(sT)|\cdot\\&\quad e^{(s+t)/2}dsdydw.\end{aligned}\tag{7}$$

Puisque x parcourt l'intervalle fini J, (7) est intégrable relativement à $dtdx$ et l'on peut bien échanger l'ordre des deux premières intégrales dans (6), ce que l'on vient de chercher.

Nous arrivons ainsi à

$$\begin{aligned}\langle\pi(\psi)a_1,\phi\rangle&=\oint_{G/B}d\hat{\nu}(g)\oint_{H/H\cap g_1Bg_1^{-1}}d\hat{\nu}(h)\int_B\psi(gbg_1^{-1}h^{-1})\cdot\\&\quad\overline{\phi(g)}\Delta_{B,G}^{-1/2}(b)\chi_l(b)\overline{\chi_f(h)}\Delta_G^{-1}(h)\Delta_{H,G}^{-1/2}(h)db.\end{aligned}$$

Donc

$$\begin{aligned}(\pi(\psi)a_1)(g)&=\oint_{H/H\cap g_1Bg_1^{-1}}d\hat{\nu}(h)\int_B\psi(gbg_1^{-1}h^{-1})\cdot\\&\quad\Delta_{B,G}^{-1/2}(b)\chi_l(b)\overline{\chi_f(h)}\Delta_G^{-1}(h)\Delta_{H,G}^{-1/2}(h)db\\&=\oint_{H/H\cap g_1Bg_1^{-1}}d\hat{\nu}(h)\oint_{B/B\cap g_1^{-1}Hg_1}\overline{\chi_f(h)}\chi_l(b)\Delta_G^{-1}(h)\Delta_{H,G}^{-1/2}(b)d\hat{\nu}(b)\cdot\\&\quad\int_{B\cap g_1^{-1}Hg_1}\psi(gbb_0^{-1}g_1^{-1}h^{-1})\Delta_{B,G}^{1/2}(b_0)\overline{\chi_l(b_0)}\cdot\\&\qquad\Delta_{B\cap g_1^{-1}Hg_1,B}(b_0)\Delta_{B\cap g_1^{-1}Hg_1}(b_0)db_0.\end{aligned}$$

Des arguments tout à fait analogues à ceux employés plus haut constatent que les deux premières intégrales sont échangeables. Finalement,

$$(\pi(\psi)a_1)(g) = \oint_{B/B\cap g_1^{-1}Hg_1} \chi_l(b)\Delta_{B,G}^{-1/2}(b)d\hat{\nu}(b)\cdot$$
$$\oint_{H/H\cap g_1Bg_1^{-1}} \overline{\chi_f(h)}\Delta_G^{-1}(h)\Delta_{H,G}^{-1/2}(h)d\hat{\nu}(h)\cdot$$
$$\int_{B\cap g_1^{-1}Hg_1} \psi(gbb_0^{-1}g_1^{-1}h^{-1})\overline{\chi_l(b_0)}\cdot$$
$$\Delta_{B,G}^{1/2}(b_0)\Delta_{B\cap g_1^{-1}Hg_1,B}(b_0)\Delta_{B\cap g_1^{-1}Hg_1}^{-1}(b_0)db_0,$$

dont la dernière intégrale est égale à

$$\int_{H\cap g_1Bg_1^{-1}} \psi(gbg_1^{-1}b'^{-1}h^{-1})\chi_{g_1\cdot l}(b')\cdot$$
$$\Delta_{g_1Bg_1^{-1},G}^{1/2}(b')\Delta_{g_1Bg_1^{-1}\cap H,\, g_1Bg_1^{-1}}(b')\Delta_{g_1Bg_1^{-1}\cap H}^{-1}(b')db'.$$

Pourvu que l'égalité

$$\Delta_H^{1/2}(b')\Delta_{g_1Bg_1^{-1}\cap H}(b') = \Delta_{g_1^{-1}Bg_1}(b') \quad (b' \in g_1Bg_1^{-1} \cap H)$$

s'établisse, ce qui aisé à constater dans notre cas, on aurait

$$\Delta_{g_1Bg_1^{-1},G}^{1/2}(b')\Delta_{g_1Bg_1^{-1}\cap H,g_1Bg_1^{-1}}(b')\Delta_{g_1Bg_1^{-1}\cap H}^{-1}(b')$$
$$= \Delta_H^{-1}(b')\Delta_{H,G}^{1/2}(b')\Delta_{g_1Bg_1^{-1}\cap H,H}^{-1}(b'),$$

et enfin

$$(\pi(\psi)a_1)(g) = \oint_{B/B\cap g_1^{-1}Hg_1} \Delta_{B,G}^{-1/2}(b)\chi_l(b)\hat{\nu}(b)\cdot$$
$$\int_H \psi(gbg_1^{-1}h)\chi_f(h)\Delta_{H,G}^{-1/2}(h)dh$$
$$= \oint_{B/B\cap g_1^{-1}Hg_1} \psi_H^f(gbg_1^{-1})\chi_l(b)\Delta_{B,G}^{-1/2}(b)d\hat{\nu}(b).$$

De même façon,

$$(\pi(\psi)a_2)(g) = \oint_{B/B\cap g_2^{-1}Hg_2} \psi_H^f(gbg_2^{-1})\chi_l(b)\Delta_{B,G}^{-1/2}(b)d\hat{\nu}(b).$$

De tout ce qui précède,

$$\langle \pi(\psi)a_1, a_1\rangle = \oint_{H/H\cap g_1 B g_1^{-1}} \chi_l(h)\Delta_{H,G}^{-1/2}(h)d\hat{\nu}(h)\cdot$$
$$\oint_{B/B\cap g_1^{-1}Hg_1} \psi_H^f(hg_1bg_1^{-1})\chi_l(b)\Delta_{B,G}^{-1/2}(b)d\hat{\nu}(b)$$
$$= \int_{\mathbf{R}} e^{it(\alpha_0-\alpha)+t/2}dt\cdot$$
$$\int_{\mathbf{R}} \psi_H^f(\exp(e^t X)\exp(xT)\exp(yY)\exp(-X))e^{i\alpha x - x/2}dxdy$$
$$= \int_{\mathbf{R}} e^{i(t-x)(\alpha_0-\alpha)+(t+x)/2}dt\cdot$$
$$\int_{\mathbf{R}^2} \psi_H^f(\exp((e^t - e^x)X)\exp(yY))e^{-iye^x}dxdy.$$

En effectuant le changement de variables $t \mapsto t + x$, $e^x = s$, on obtient

$$\langle \pi(\psi)a_1, a_1\rangle$$
$$= \int_{\mathbf{R}} e^{it(\alpha_0-\alpha)+t/2}dt \int_{\mathbf{R}\times\mathbf{R}_+} \psi_H^f(\exp(s(e^t - 1)X)\exp(yY))e^{-iys}dyds.$$

D'une façon analogue,

$$\langle \pi(\psi)a_2, a_2\rangle = \oint_{H/H\cap g_2 B g_2^{-1}} \chi_f(h)\Delta_{H,G}^{-1/2}(h)d\hat{\nu}(h)\cdot$$
$$\oint_{B/B\cap g_2^{-1}Hg_2} \psi_H^f(hg_2bg_2^{-1})\chi_l(b)\Delta_{B,G}^{-1/2}(b)d\hat{\nu}(b)$$
$$= \int_{\mathbf{R}} e^{it(\alpha_0-\alpha)+t/2}dt \int_{\mathbf{R}\times\mathbf{R}_+} \psi_H^f(\exp(s(1 - e^t)X)\exp(yY))e^{iys}dyds.$$

Ces calculs se terniment donc à

$$\langle \pi(\psi)a_1, a_1\rangle + \langle \pi(\psi)a_2, a_2\rangle$$
$$= \int_{\mathbf{R}} e^{it(\alpha_0-\alpha)+t/2}dt\cdot$$
$$\int_{\mathbf{R}^2} \psi_H^f(\exp(s(1-t)X)\exp(yY))e^{iys}dyds.$$

Si l'on y pose

$$\Psi(t) = \int_{\mathbf{R}^2} \psi_H^f(\exp(s(1 - e^t)X)\exp(yY))e^{iys}dyds,$$

cette fonction est infiniment différentiable pour t non nulle car, dans ce cas, l'intégrale serait effectuée sur un compact. D'ailleurs, la fonction $\mathbf{R}^2 \ni (s,y) \mapsto \psi_H(\exp(sX)\exp(yY))$ appartenant à $C_c^\infty(\mathbf{R}^2)$, dont on fait des L^1-approximations par des fonctions de la forme $\sum_j \xi_j(s)\eta_j(y)$ $(\xi_j, \eta_j \in C_c^\infty(\mathbf{R}))$, pour un nombre positif ε quelconque, on peut choisir des ξ_j, η_j de telle manière que

$$|\Psi(t) - (2\pi)^{1/2} \sum_j \int_{\mathbf{R}} \xi_j(s(1-e^t))(\eta_j)^\wedge(s)ds| < \varepsilon$$

quel que soir $t \in \mathbf{R}$, où $(\eta_j)^\wedge$ désigne la transformée de Fourier inverse de η_j. Compte tenu du fait que la limite, lorsque t tend vers zéro, de

$$\int_{\mathbf{R}} \xi_j(s(1-e^t))(\eta_j)^\wedge(s)ds$$

est égale à $(2\pi)^{1/2}\xi_j(0)\eta_j(0)$, $\Psi(t)$ est continue même en $t = 0$.

Nous considérons à la fin la fonction $e^{t/2}\Psi(t)$, lorsque $t \mapsto -\infty$, elle décroît rapidement grâce au facteur $e^{t/2}$. Examinons son comportement quand t tend vers $+\infty$. Un changement de variables mène à

$$e^{t/2}\Psi(t) = e^{t/2}(1-e^t)^{-1} \int_{\mathbf{R}^2} \psi_H^f(\exp(sX)\exp(yY))e^{iys(1-e^t)^{-1}}dsdy,$$

et à

$$|e^{t/2}\Psi(t)| \leq e^{t/2}(1-e^t)^{-1} \int_{\mathbf{R}^2} |\psi_H^f(\exp(sX)\exp(yY))|dsdy,$$

ce qui prouve que $e^{t/2}\Psi(t)$ est à décroissance rapide, t tendant vers $+\infty$.

Il en découle que la formule d'inversion de Fourier peut amener à la notre formule de Plancherel concrète pour la représentation monomiale $\tau = \mathrm{ind}_H^G \chi_f$:

$$(2\pi)^{-2} \int_{\mathbf{R}} (\langle \pi(\psi)a_1(\alpha), a_1(\alpha)\rangle + \langle \pi(\psi)a_2(\alpha), a_2(\alpha)\rangle d\alpha$$
$$= (2\pi)^{-1} \int_{\mathbf{R}} \psi_H^f(\exp(yY))e^{iys}dsdy = \psi_H^f(e).$$

Références

[1] L. Auslander and C.C. Moore, *Unitary representations of solvable Lie groups.* Mem. Amer. Soc. No **62**, 1966.

[2] Y. Benoist, *Analyse harmonique sur les symétriques nilpotents*, J. Func. Anal. **59** (1984), 211–253.
[3] Y. Benoist, *Multiplicité un pour les espaces symétriques exponentiels*, Mém. Soc. Math. France **15** (1985), 1–37.
[4] P. Bernat et al., "Représentations des groupes de Lie résolubles," Dunod, Paris, 1972.
[5] P. Bonnet, *Transformation de Fourier des distributions de type positif sur un groupe de Lie unimodulaire*, J. Func. Anal. **55** (1984), 220–246.
[6] L. Corwin, F.P. Greenleaf and G. Grélaud, *Direct integral decompositions and multiplicities for induced representations of nilpotent Lie groups*, Trans. Amer. Math. Soc. **304** (1987), 549–583.
[7] H. Fujuwara, *Représentations monomiales des groupes de Lie nilpotents*, Pacific J. Math. **127** (1987), 329–352.
[8] H. Fujuwara et S. Yamagami, *Certaines représentations monomiales d'un groupe de Lie résoluble exponentiel*. à paraître dans Advanced Studies in Pure Math.
[9] G. Grélaud, *Désintégration des représentations induites des groupes de Lie résolubles exponentiels*. Thèse de 3e cycle, Univ. de Poitiers, 1973.
[10] G. Grélaud, *Sur les représentations des groupes de Lie résolubles*. Thèse, Univ. de Poitiers, 1984.
[11] A.A. Kirillov, *Représentations unitaires des groupes de Lie nilpotents*, Uspekhi Mat. Nauk **17** (1962), 57–110.
[12] R. Lipsman, *Orbital parameters for induced and restricted representations*. à paraître dans Trans. Amer. Math.
[13] R. Lipsman, *Harmonic analysis on exponential solvable homogeneous spaces: The algebraic or symmetric cases*. à paraître.
[14] R. Lipsman, *Induced representations of completely solvable Lie groups*. à paraître.
[15] R. Penney, *Abstract Plancherel theorems and a Frobenius reciprocity theorem*, J. Func. Anal. **18** (1975), 177–190.
[16] S.R. Quint, *Decomposition of induced representations of solvable exponential Lie groups*. Dissertation, Univ. of California, Berkeley, 1973.
[17] S. Sternberg, *Lectures on Differential Geometry*. Prentice-Hall, 1964.
[18] O. Takenouchi, *Sur la facteur-représentation d'un groupe de Lie résoluble de type* (E), Math. J. Okayama Univ. **7** (1957), 151–161.
[19] M. Vergne, *Étude de certaines représentations induites d'un groupe de Lie résoluble exponentiel*, Ann. Sci. Ec. Norm. Sup. **3** (1970), 353–384.

Faculté de Technologie
Université de Kinki à Kyushu
11–6, Kayanomori, Iizuka
820, Japon

The Surjectivity Theorem, Characteristic Polynomials and Induced Ideals

ANTHONY JOSEPH

1. INTRODUCTION

Throughout we follow almost faithfully the notation of [9] as supplemented by [17].

1.1. Let $\underline{g}$ be a complex semisimple Lie algebra and $U(\underline{g})$ its enveloping algebra. One has now a very explicit description of the set $\operatorname{Prim} U(\underline{g})$ of primitive ideals of $U(\underline{g})$ and one could hope to say how the subset $\operatorname{Prim}_c U(\underline{g})$ of completely prime, primitive ideals lies in $\operatorname{Prim} U(\underline{g})$. When $\underline{g}$ is of type A_n (or even just has A_n factors) a theorem of C. Moeglin [26] asserts that if $I \in \operatorname{Prim}_c U(\underline{g})$ then I is induced from a character of a parabolic subalgebra of $\underline{g}$. (The converse assertion which holds for all $\underline{g}$ is an easier and older result due to N. Conze [8], 3.1). Although this is very satisfactory it does not merge well with the Goldie rank picture. Thus one cannot say which coherent families of primitive ideals have a completely prime member - equivalently which Goldie rank polynomials take the value 1. Conversely in type A_n the Goldie ranks are completely known at least implicitly; but this does not allow us to extract Moeglin's theorem.

1.2. Moeglin's theorem should have the following generalization. Identify $\underline{g}$ with its dual $\underline{g}^*$ through the Killing form and let G denote the adjoint group of $\underline{g}$. Take $J \in \operatorname{Prim} U(\underline{g})$. By [18] its associated variety $\mathcal{V}(J)$ is the Zariski closure $\bar{\mathcal{O}}$ of a nilpotent orbit $\mathcal{O}$ in $\underline{g}$. As in [4] we can write $\mathcal{O} = \operatorname{Ind}_{\underline{r}}^{\underline{g}} \mathcal{O}'$ for some nilpotent orbit $\mathcal{O}' \subset \underline{r}$ where $\underline{r}$ is a Levi factor of $\underline{g}$. We call $\mathcal{O}$ rigid if for every such presentation $\underline{r} = \underline{g}$ and $\mathcal{O}' = \mathcal{O}$. In particular we can take $\mathcal{O}'$ to be rigid in the above without loss of generality. Now we should like to show that if $J \in \operatorname{Prim}_c U(\underline{g})$, then it is induced from $J' \in \operatorname{Prim}_c U(\underline{r})$ satisfying $\mathcal{V}(J') = \bar{\mathcal{O}}'$. Notice, however, that except in type A_n the pair $(\underline{r}, \mathcal{O}')$ is not uniquely determined (up to conjugation). Moreover, it can happen that $\operatorname{Ind}_{\underline{r}_1}^{\underline{g}} \{0\} = \operatorname{Ind}_{\underline{r}_2}^{\underline{g}} \{0\}$ whilst the ideals induced from the augmentation ideals of $U(\underline{r}_i) : i = 1, 2$ are distinct. (For example [3], 7.6, in type B_2). It can also happen that a rigid orbit can give rise to more than one $J \in \operatorname{Prim}_c U(\underline{g})$. (For example [23], Thm. 4.10, in type G_2). One would expect that these diseases would cause serious obstructions to any proof.

1.3. Under a mild technical restriction which can in principle be eliminated by other means we shall establish this generalization for maximal ideals. Actually we prove the slightly surprising fact that if $J \in \operatorname{Max} U(\underline{g})$ is not induced from a proper Levi factor, then $\mathcal{V}(J)$ is the closure of a rigid orbit (under the above technical restriction). The proof uses the surjectivity theorem established in [20].

1.4. Of course 1.3 only makes a small scratch at the question in 1.2 and one can ask if Goldie rank polynomials can be used more effectively. In ([21], Sect. 8) we established a positivity property for these polynomials which was used in ([21], 8.7) to determine that part of $\operatorname{Prim}_c U(\underline{g})$ corresponding to regular integral central characters. Here we note a further consequence of this positivity towards analysing $\operatorname{Prim}_c U(\underline{g})$.

2. Fibre type

2.1. Let $\underline{g} = \underline{n}^- \oplus \underline{h} \oplus \underline{n}$ be a triangular decomposition for $\underline{g}$ and R (resp. R^+) the corresponding system of non-zero (resp. positive) roots, with $B \subset R^+$ the set of simple roots, $P(R)$ the lattice of weights and W the Weyl group. Let $\alpha^\vee = 2\alpha/(\alpha,\alpha)$ denote the coroot corresponding to $\alpha \in R$ and $s_\alpha \in W$ the reflection $x \mapsto x - (\alpha^\vee, x)\alpha$. Let ρ denote the half sum of the positive roots.

2.2. For each $\lambda \in \underline{h}^*$, set $R_\lambda = \{\alpha \in R \mid (\alpha^\vee, \lambda) \in \mathbb{Z}\}$, $R_\lambda^0 = \{\alpha \in R \mid (\alpha^\vee, \lambda) = 0\}$. Clearly, $R_\lambda^0 \subset R_\lambda$ and both are root systems with Weyl groups which we shall denote by W_λ^0, W_λ. Furthermore, $R_\lambda^+ := (R_\lambda^0 \cap R^+) \cup \{\alpha \in R_\lambda \mid (\alpha^\vee, \lambda) > 0\}$ is a positive system for R_λ and we shall denote the corresponding set of simple roots by B_λ. We call λ regular if $R_\lambda^0 = 0$ and dominant if $R_\lambda^+ \subset R^+$.

2.3. Given $\lambda \in \underline{h}^*$ we set $\Lambda = \lambda + P(R)$ which we may also consider as its representative in $\underline{h}^*/P(R)$. Let Λ^+ denote the set of dominant elements of Λ. Obviously $R_\mu = R_\lambda$, $\forall \mu \in \Lambda$. Taking μ regular, R_μ^+ and so B_μ are uniquely determined and we can take $B_\lambda = B_\mu$ without loss of generality. Since W acts on $\underline{h}^*$, so does the affine Weyl group $W_{\text{aff}} = W \ltimes P(R)$. We use $\hat{\Lambda}$ to denote the representative of Λ in $\underline{h}^*/W_{\text{aff}}$. Since $R_\mu = R_\lambda$ for $\mu = w\lambda : w \in W_\lambda$ we can always pick $\mu \in \hat{\Lambda}$ dominant and regular. Yet different dominant, regular representatives do not necessarily give the same R_μ^+, B_μ, rather they are conjugated by the action of W. In type A_n one may always conjugate B_μ into B; but this is false in general. Again the simple roots B_λ^0 corresponding to $R_\lambda^0 \cap R^+$ lie in B so we cannot always find $\mu \in \Lambda$ vanishing on arbitrarily many roots in B_λ, though we can always do this one root at a time ([14], p. 47). Given $B' \subset B$, we let $W_{B'}$ denote the subgroup of W generated by the $s_\alpha : \alpha \in B'$.

2.4. Let $L(\lambda) : \lambda \in \underline{h}^*$ denote the unique simple highest weight module with highest weight $\lambda - \rho$. Set $J(\lambda) = \operatorname{Ann} L(\lambda)$. By Duflo's theorem ([10], II, Thm. 1) the map $\lambda \mapsto J(\lambda)$ is surjection of $\underline{h}^*$ onto $\operatorname{Prim} U(\underline{g})$. Identify $Z(\underline{g})$ with $\underline{h}^*/W$ via the Harish-Chandra isomorphism. Then the map $\lambda \mapsto J(\lambda) \cap Z(\underline{g}) \in \operatorname{Max} Z(\underline{g})$ is just the map sending λ to its orbit $\hat{\lambda} := W\lambda$ under W. The (finite) subset $X^{\underline{g}}_{\hat{\lambda}} := \{J(\mu) \mid \mu \in W\lambda\}$ is called the primitive fibre over $\hat{\lambda}$. Assume λ dominant and set $D_\lambda = \{x \in W \mid xB_\lambda \subset R^+\}$. Then each $w \in W$ can be written uniquely in the form $w = xy : x \in D_\lambda,\ y \in W_\lambda$ and one has ([15], lemma 6)

$$(*) \qquad J(xy\lambda) = J(y\lambda)\,,\ \forall x \in D_\lambda\,,\ y \in W_\lambda\,.$$

2.5. Each $J \in \operatorname{Prim} U(\underline{g})$ defines a unique nilpotent orbit $\mathcal{O}_J \subset \underline{g}^*$ whose closure is just the associated variety $\mathcal{V}(J)$ of J. This gives a map $J \mapsto \mathcal{O}_J$ of $\operatorname{Prim} U(\underline{g})$ into the set $\mathcal{S}^{\underline{g}}$ of nilpotent orbits in $\underline{g}^*$. For each $\hat{\lambda} \in \underline{h}^*/W$, let $\mathcal{S}^{\underline{g}}_{\hat{\lambda}}$ denote the image of its restriction to $X^{\underline{g}}_{\hat{\lambda}}$. Using the translation principle (see [14], 17.13) one shows that $\mathcal{S}^{\underline{g}}_{\hat{\lambda}}$ is independent of a regular representative of $\hat{\lambda}$ in $\underline{h}^*/W_{\text{aff}}$ and becomes increasingly smaller as R^0_λ increases. In an essentially trivial fashion this definition carries over to $\underline{g}$ reductive and furthermore $\mathcal{S}^{\underline{g}}_{\hat{\lambda}}$ only depends on the restriction of λ to the semisimple part of $\underline{g}$.

2.6. Given $\mathcal{O} \in \mathcal{S}^{\underline{g}}_{\hat{\lambda}} : \lambda \in \underline{h}^*$ we say that $\mathcal{O}$ is $\hat{\lambda}$ rigid if there is no proper subset $B_1 \subsetneq B$ such that $\mathcal{O} = \operatorname{Ind}^{\underline{g}}_{\underline{r}} \mathcal{O}_1$ where $\underline{r}$ is the Levi factor corresponding to B_1 and $\mathcal{O}_1 \in \mathcal{S}^{\underline{r}}_{\hat{\mu}}$ with $\lambda - \mu \in P(R)$. (This makes sense due to 2.4(*)). A priori this is weaker than saying that $\mathcal{O}$ is rigid; but hopefully these notions are equivalent. We remark that $\mathcal{S}^{\underline{g}}_{\hat{\rho}}$ is just the set of special orbits in the sense of [1]. A special orbit is $\hat{\rho}$-rigid exactly if it cannot be induced from a *special* orbit of a proper Levi factor. Even in type B_3 it can happen that a special orbit is induced from a non-special orbit from a type B_2 subalgebra - but this orbit is also induced from a special orbit from a type $A_1 \times A_1$-subalgebra, a phenomenon which has some curious consequences (see [20], 8.7).

2.7. The problem raised in 2.6 can be reduced by standard methods to the case $\operatorname{card} B_\lambda = \operatorname{card} B$. Indeed ([13], p. 118) we can find $B' \subset B$ such that $\operatorname{card} B' = \operatorname{card} B_\lambda$ and such that $R_\lambda \subset R' := \mathbb{Z}B' \cap R$ up to conjugation by W. It was already noted by Jantzen ([13], p. 6) that Verma module multiplicities can then be computed from the corresponding multiplicities taken with respect to the Levi factor $\underline{r}_{B'}$ defined by B'.

Essentially all questions about Prim $U(\underline{g})$ can be similarly reduced. Set $I(R) = \{\lambda \in \underline{h}^* \mid \text{card}\, B_\lambda = \text{card}\, B\}$. Then $I(R)/W_{\text{aff}}$ is a finite set to which one may limit considerations. In particular if $\underline{g}$ has only type A_n factors, this set has cardinality one and it follows that $\mathcal{O}$ is $\hat{\lambda}$ rigid only if $\mathcal{O}$ is rigid and so only if $\mathcal{O} = \{0\}$.

3. A CRITERION FOR A MAXIMAL IDEAL TO BE INDUCED

3.1. Let $j : \underline{g} \to \underline{g} \times \underline{g}$ be the diagonal map. Set $\underline{k} = j(\underline{g})$, $\underline{t} = j(\underline{h})$, $\underline{u} = j(\underline{n}^-)$, $\underline{u}^- = j(\underline{n})$. Given M a $U(\underline{g})$ module, we let $F(M)$ (resp. $A(M)$) denote the largest subspace of $\text{End}_{\mathbf{C}}\, M$ on which $\underline{k}$ (resp. $\underline{u}$) acts locally finitely (resp. nilpotently). Both are obviously $U(\underline{g})$ bisubmodules of $\text{End}_{\mathbf{C}}\, M$ and furthermore $F(M)$ is a submodule of $A(M)$. If $M = L(\lambda)$ we shall simply denote these spaces by $F(\lambda)$, $A(\lambda)$.

3.2. Consider $A(\lambda)$ as a $U(\underline{k})$ module. We showed in ([20], 2.3, 3.1) that $A(\lambda)$ admits primary decomposition with respect to Max $Z(\underline{k})$ and that its primary factors lie in the $\underline{O}$ category with respect to $\underline{k}, \underline{t}, \underline{u}$ so in particular have finite length. It follows that $A(\lambda)$ admits a maximal submodule $G(\lambda)$ (necessarily unique) with no finite dimensional quotients. Actually it is convenient to describe $G(\lambda)$ in another fashion. Let M be a $\underline{k}$ module. Then $H_0(\underline{u}, M)^* \xrightarrow{\sim} H^0(\underline{u}^-, M^*)$ where duality is defined with respect to the Chevalley antiautomorphism interchanging $\underline{u}$ and $\underline{u}^-$. Thus for $M \in Ob\underline{O}$ one has $M \neq \underline{u}M$ if and only if M has a finite dimensional quotient. Using the $Z(\underline{k})$ primary decomposition we conclude that $G(\lambda) = \underline{u}G(\lambda)$ and further use of primary decomposition and the above result (the $\underline{u}$-adic Artin-Rees lemma fails here) gives

$$G(\lambda) = \bigcap_{i \in \mathbb{N}} \underline{u}^i A(\lambda).$$

This description gives the following result an easy induction proof which we omit.

LEMMA. *$G(\lambda)$ is a two-sided ideal in $A(\lambda)$.*

3.3. It was claimed in ([20], 7.1) that $A(\lambda)$ is a Goldie ring (and necessarily prime).Actually there was a slight flaw in the proof; but a more careful growth rate estimate established this result [22]. Set $U(\lambda) = U(\underline{g})/J(\lambda)$. It is easy to show [22] that $U(\lambda)$ and $A(\lambda)$ have the same Gelfand-Kirillov dimension. Hence by Goldie's theorem ([11], Chap. 4) and ([5], 3.4) one concludes that $G(\lambda) \cap U(\lambda) = 0$ implies $G(\lambda) = 0$ which is easily seen to further imply that $F(\lambda) = A(\lambda)$. Now suppose that $U(\lambda) \subset G(\lambda)$. This means (in the precise sense of [20], 5.1) that the identity of $U(\lambda)$ which we can identify with the unique trivial $\underline{k}$ type in $A(\lambda)$, occurs in $G(\lambda)$

as a non-split submodule. Now let $A_{\hat{\rho}}$ denote the primary component of $A(\lambda)$ corresponding to the augmentation ideal of $Z(\underline{k})$. Examining the $\underline{t}$ weights of $\mathrm{End}_{\underline{n}^-}(L(\lambda)) = A(\lambda)^{\underline{u}}$ one easily checks that $A_{\hat{\rho}}$ is a subring of $A(\lambda)$ - even though the generators of $Z(\underline{k})$ do *not* act by derivations (see also [20], 2.4). Thus $A_{\hat{\rho}}$ is an $\underline{O}$ ring. Its nilradical N is therefore $\underline{k}$ invariant and nilpotent ([19], 4.4), hence zero. The structure theorem ([19], 5.6) on semisimple $\underline{O}$ rings then shows that $A_{\hat{\rho}}$ is a simple ring (recall it has a unique trivial $\underline{k}$ type) and moreover isomorphic as a $\underline{k}$ module to the $\underline{O}$ dual $\delta M_{B'}(\rho)$ of the module induced from the trivial character of the parabolic subalgebra $\underline{p}_{B'}$ defined by a $B' \subset B$. Furthermore $B' \subsetneq B$ for otherwise $A_{\hat{\rho}}$ reduces to scalars contradicting that the trivial $\underline{k}$ type in $A(\lambda)$ is non-split. This further implies ([20], Thm. 3.3) that $L(\lambda)$ is induced from a simple highest weight $\underline{p}_{B'}$ module $L_{B'}(\lambda)$. Set $J_{B'}(\lambda) = \mathrm{Ann}\, L_{B'}(\lambda)$. We have proved the

Proposition. *Suppose $J(\lambda)$ is a maximal two-sided ideal. Then either $F(\lambda) = A(\lambda)$ or there exists $B' \subsetneq B$ and a simple $\underline{p}_{B'}$ highest weight module $L_{B'}(\lambda)$ such that $J(\lambda) = \mathrm{Ind}_{\underline{p}_{B'}}^{\underline{g}} J_{B'}(\lambda) := \mathrm{Ann}\, U(\underline{g}) \otimes_{U(\underline{p}_{B'})} L_{B'}(\lambda)$, that is $J(\lambda)$ is non-trivially induced.*

Remark: Note that $\mathrm{Ann}\, L_{B'}(\lambda)$ is necessarily a maximal ideal of $U(\underline{r}_{B'})$. This is because induction preserves inclusion ([6], Sect. 3) and augments Gelfand-Kirillov dimension uniformly by $2 \dim \underline{m}_{B'}$. Note also ([6], 3.7 a)) that induction is a well-defined operation on ideals and not just on modules.

3.4. Before going further let us note the relevance of $\hat{\lambda}$-rigidity to discussing induced ideals.

Lemma. *Fix $\lambda \in \underline{h}^*$ and suppose that $J(\lambda)$ is the annihilator of a simple highest weight module M (not necessarily $L(\lambda)$) induced from a simple highest weight module N of a proper parabolic subalgebra $\underline{p}_{B'}$ of $\underline{g}$. Then $\mathcal{O}_{J(\lambda)}$ is not $\hat{\lambda}$-rigid.*

Let $\underline{r}$ denote the reductive part of $\underline{p}_{B'}$ and set $J_1 = \mathrm{Ann}_{U(\underline{r})} N$. Then $\mathcal{O}_{J(\lambda)} = \mathrm{Ind}_{\underline{r}}^{\underline{g}} \mathcal{O}_{J_1}$ via ([18], 3.11) and ([17], 7.10). Suppose N has highest weight $\mu - \rho$ as an $\underline{r}$-module. Then $M = L(\mu)$. Since $J(\lambda) = J(\mu)$, there exists $w \in W$ such that $\mu = w\lambda$. By 2.4(*) we can assume $w \in W_\lambda$ without loss of generality and then $\mu - \lambda \in P(R)$. This shows that $\mathcal{O}_{J(\lambda)}$ is not $\hat{\lambda}$-rigid.

3.5. We now prove the converse to 3.4 for maximal ideals.

Theorem. *Fix $\lambda \in \underline{h}^*$. Suppose that $J(\lambda) \in \mathrm{Max}\, U(\underline{g})$ and $\mathcal{O}_{J(\lambda)}$ is not $\hat{\lambda}$-rigid. Then $J(\lambda)$ is induced from a maximal ideal in the enveloping*

algebra of a proper Levi factor of $\underline{g}$.

By hypothesis we can write $\mathcal{O}_{J(\lambda)} = \mathrm{Ind}_{\underline{r}}^{\underline{g}} \mathcal{O}_{J_1}$ with $\underline{r}$ a Levi factor of $\underline{g}$ defined by $B_1 \subsetneq B$ and J_1 the annihilator of the simple highest weight module $L_{B_1}(\mu)$ for $\underline{r}$ of highest weight $\mu - \rho$. By ([20], 5.7) we can choose $\nu \in P(R)^+ \cap B_1^\perp$ sufficiently large such that the module induced from $L_{B_1}(\mu) \otimes \mathbf{C}_{-\nu}$ is simple, hence isomorphic to $L(\mu') : \mu' = \mu - \nu$. Now $\mathcal{O}_{J(\mu')} = \mathrm{Ind}_{\underline{r}}^{\underline{g}} \mathcal{O}_{J_1} = \mathcal{O}_{J(\lambda)}$. By (*) of 2.4 and the hypothesis we can choose $\lambda' \in W\lambda$ such that $J(\lambda) = J(\lambda')$ and $\lambda' - \mu \in P(R)$, so then $\lambda' - \mu' \in P(R)$. Then by ([13], 2.11) and self-duality of simples in the $\underline{O}$ category we can choose $\mu_1, \mu_2 \in \Lambda$ regular with $\mu_1 \in W_\lambda \mu_2$ and finite dimensional simple $\underline{g}$ modules E_1, E_2 such that $L(\mu')$ (resp. $L(\lambda')$) is a submodule of $E_1 \otimes L(\mu_1)$ (resp. $E_2 \otimes L(\mu_2)$). Then by ([18], 3.11) and ([17], 4.3) $\bar{\mathcal{O}}_{J(\mu_1)} = \mathbf{G}\mathcal{V}(L(\mu_1)) = \mathbf{G}\mathcal{V}(L(\mu')) = \bar{\mathcal{O}}_{J(\mu')} = \bar{\mathcal{O}}_{J(\lambda')} = \mathbf{G}\mathcal{V}(L(\lambda')) = \mathbf{G}\mathcal{V}(L(\mu_2)) = \bar{\mathcal{O}}_{J(\mu_2)}$. Writing $\mu_i = w_i\xi$, $w_i \in W_\lambda$: $i = 1, 2$, ξ is dominant (and regular), this last equality just means that w_1, w_2 belong to the same two-sided cell $\mathcal{DC}$ and W_λ by ([18], 3.10) and say the discussion in ([17], Sect. 5). Choose $y \in \mathcal{DC}$ in the intersection of the left cell containing w_2 and the right cell containing w_1 which is non-empty by ([25], 12.16). Then by ([12], 3.8) we can find a finite dimensional simple $\underline{g}$ module E_3 such that $L(w_1\xi)$ is a submodule of $E_3 \otimes L(y\xi)$. On the other hand $J(\mu_2) = J(w_2\xi) = J(y\xi)$. Then by ([13], 2.11) we can choose a finite dimensional simple $\underline{g}$ module E_4 such that $L(\lambda'')$ is a submodule of $E_4 \otimes L(y\xi)$ and $J(\lambda'') = J(\lambda')$. Using that simple modules are self-dual in the $\underline{O}$ category, we conclude that $L(\lambda'')$ is a simple submodule of some $E \otimes L(\mu')$. Since $L(\mu')$ is induced, this means that $L(\lambda'')$ is not rigid in the sense of ([20], 1.2) and so $F(\lambda'') \subsetneq A(\lambda'')$ by ([20], 7.6). The conclusion of the theorem then results from 3.3.

REMARK: Note that the proof does not require any connection between B' in 3.3 and B_1 above, that is $L(\lambda'')$ and $L(\mu')$ do not have to be induced from the same Levi factor.This is as it should be and indicates that an important technical obstacle is overcome.

3.6. Taking account of our remarks in 2.7, the above already gives in type A_n the following result which appears to be new.

COLLARY. *Suppose $\underline{g}$ has only factors of type A_n. Then $J \in \mathrm{Max}\, U(g)$ is induced from a primitive ideal of finite codimension in the enveloping algebra of a Levi factor.*

4. A SUPPORT LEMMA

4.1. Let $\mathcal{O}$ be a nilpotent orbit and C an irreducible component of $\mathcal{O} \cap \underline{n}$. We call C an orbital variety in $\underline{g}$. Recall the definition ([17], 2.4) of the characteristic polynomial $p_C \in S(\underline{h})$ of C. By ([21], 8.3) p_C can be written as sum with rational coefficients ≥ 0 of products of simple roots. We define $\operatorname{Supp} p_C$ to be that subset of simple roots which appear in this expression for P_C.

4.2. Let $\underline{r} = \underline{r}_{B'}$ be the Levi factor of $\underline{g}$ defined by $B' \subset B$. Set $\underline{n}_{B'} = \underline{n} \cap \underline{r}$ and let $\underline{m}_{B'}$ denote the $\underline{h}$ stable complement of $\underline{n}_{B'}$ in $\underline{n}$. Let $\mathcal{O}'$ be a nilpotent orbit in $\underline{r}$ and C' a component of $\underline{n}_{B'} \cap \mathcal{O}'$. We define $\operatorname{Ind}_{\underline{r}}^{\underline{g}} C'$ to be the unique component of $\underline{n} \cap \operatorname{Ind}_{\underline{r}}^{\underline{g}} \mathcal{O}'$ with the same Zariski closure as $C' + \underline{m}_{B'}$. (Recall here that by definition $\operatorname{Ind}_{\underline{r}}^{\underline{g}} \mathcal{O}'$ has the same Zariski closure as $\mathbf{G}(C' + \underline{m}_{B'})$. One may remark ([17], 9.6 (i)) that every component C' above has the same Zariski closure as $\mathbf{B}'(\underline{n}_{B'} \cap {}^w\underline{n}_{B'})$, for some $w \in W_{B'}$, where $\mathbf{B}'$ is the Borel subgroup of $\mathbf{G}' \subset \mathbf{G}$ respectively corresponding to the subalgebras $\underline{b}' := \underline{h} + \underline{n}_{B'}$, $\underline{r}_{B'}$ of $\underline{g}$. Now let $\mathbf{M}$, $\mathbf{B}$ be the subalgebras of $\mathbf{G}$ corresponding to $\underline{m}_{B'}$, $\underline{b} := \underline{h} + \underline{n}$. One has $\mathbf{B} = \mathbf{M}\mathbf{B}'$. Now $\underline{n} \cap {}^w\underline{n} = (\underline{n}_{B'} \cap {}^w\underline{n}_{B'}) \oplus \underline{m}_{B'}$. Both components in this direct sum are $\mathbf{B}'$ stable and since $[\underline{m}_{B'}, \underline{n}] \subset \underline{m}_{B'}$, we conclude that

$$\mathbf{B}(\underline{n} \cap {}^w\underline{n}) = \mathbf{B}'(\underline{n}_{B'} \cap {}^w\underline{n}_{B'}) + \underline{m}_{B'} .$$

Thus $\operatorname{Ind}_{\underline{r}}^{\underline{g}} C'$ is just the unique component of $\underline{n} \cap \operatorname{Ind}_{\underline{r}}^{\underline{g}} \mathcal{O}'$ with the same Zariski closure as $\mathbf{B}(\underline{n} \cap {}^w\underline{n})$. However, it should be noted that there is considerable non-uniqueness of w in this description.

4.3. Retain the above notation.

LEMMA. *Let C be an orbital variety in $\underline{g}$. Set $B' = \operatorname{Supp} p_C$. Then $C = \operatorname{Ind}_{\underline{r}_{B'}}^{\underline{g}} C'$ for some orbital variety C' in $\underline{r}_{B'}$. Moreover $\bar{C} \cap \underline{n}_{B'} = \bar{C}'$.*

We may, of course, assume $B' \subsetneq B$. We claim that the support condition implies that every component of $\bar{C} \cap \underline{n}_{B'}$ has codimension $\dim \underline{m}_{B'}$. This follows by the reasoning in ([21], 8.3). Take an ordering of R^+ so that the roots from $R^+ \setminus R'^+$ (where $R' = \mathbb{Z}B' \cap R$ and $R'^+ = R' \cap R^+$) are selected first and set $\underline{m}_i = \underline{m}_{\alpha_i}$ (notation [21], 8.2). In any chain $\bar{C} = C_1, C_2, \ldots, C_n$ of closed irreducible subvarieties of $\underline{n}$ obtained by taking C_{i+1} to be an irreducible component of $\underline{m}_i \cap C_i$, we claim that $C_{i+1} \subsetneq C_i$ (and consequently $\dim C_{i+1} = \dim C_i - 1$) whenever $\alpha_i \in R^+ \setminus R'^+$. Otherwise by ([21], 8.3) α_i occurs in the expression for p_C contradicting the support condition.

Now suppose C has the same Zariski closure as $\mathbf{B}(\underline{n} \cap {}^y\underline{n}) : y \in W$. We can write $\underline{n} \cap {}^y\underline{n} = \underline{n}_{B'} \cap {}^y\underline{n} + \underline{m}$, for some $\underline{m} \subset \underline{m}_{B'}$. Now $\underline{n}_{B'} \cap {}^y\underline{n}$ is the root subspace of $\underline{n}_{B'}$ spanned by the roots $S := \{\alpha \in R'^+ \mid y^{-1}\alpha \in R^+\}$. If $\alpha, \beta \in S$ (resp. $\alpha, \beta \notin S$) and $\alpha + \beta \in R'^+$, then obviously $\alpha + \beta \in S$

(resp. $\alpha+\beta \notin S$). It is well-known that this implies that there exists $w \in W_{B'}$ such that $S = R'^{+} \cap w^{-1}R'^{+}$, and then $\underline{n}_{B'} \cap {}^{y}\underline{n} = \underline{n}_{B'} \cap {}^{w}\underline{n}_{B'} := \underline{q}$. Let C' be the orbital variety in $\underline{r}_{B'}$ with the same Zariski closure as $\mathbf{B}'_{\underline{q}}$.

Now $\underline{n}_{B'} \cap \mathbf{B}(\underline{n} \cap {}^{y}\underline{n}) = \underline{n}_{B'} \cap \mathbf{M}(\mathbf{B}'\underline{q} + \mathbf{B}'\underline{m}) = \mathbf{B}'\underline{q}$. On the other hand $\mathbf{B}(\underline{n} \cap {}^{y}\underline{n})$ is locally closed in $\underline{n}$, so quasi-projective and we conclude from ([27], p. 59, Cor. 1) that $\dim C' = \dim \mathbf{B}'\underline{q} \geq \dim \mathbf{B}(\underline{n} \cap {}^{y}\underline{n}) - \dim \underline{m}_{B'} = \dim C - \dim \underline{m}_{B'} = \dim(\bar{C} \cap \underline{n}_{B'})$. Since $\bar{C} \supset \mathbf{B}(\underline{n} \cap {}^{y}\underline{n})$, we conclude from this and the previous calculation, an equality in the above and in particular that $\bar{C}'$ is an irreducible component of $\bar{C} \cap \underline{n}_{B'}$.

Finally $\underline{n} \cap {}^{y}\underline{n} \subset \underline{n}_{B'} \cap {}^{y}\underline{n} + \underline{m}_{B'} \cap {}^{w}\underline{n}_{B'} + \underline{m}_{B'} = \underline{n} \cap {}^{w}\underline{n}$, since $w \in W_{B'}$. Consequently $\overline{\operatorname{Ind} C'} \supset \bar{C}$. On the other hand $\dim(\operatorname{Ind} C') = \dim C' + \underline{m}_{B'} = \dim C$, by the above. Hence $\operatorname{Ind} C'$ and C are orbital varieties with the same Zariski closure and so must coincide. The identity $\overline{\operatorname{Ind} C'} = \bar{C}' + \underline{m}_{B'}$, implies $\bar{C} \cap \underline{n}_{B'} = \overline{\operatorname{Ind} C'} \cap \underline{n}_{B'} = \bar{C}'$.

REMARKS: Let $I(\bar{C})$ denote the ideal of definition of $\bar{C}$ in $S(\underline{n}^{-})$. If $I(\bar{C})$ has codimension C generators which are $\underline{h}$ weight vectors, the assertion of the lemma follows easily from ([7], 4.15).

It is not obvious in general if $\bar{C}'$ (as defined above) has the same dimension as $\bar{C} \cap \underline{n}_{B'}$. One could ask if the components of $\bar{C} \cap \underline{n}_{B'}$ are closures of orbital varieties in $\underline{r}_{B'}$. By ([17], 7.3) it would be enough to show that $\bar{C} \cap \underline{n}_{B'}$ is involutive. Whilst $\langle I(\bar{C}) + \underline{m}_{B'} \rangle$ is closed under the Poisson bracket on $S(\underline{n}^{-})$, it is not obvious that this holds for its radical $I(\bar{C} \cap \underline{n}_{B'})$.

4.4. Take $\lambda \in \underline{h}^{*}$ dominant and regular. Recall ([16], II, 5.1) that for each $w \in W_{\lambda}$, there exists a polynomial p_w on $\underline{n}^{*}$ such that $p_w(\mu) = \operatorname{rk} U(\underline{g})/J(w\mu)$, $\forall \mu \in \lambda^{+}$, where rk denotes Goldie rank. More precisely p_w is a positive sum of products of roots in R_{λ}^{+}. If λ is not integral, we may find $x \in W \setminus \{1d\}$ such that $x\lambda$ is still dominant. Then $B_{x\lambda} = xB_{\lambda}$. Yet by 2.4.(*) the ideals obtained on applying the Duflo map to $W_{\lambda}\lambda$ is in natural correspondence with those obtained from $W_{x\lambda}(x\lambda) = xW_{\lambda}\lambda$. Consequently the Goldie rank polynomial p_w is unchanged, except that if we wish to compute Goldie rank with respect to $x\lambda$ then we must transport p_w to $x.p_w$ (so then $(x.p_w)(x\lambda) = p_w(\lambda)$) which exactly replaces each $\alpha \in B_{\lambda}$ occurring in p_w by $x\alpha \in B_{x\lambda}$. Notice, however, that this action changes both the simple modules and their associated varieties (see [17], Sect. 10 for example). With this understood we now announce the

THEOREM. *Assume $\lambda \in \underline{h}^{*}$ dominant and regular. Suppose $B_1 := \operatorname{Supp} p_w \subsetneq B_{\lambda}$. Then there exists $B' \subsetneq B$ such that $J(w\lambda)$ is induced* *from some* $J \in \operatorname{Prim} U(\underline{r}_{B'})$.

Choose a basis $\{v_1, v_2, \ldots, v_\ell\}$ for $\mathbb{Q}R$ such that $\{v_{r+1}, \ldots, v_\ell\} = B_\lambda$ and $\{v_{s+1}, \ldots, v_\ell\} = B_1$. Then the lexicographic ordering corresponding to this basis defines a positive system $xR^+ : x \in W$. Moreover by this choice there exist subsets $B' \subset B'' \subset B$ such that $B_1 \subset \mathsf{N}(xB')$, $B_\lambda \subset \mathsf{N}(xB'')$ and $\operatorname{card} B_1 = \operatorname{card} B'$, $\operatorname{card} B_\lambda = \operatorname{card} B''$. Set $\tilde{B}_1 = x^{-1}B_1$. Since $B_{x^{-1}\lambda} = x^{-1}B_\lambda \subset R^+$, it follows that $x^{-1}\lambda$ is dominant (and regular). Thus we may replace λ by $\tilde{\lambda} := x^{-1}\lambda$ and then p_w is replaced by $\tilde{p}_w := x^{-1}p_w$. Moreover, $\operatorname{Supp} x^{-1}p_w = \tilde{B}_1 \subset \mathsf{N}B'$. Yet by definition of Supp, this just means that $\operatorname{Supp} x^{-1}p_w = B'$, if one takes account of the condition $\operatorname{card} \tilde{B}_1 = \operatorname{card} B'$, which also forces B' to be a strict subset of B. Since $x^{-1}B_\lambda \subset R^+$, we have by 2.4.(*) that $J(w\lambda) = J(x^{-1}w\lambda) = J(\tilde{w}\tilde{\lambda})$ where $\tilde{w} = x^{-1}wx \in x^{-1}W_\lambda x = W_{\tilde{\lambda}}$. Again $p_w(\lambda) = (x^{-1}.p_w)(\tilde{\lambda}) = \tilde{p}_w(\tilde{\lambda})$. Since $p_w = p_{\tilde{w}}$, we also have $\tilde{p}_w = \tilde{p}_{\tilde{w}}$, and we can work from now on in the tilde variables. For convenience we drop the tilde.

Now consider the characteristic polynomial $p_{\mathcal{S}(L(w\lambda))}$ attached to the vanishing cycle $\mathcal{S}(L(w\lambda))$ of $L(w\lambda)$, by the procedure in ([17], Sect. 5). Up to a scalar it equals $p_{w^{-1}}$. Here we could have assumed that w is an involution (via [16], 3.4) without changing $J(w\lambda)$ and this we shall now do.

One has (up to an overall scalar) that

$$(*) \qquad p_w = p_{\mathcal{S}(L(w\lambda))} = \sum_i k_i p_{C_i}$$

where the k_i are positive integers (depending on λ, which we consider fixed) determining the multiplicity of each component C_i of $\mathcal{S}(L(w\lambda))$ in $\underline{n}$. Each C_i is the closure of an orbital variety (corresponding to the fixed nilpotent orbit $\mathcal{O}_{J(w\lambda)}$). The positivity condition ([21], 8.3) implies that $\operatorname{Supp} p_{C_i} \subset B'$, for all i. By 4.3, the $C_i' = C_i \cap \underline{n}_{B'}$ are closures of orbital varieties in $\underline{r}_{B'}$. Moreover $C_i = C_i' + \underline{m}_{B'}$.

Now let L' denote the $U(\underline{r}_{B'})$ submodule of $L := L(w\lambda)$, generated by the (unique up to scalars) highest weight vector e of L. It is clear L' is a highest weight module (with highest weight $w\lambda - \rho$) and one checks that the non-degenerate contravariant form on L restricts to a non-degenerate contravarient form on L' (using weight space decomposition). Thus L' is simple. Now $\underline{m}_{B'}e = 0$, so L is a $U(\underline{p}_{B'})$ module.

Define $\operatorname{Ind} L' := U(\underline{g}) \otimes_{U(\underline{p}_{B'})} L'$. Universality gives a surjection $\operatorname{Ind} L' \twoheadrightarrow L$. We claim that $C := \mathcal{V}(L) = \mathcal{V}(\operatorname{Ind} L')$, that is the associated varieties of L and $\operatorname{Ind} L'$ on $\underline{n}$ (but not necessarily their vanishing cycles) coincide. Now $C = C \cap \underline{n}_{B'} + \underline{m}_{B'}$, so $\sqrt{\operatorname{gr} \operatorname{Ann}_{U(\underline{n}^-)} e}$ is generated by its intersection with $S(\underline{n}_{B'}^-)$. Take x in this common intersection. We can assume x homogeneous without loss of generality, hence of the form $\operatorname{gr} y$ with $y \in U(\underline{n}_{B'}^-)$. Take the canonical filtrations in $U(\underline{n}^-)$, $U(\underline{n}_{B'}^-)$, $L = U(\underline{n}^-)e$ and assume y to be of filtration degree d. By Bernstein's criterion ([2], Prop. 1.4(2)) for each $k \in \mathbb{N}$, the element $y^k e$ has filtration degree $\varphi(k) \leq kd$, where $kd - \varphi(k) \to \infty$ as $k \to \infty$. However, by Bernstein's criterion again, this just means that $x = \operatorname{gr} y \in \sqrt{\operatorname{gr} \operatorname{Ann}_{U(\underline{n}_{B'}^-)} e}$. Yet $1 \otimes e$ is the canonical generator of $\operatorname{Ind} L'$ and of course $\operatorname{Ann}_{U(\underline{n}_{B'}^-)}(1 \otimes e) = \operatorname{Ann}_{U(\underline{n}_{B'}^-)} e$, so this proves the required assertion.

Set $C' = \mathcal{V}(L')$, $J' = \operatorname{Ann} L'$. Since $\mathbf{G}'C' = \bar{\mathcal{O}}_{J'}$ and by the above $C' = C \cap \underline{n}_{B'} = \cup C_i'$, we conclude that $\mathbf{G}'C_i' = \bar{\mathcal{O}}_{J'}$. That is the C_i' all generate the same $\mathbf{G}'$ orbit and so the $p_{C_i'}$ all generate the same $W_{B'}$ module M, which is simple. (It is the Springer representation attached to $\mathcal{O}_{J'}$. For further details see [21], introduction). Since $P_{C_i} = p_{C_i'}$ we conclude that $p_w \in M$ by $(*)$.

The Goldie rank polynomial attached to J' is a non-zero vector in M. Then by ([16], II, 5.5) amongst the ideals in the primitive fibre $X_{\hat{\lambda}}^{\underline{r}_{B'}} \subset \operatorname{Prim} U(\underline{r}_{B'})$ over $\hat{\lambda}$ there is a subset whose associated Goldie rank polynomials $\{p_i\}$ give a basis for M. Let $J_{B'}(\mu) : \mu \in \underline{h}^*$ denote the annihilator of the $U(\underline{r}_{B'})$ simple highest weight module with highest weight $\mu - \rho$ and viewed as a $\underline{p}_{B'}$ module in the usual fashion. Let w_λ denote the unique longest element in W. Then we can take this set of primitive ideals to be $\{J_{B'}(w_i w_\lambda \lambda) : w_i \in S\}$ where S is an appropriate subset of W_{B_1}. Let $M_{B'}(x w_\lambda \lambda) : x \in W_{B'}$, denote the Verma module for $\underline{r}_{B'}$ with highest weight $x w_\lambda \lambda - \rho$ and $L_{B'}(x w_\lambda \lambda)$ its unique simple quotient. A comparison of the formal character of $U(\underline{g}) \otimes_{U(\underline{p}_{B'})} L_{B'}(w_i w_\lambda \lambda)$ with that of its simple quotient $L(w_i w_\lambda \lambda)$ using the fact that by the truth of the Kazhdan-Lusztig conjectures one has $[M_{B'}(x w_\lambda \lambda) : L_{B'}(y w_\lambda \lambda)] = [M(x w_\lambda \lambda) : L(y w_\lambda \lambda)]$, $\forall x, y \in W_{B'}$ shows that the induced module is simple and so one has $J(w_i w_\lambda \lambda) = \operatorname{Ind}_{\underline{p}_{B'}}^{\underline{g}}(J_{B'}(w_i w_\lambda \lambda))$. Moreover, the Goldie rank polynomials attached to the induced ideals are again the p_i by say combining ([20], 6.1, 6.8) with ([16], I, 5.12(iii)). Since $p_w \in M$ it follows from ([16], II, Thm. 5.5) that p_w must in fact be one of the p_i and then $J(w\lambda)$ must be the corresponding induced ideal.

REMARKS: Even the fact noted above that p_w generates a simple W_{B_1}

module appears to be new. The same reasoning shows that if card $B_\lambda <$ card B, then $J(w\lambda)$ is induced for all $w \in W$; but this was already noted in 2.7. For the conclusion of the theorem, we do not need that λ be regular (exept that this makes the choice of w unambiguous). Indeed if $J(w\lambda)$ is induced for λ regular, then the translated ideals $J(w\mu) : \mu \in \Lambda^+$ are similarly induced by an easy application of the translation principle. (Use $E \otimes \text{Ind}\, L' \xrightarrow{\sim} \text{Ind}(L' \otimes E)$ to show that the summand of the left hand side defined in ([13], 2.11) and used in ([6], 2.9) is an induced module).

5. The integral case

5.1. We deduce from 3.5 and 4.4 some consequences for $\text{Prim}_c\, U(\underline{g})$ in the integral case. In principle the arguments carry over to the non-integral case; but there are some subtle technicalities arising from difficulties in translating to the walls which we have not had time to analyze. Assume from now on that $\lambda \in P(R)^+$. Let $P(R)^{++}$ denote the regular elements of $P(R)^+$. We say that $J \in \text{Prim}\, U(\underline{g})$ is rigid if it cannot be written in the form $\text{Ind}_{\underline{p}}^{\underline{g}} J'$ for some $J' \in \text{Prim}\, U(\underline{p})$ with $\underline{p}$ a proper parabolic subalgebra of $\underline{g}$.

5.2. For completion we recall without proof the result noted in ([21], 8.7).

Theorem. *Assume $\lambda \in P(R)^{++}$, take $w \in W$ and set $B' = \text{Supp}\, p_w$. Suppose $J(w\lambda) \in \text{Prim}_c\, U(\underline{g})$. Then $J(w\lambda) = J(w_B w_{B'} \lambda)$ and is the annihilator of the module induced from the one dimensional $\underline{p}_{B''} : B'' = w_B B'$ module with highest weight $w_B w_{B'} \lambda - \rho$.*

5.3. Given $w \in W$, set $\tau(w) = \{\alpha \in B \mid w\alpha \in R^+\}$. After Borho-Jantzen ([6], 2.14) one knows that $J(w\lambda) : \lambda \in P(R)^{++}$ degenerates on translation to the α-wall if and only if $\alpha \in \tau(w)$. It was a consequence of this that $J(w\lambda) \neq J(w'\lambda)$ if $\tau(w) \neq \tau(w')$. A result of a similar nature was discovered independently by Duflo ([10], Prop. 11). One calls τ the Borho-Jantzen-Duflo τ-invariant. The ideals $J(w_B w_{B'} \lambda) : B' \subset B$ have τ-invariant B' and are the minimal ideals in $X_\lambda^{\underline{g}}$ with this value of the τ-invariant. For λ integral, these are always induced and many properties of them can be explicitly determined (for example the Goldie rank polynomial is the product of the positive roots generated by B', normalized to take value 1 at $\lambda = \rho$).

At the other extreme one may consider the ideals in $X_\lambda^{\underline{g}}$ which are maximal for a given value of τ. By ([6], 2.17) these take the form $J(w_{B'}\lambda)$, $B' \subset B$. In this case one does not have an explicit formula for the Goldie rank polynomial and relatively little idea of when such ideals are induced. We prove the

THEOREM. *Take $\lambda \in P(R)^+$ and $B' \subset B$. Suppose $J = J(w_{B'}\lambda) \in \operatorname{Prim}_c U(\underline{g})$. Then the following two assertions are equivalent.*

(i) *J is rigid.*

(ii) *$\mathcal{O}_J$ is $\hat{\lambda}$-rigid.*

By 3.4, we have (ii) $\Rightarrow$ (i). Set $p = p_{w_{B'}}$. To prove the opposite assertion it is enough by 4.4 to consider the case $\operatorname{Supp} p = B$. Now take $\mu \in P(R)^+$ and set $n_\alpha^\mu = (\alpha^\vee, \mu) : \alpha \in B$ which are non-negative integers. By the positive property ([21], 8.6) for p it follows that $p(\mu) \leq p(\lambda)$ if $n_\alpha^\mu \leq n_\alpha^\lambda$, $\forall \alpha \in B$ with a strict inequality if $n_\alpha^\mu < n_\alpha^\lambda$ for some $\alpha \in B$. We claim that $p(\mu) \geq 1$ if $n_\alpha^\mu > 0$ for all $\alpha \in B \backslash B'$. Otherwise $p(\mu) = 0$ and $J(w_{B'}\mu)$ has degenerated at some $\alpha \in B'$ wall, contradicting ([6], 2.14) – see also the remarks in ([16], III, 1.1). We conclude that

$$(*) \qquad (\alpha^\vee, \lambda) = \begin{cases} 0: & \alpha \in B'. \\ 1: & \alpha \in B \setminus B'. \end{cases}$$

However by the first of these equalities $J(w_{B'}\lambda) = J(\lambda)$ is a maximal ideal in $X_{\hat{\lambda}}^{\underline{g}}$ and so the assertion follows from 3.5.

REMARKS: We emphasize that we have established the additional information $(*)$ above (under the hypothesis that $\operatorname{Supp} p = B$).Curiously this condition is not sufficient to ensure that $J \in \operatorname{Prim}_c U(\underline{g})$. In ([24], table) we gave a complete list of the Goldie ranks of these ideals in the case when $B' = B \setminus \{\alpha\}$ and $(\lambda, \beta) = 0$, $\forall \beta \in B'$ (which we called an α-corner). All such ideals were found to be induced, that is the associated orbits are not rigid.

REFERENCES

[1] D. Barbasch and D.A. Vogan, *Unipotent representations of complex semisimple groups*, Annals of Math. **121** (1985), 41–110.

[2] I.N. Bernstein, *Modules over a ring of differential operators*, Func. Anal. Appl. **5** (1971), 89–101.

[3] W. Borho, *Primitive vollprime Ideale in der Eienhüllenden von* $\underline{so}(5, \mathbb{C})$, J.Algebra **43** (1976), 619–654.

[4] W. Borho, *Über Schichten halbeinfacher Lie-Algebren*, Invent. Math **65** (1981), 283–317.

[5] W. Borho and H. Kraft, *Über die Gelfand-Kirillov-Dimension*, Math. Ann. **220** (1976), 1–24.

[6] W. Borho and J.C. Jantzen, *Über primitive Ideale in der Einhüllenden einer halbeinfachen Lie-Algebra*, Invent. Math. **39** (1977), 1–53.

[7] W. Borho, J.-L. Brylinski and R. MacPherson, *Equivariant K-theory approach to nilpotent orbits*, preprint, IHES, 1986.

[8] N. Conze, *Algèbres d'opératerurs differentiels et quotients des algèbres enveloppantes*, Bull. Soc. Math. France **102** (1974), 379–415.

[9] J. Dixmier, "Algèbres enveloppantes," Cahiers scientifiques XXXVII, Gauthier-Villars, Paris, 1974.

[10] M. Duflo, *Sur la classification des idéaux primitifs dans l'algèbre enveloppante d'une algèbre de Lie semisimple*, Annals of Math. **105** (1977), 107–120.

[11] I.N. Herstein, "Topics in ring theory," Chicago Lectures in Mathematics, University of Chicago Press, Chicago, 1969.

[12] O. Gabber and A. Joseph, *On the Bernstein-Gelfand-Gelfand resolution and the Duflo sum formula*, Compos. Math **43** (1981), 107–131.

[13] J.C. Jantzen, "Moduln mit einen höchsten Gewicht," Lecture Notes in Mathematics 750, Springer-Verlag, Berlin, Heidelberg, New York, Tokyo, 1979.

[14] J.C. Jantzen, "Einhüllenden Algebren haleinfacher Lie-Algebren," Springer-Verlag, Berlin, Heidelberg, New York, Tokyo, 1983.

[15] A. Joseph, *A characteristic variety for the primitive spectrum of a semisimple Lie algebra*, in "Non-commutative harmonic analysis," Lecture Notes in Mathematics 587 (Ed. J. Carmona and M. Vergne), Springer-Verlag, Berlin, Heidelberg, New York, Tokyo, 1977, pp. 102–118.

[16] A. Joseph, *Goldie rank in the enveloping algebra of a semisimple Lie algebra, I*, J. Algebra **66** (1980), 269–283; *II*, J. Algebra **66** (1980), 284–306; *III*, **73** (1981), 295–326.

[17] A. Joseph, *On the variety of a highest weight module*, J. Algebra **88** (1984), 238–278.

[18] A. Joseph, *On the associated variety of a primitive ideal*, J. Algebra **93** (1985), 509–523.

[19] A. Joseph, *Rings which are modules in the Bernstein-Gelfand-Gelfand $\mathcal{O}$ category*, J. Algebra **113** (1988), 110-126.

[20] A. Joseph, *A surjectivity theorem for rigid highest weight modules*, Invent. Math. **92** (1988), 567–596.

[21] A. Joseph, *On the characteristic polynomials of orbital varieties*, Ann. Sci. Ec. Norm. Sup. (to appear).

[22] A. Joseph, *Rings of b-finite endomorphisms of highest weight modules are Goldie*, in "Proceedings of Amitsur conference," Bar-Ilan University, 1989 (to appear).

[23] A. Joseph, *Kostant's problem and Goldie rank*, in "Non-commutative harmonic analysis," Lecture Notes in Mathematics 880 (Ed. J. Carmona and M. Vergne), Springer-Verlag, Berlin, Heidelberg, New York, Tokyo, 1981, pp. 249–266.

[24] A. Joseph, *Multiplicity of the adjoint representation in simple quotients of the enveloping algebra of a simple Lie algebra*, Trans. Amer. Math. Soc. (to appear).

[25] G. Lusztig, *Characters of reductive groups over finite fields*, "Annals of Math. studies 107," Princeton, New Jersey, 1984.

[26] C. Moeglin, *Idéaux complétement premiers de l'algébre enveloppante de* $\underline{gl}_n(\mathbb{C})$, J. Algebra **106** (1987), 287–366.

[27] I.R. Shafarevitch, "Basic algebraic geometry," Springer-Verlag, Berlin, Heidelberg, New York, Tokyo, 1977.

A. JOSEPH

The Donald Frey Professorial Chair
Department of Theoretical Mathematics
The Weizmann Institute of Science
Rehovot 76100
ISRAEL

and

Laboratoire de Mathématiques Fondamentales
Equipe de recherche associée au CNRS
Université de Pierre et Marie Curie
FRANCE

A Formula of Gauss–Kummer and the Trace of Certain Intertwining Operators

BERTRAM KOSTANT[1]

§ 1. INTRODUCTION

1.1. In our opinion one of the most interesting aspects of the orbit method is the occurrence of "independence of polarization". From the perspective of geometric quantization, given two transverse polarizations, one may set up a formal kernel operator which then intertwines the two corresponding quantizations. This has been referred to as the BKS kernel. See e.g. [Sn], § 5. For the case of real polarizations of a coadjoint orbit of a nilpotent Lie group this more or less comes down to the Fourier transform. However for hyperbolic coadjoint orbits of a semisimple Lie group one may easily relax the condition of transversality and then the transforms become the well-known and well-studied intertwining operators associated to the spherical principal series. As one knows these operators are parameterized by elements of the Weyl group. The particular case when the polarizations are transverse corresponds to the long element of the Weyl group. In this paper we find that an analytic continuation of this operator is traceable and an expression for the trace relates to a known formula of Gauss–Kummer for hypergeometric series.

A result of Gauss for the hypergeometric series

$$F(a,b;c;z)=\sum_{k=0}^{\infty}\frac{(a)_k\cdot(b)_k}{(c)_k\cdot(1)_k}\cdot z^k \tag{.1}$$

when $\mathrm{Re}(c-a-b)>0$ is the famous equation

$$F(a,b;c;1)=\frac{\Gamma(c-a-b)\cdot\Gamma(c)}{\Gamma(c-a)\cdot\Gamma(c-b)}\,. \tag{.2}$$

Using (.1) together with certain "quadratic transformations" of F due to Gauss and Kummer one obtains the also well known equation

$$F(a,b;1+a-b;-1)=2^{-a}\cdot\frac{\Gamma(1+a-b)\cdot\Gamma(1/2)}{\Gamma(1-b+a/2)\cdot\Gamma((a+1)/2)} \tag{.3}$$

of Kummer.

[1]Supported in part by NSF Grant No. DMS–87–03278

Putting $1+a-b=\tau$, $a=n-1$ so that $b=-\tau+n$ Kummer's equation (.3) may be written

$$(.4)\qquad F(n-1,-\tau+n;\tau;-1)=2^{-n+1}\cdot\frac{\Gamma(\tau)\cdot\Gamma(1/2)}{\Gamma(n/2)\cdot\Gamma(\tau-(n-1)/2)}\,.$$

But if $n\geq 2$ is an integer then, as a function of τ, one recognizes that 2 times the right side of (.4) is the inverse of Harish–Chandra's c-function (suitably parameterized) for the generalized Lorentz group $SO(n,1)$.

In this paper it will be shown that equation (.4), for suitable values of the parameters, is a special case of a much more general theorem. The theorem is a statement, for an "arbitrary" semisimple Lie group G, about the c-function, the traceability (which we show exists) of an intertwining operator, $\mathcal{A}$, and the trace of $B\mathcal{A}$ for a certain bounded operator B. The generalization of the sum on the left side of (.4) arises since $\operatorname{tr} B\mathcal{A}$ can be computed as a sum of traces of the restriction of $B\mathcal{A}$ to all suitable K-types. The generalization of the right side of (.4) is the inverse of the c-function and the theorem, among other things, asserts that the latter equals $\operatorname{tr} B\mathcal{A}$. The equality in (.4), (with suitable values of the parameters), follows from the theorem when $G=SO(n,1)$.

1.2. In more detail assume G is a (not necessarily connected) semisimple group with finite center. If not connected we assume at least that (1) in § 2.1 is satisfied. Adopting standardized notation let $G=KAN$ be an Iwasawa decomposition of G and let M be the centralizer of A in K. We use the corresponding lower case German letter to denote the Lie algebra of each of these groups. Let W be the restricted Weyl group operating in $\mathfrak{a}$ and its dual $\mathfrak{a}^*$ and let $\kappa\in W$ be the longest element. Let $D\subseteq\mathfrak{a}^*$ be the open Weyl chamber and let $D_{\mathbf{C}}=\{\nu\in\mathfrak{a}^*_{\mathbf{C}}\mid \operatorname{Re}\nu\in D\}$ so that $D_{\mathbf{C}}$ is an open cone in $\mathfrak{a}^*_{\mathbf{C}}$. Let $\rho\in D$ have its usual meaning. For any $v\in W$ and $\mu\in\rho+D_{\mathbf{C}}$ let

$$A(v,\mu):L_2(K)\longrightarrow L_2(K)^M$$

be the corresponding intertwining operator associated with spherical representations of G. (See e.g. [HC], [H_1], [KN-S], [KU-S], [S].) The indicated action of M is by right translations. Also for any $v\in W$ let T_v be the unitary operator on $L_2(K)^M$ induced by a right translation associated to v.

By showing that $T_\kappa A(\kappa,\mu)$, for μ in a suitable translate of the open set $D_{\mathbf{C}}$, is given by right convolution by a sufficiently smooth (although not C^∞ in general) function on K we prove

THEOREM A. *There exists an element $\rho_0 \in \rho + D$ such that for any $\mu \in \rho_0 + D_{\mathbf{C}}$ the intertwining operator $A(\kappa, \mu)$ on $L_2(K)$ is of trace class. Furthermore*

$$\operatorname{tr} T_v A(\kappa, \mu) = 0 \text{ if } v \neq \kappa \tag{.51}$$

and

$$\operatorname{tr} T_\kappa A(\kappa, \mu) = 1\,. \tag{.52}$$

1.3. If $\gamma \in K^\wedge$ and $\mu \in \mathfrak{a}_{\mathbf{C}}^*$ then we have defined a $l(\gamma) \times l(\gamma)$ matrix $P^\gamma(\mu)$ in [K] where for $V_\gamma \in \gamma$

$$l(\gamma) = \dim(V_\gamma^*)^M\,.$$

(The definition of $P^\gamma(\mu)$ here is slightly different from that in [K].) We have proved in [K] that $P^\gamma(\mu)$ is invertible if μ is in the closure of $\rho + D_{\mathbf{C}}$ and have noticed here that for such μ and any $\nu \in \mathfrak{a}_{\mathbf{C}}^*$ we can invariantly regard

$$P^\gamma(\nu) \cdot P^\gamma(\mu)^{-1} \in \operatorname{End}(V_\gamma^*)^M\,.$$

On the other hand the action of W on $(V_\gamma^*)^M$ defines for any $v \in W$ an operator

$$Q_v^\gamma \in \operatorname{Aut}(V_\gamma^*)^M\,.$$

Now for $\mu \in \mathfrak{a}_{\mathbf{C}}^*$ let $c(\mu)$ be Harish–Chandra's c-function for G. Our parameterization is somewhat different from, say, that in [H]. See (67.5) here.

By restricting the operators on the left side of (.51) and (.52) to the various K-types and using a result on such restrictions of $A(\kappa, \mu)$ – (a more general restriction result was found independently by Wallach. See Theorem 18 or [W]) – we obtain as a consequence of Theorem A.

THEOREM B. *Let $\gamma \in K^\wedge$, $v \in W$ and $\mu \in \mathfrak{a}_{\mathbf{C}}^*$. Let ρ_0 be as in Theorem A. (See (105).) Assume $\mu \in \rho_0 + D_{\mathbf{C}}$. Then*

$$c(\mu) \neq 0$$

and

$$\sum_{\gamma \in K^\wedge} \dim\gamma \cdot \operatorname{tr} Q_\kappa^\gamma \cdot P^\gamma(\kappa\mu + 2\rho) \cdot (P^\gamma(\mu))^{-1} = c(\mu)^{-1} \tag{.61}$$

is an absolutely convergent sum. Furthermore if $v \neq \kappa$ then

$$\sum_{\gamma \in K^\wedge} \dim\gamma \cdot \operatorname{tr} Q_v^\gamma \cdot P^\gamma(\kappa\mu + 2\rho) \cdot (P^\gamma(\mu))^{-1} = 0 \tag{.62}$$

and in particular, putting $v = 1$,

$$\sum_{\gamma \in K^\wedge} \dim \gamma \cdot \operatorname{tr} P^\gamma(\kappa\mu + 2\rho) \cdot (P^\gamma(\mu))^{-1} = 0$$

is also an absolutely convergent sum.

1.4. For convenience assume $\operatorname{Ad} G = G_{\mathbb{R}}$ using the notation of [K] (that is, $\operatorname{Ad} G$ is as large as possible). In the split rank 1 case the equations (.61) and (.62) become very explicit since in the general case we have determined $\det P^\gamma(\mu)$ (in [K]) but in the split rank 1 case (as shown in [K]) one always has $l(\gamma) \leq 1$. To apply Theorem B we need only concern ourselves with those γ for which $l(\gamma) = 1$. Such γ have been determined in [K]. (A more explicit description is given in [J] and [J-W]). To each such γ there corresponds a pair of integers i, j. Let $d(i,j)$ be the sum of $\dim \gamma$ for those γ which correspond to i, j. (The map $\gamma \longrightarrow i, j$ is one–one in case $\mathfrak{g}$ is not isomorphic to Lie $SU(n,1)$. If $\mathfrak{g} \cong \operatorname{Lie} SU(n,1)$ then the map is one–one if $j = 0$ and two to one when $j > 0$.)

In the split rank 1 case Theorem B and Theorem 2.8.8 of [K] yields the following "2 parameter" generalization of the Kummer equation (.4) – at least for certain restrictions on the variables.

THEOREM C. *Let $\mathfrak{g}$ be simple real Lie algebra of split rank 1. Let t be the dimension of the root space corresponding to a simple restricted root α and let s be the dimension of the root space corresponding to 2α so that s takes the possible values*

$$s = 0, 1, 3 \text{ and } 7.$$

Then if $\tau \in \mathbf{C}$ and $\operatorname{Re} \tau > 4s + 3t + 2$ one has

$$\frac{2^{\tau - 2s - t} \cdot \Gamma(1/2 \cdot (1 + \tau - s)) \cdot \Gamma(\tau/2)}{\Gamma(1/2 \cdot (s + t + 1)) \cdot \Gamma(\tau - (s + 1/2 \cdot t))} \tag{.71}$$

$$= \sum_{i,j} (-1)^{2i+j} \cdot d(i,j) \cdot \varphi_{i,j}(\tau) \tag{.72}$$

where for $\xi = -\tau + 2s + t$

$$\varphi_{i,j}(\tau) = \frac{[\xi(\xi+2)\ldots(\xi+2(i+j)-2)] \cdot [(\xi+1-s)(\xi+3-s)\ldots(\xi+2i-1-s)]}{[\tau(\tau+2)\ldots(\tau+2(i+j)-2)] \cdot [(\tau+1-s)(\tau+3-s)\ldots(\tau+2i-1-s)]}\,. \tag{.73}$$

The sum (.72) is absolutely convergent and is over all $i, j \in \mathbf{Z}_+$ if $s \neq 0$ but j is restricted to $0, 1$ if $s = 0$. The sum (.71) is a non-zero complex number and is the inverse of the value of the c-function at $\tau\alpha$.

On the other hand if we drop the factor $(-1)^{2i+j}$ *in (.72) then one has the absolutely convergent sum*

$$\sum_{i,j} \cdot d(i,j) \cdot \varphi_{i,j}(\tau) = 0\,. \tag{.74}$$

REMARK D: The alternation of signs in (.72) is due to the rank 1 operator Q^{γ}_{κ}. In particular for $SO(n,1)$ *the alternation of signs in Kummer's formula* (.4) *(that is, taking the special value of the hypergeometric function at* $z = -1$) is, from our point of view, due to the presence of Q^{γ}_{κ} in (.61).

The verification that (.72), when $s = 0$, reduces to Kummer's equation (.4) for n an integer ≥ 2 and $\operatorname{Re}\tau$ sufficiently large is Theorem 36 in this paper.

REMARK E: After learning of Theorem C, T. Koornwinder, in a private communication, has demonstrated to me that he can obtain Theorem C using a degenerate case of the addition formula of Jacobi functions as defined in [Koo]. Furthermore the statement is true for continuous values of the parameters s and t. His result uses Theorem 8.1 in [Koo]. This of course raises the unsolved question as to whether there is some sort of extension of (the higher rank) Theorem B involving continuous values of the parameters.

1.5. The results (.52), (.61) and (.71) have been known to us when [K] was written and are alluded to in the introduction of [K]. On the other hand the vanishing results (.51), (.62) and (.74), are relatively recent.

§ 2. PRELIMINARIES

2.1. Let G be a (not necessarily connected) semisimple Lie group with finite center. Let $\mathfrak{g}$ be the Lie algebra of G. In this paper we shall need to make use of results in [K]. However we would like to use a simpler notation here and to be somewhat more general. In [K] the main results were about $G_{\mathbb{R}} = \{g \in \operatorname{Ad} \mathfrak{g}_{\mathbb{C}} | g \text{ stabilizes } \mathfrak{g}\}$, using the notation of [K]. For our purposes here it suffices to assume (and we do) that the adjoint representation Ad of G on $\mathfrak{g}_{\mathbb{C}}$ factors through $G_{\mathbb{R}}$. That is

$$\operatorname{Ad} : G \longrightarrow G_{\mathbb{R}}\,. \tag{1}$$

Of course if G is connected then (1) is automatically satisfied.

Let $\mathfrak{k} \subseteq \mathfrak{g}$ be the Lie algebra of a maximal compact subgroup K of G. Let θ be the corresponding Cartan involution of G and $\mathfrak{g}$ and let

$$\mathfrak{g} = \mathfrak{k} + \mathfrak{p}$$

be the corresponding Cartan decomposition of $\mathfrak{g}$.

The situation here is more general than in [K] in that we are not assuming here that (1) is surjective. We will use the term "K_θ case"to mean we are in the case where (1) is surjective. That is, the case where $\mathrm{Ad}_{\mathfrak{g}} K = K_{\theta,\mathbb{R}}$ in the notation of [K]. See Propositions 1 and 2 in [K-R] on pages 761 and 762.

Let $\mathfrak{a}$ be a maximal abelian subalgebra of $\mathfrak{p}$ and let $A \subseteq G$ be the corresponding subgroup of G. Let M' be the normalizer of A in K and let $M \subseteq M'$ be the centralizer of A in K. Then $W = M'/M$ is the (restricted) Weyl group which we regard as operating in $A, \mathfrak{a}$ and in (by contragredience) the real $\mathfrak{a}^*$ and complex $\mathfrak{a}^*_{\mathbb{C}}$ dual spaces to $\mathfrak{a}$. Let N be a maximal unipotent subgroup of G which is normalized by MA and let $\mathfrak{n}$ be its Lie algebra so that

$$\mathfrak{g} = \mathfrak{k} + \mathfrak{a} + \mathfrak{n} \tag{2}$$

is an Iwasawa decomposition of $\mathfrak{g}$ and

$$G = KAN \tag{3}$$

is the corresponding Iwasawa decompostition of G.

2.2. Let X^∞ be the space of all C^∞ complex valued functions f of K such that $f(km) = f(k)$ for all $k \in K$ and $m \in M$. Then X^∞ is a K-module (and hence a $\mathfrak{k}$-module) with respect to a representation

$$\pi : K \longrightarrow \mathrm{End}\, X^\infty \tag{4}$$

where if $k, k' \in K$ then $\pi(k)f(k') = f(k^{-1}k')$.

Let X be the space of all K-finite elements in X^∞. Let $K^\wedge$ be the set of all equivalence classes of irreducible finite dimensional complex K-modules. Then X is a completely reducible K submodule of X^∞ and one has the direct sum

$$X = \sum_{\gamma \in K^\wedge} X^\gamma \tag{5}$$

where, for $\gamma \in K^\wedge$, X^γ is the γ-primary component in X. Now for any $\gamma \in K^\wedge$ fix $V_\gamma \in \gamma$ and let V_γ^* be the contragredient K-module. Also for any K-module V let V^M be the space of all M fixed vectors in V. Now regarding $V_\gamma \otimes (V_\gamma^*)^M$ as a K-module where K operates only on the left factor one has, by the Frobenius reciprocity theorem, an identification

$$X^\gamma = V_\gamma \otimes (V_\gamma^*)^M \tag{6}$$

as K-modules where if $k \in K$, $v \in V_\gamma$ and $v' \in (V_\gamma^*)^M$ then

(7) $$(v \otimes v')(k) = < v, k \cdot v' >$$

so that (5) becomes

(8) $$X = \sum_{\gamma \in K^\wedge} V_\gamma \otimes (V_\gamma^*)^M .$$

Now any $\mu \in \mathfrak{a}_{\mathbf{C}}^*$ defines a character χ_μ of A where if $h \in \mathfrak{a}$ then $\chi_\mu(a) = e^{<\mu,h>}$ in case $a = \exp h$. We will also write a^μ for $\chi_\mu(a)$. Now let $\mu \in \mathfrak{a}_{\mathbf{C}}^*$ and let $f \in C^\infty(K)$. Recalling the Iwasawa decomposition (3) let f_μ be the C^∞ complex valued function on G defined by putting

(9) $$f_\mu(kan) = a^{-\mu} f(k)$$

for all $k \in K$, $a \in A$, and $n \in N$. Then, depending upon μ, one defines a representation of G on X^∞

$$\pi_\mu : G \longrightarrow \mathrm{End}\, X^\infty$$

by putting

(10) $$\pi_\mu(g) f(k) = f_\mu(g^{-1}k)$$

for $g \in G$, $k \in K$ and $f \in X^\infty$. The fact that $\pi_\mu(g)f$ is indeed in X^∞ follows from the observation that if $f \in X^\infty$ then

(11) $$f_\mu(gm) = f_\mu(g)$$

for any $g \in G$ and $m \in M$. One notes also of course that

(12) $$\pi_\mu | K = \pi .$$

The representation π_μ of G defines a representation of $\mathfrak{g}$ and hence of $U = U(\mathfrak{g})$, the universal enveloping algebra of $\mathfrak{g}$, on X^∞ where if $x \in \mathfrak{g}$, $f \in X^\infty$, and $k \in K$,

(13) $$\pi_\mu(x) f(k) = d/dt\, f_\mu(\exp(-tx)k)|_{t=0} .$$

Although not a G-module one notes that X with respect to π_μ is a $(\mathfrak{g}, K)$-submodule of X^∞. If we compare this with the notation of [K] then if $G = G_{\mathbf{R}}$ (see (1)) this representation is realized in [K] by the module (see Def. 1.7.2., page 256 in [K])

(14) $$X_{\kappa\mu} = \{ f_\mu | f \in X \}$$

where κ is the longest element in W.

2.3. Now let $\Delta \subseteq \mathfrak{a}^*$ be the set of (restricted) roots for the pair $\{\mathfrak{g}, \mathfrak{a}\}$ and let $\Delta_+ \subseteq \Delta$ be the set of (positive, restricted) roots corresponding to the adjoint action of $\mathfrak{a}$ on $\mathfrak{n}$. If $\varphi \in \Delta$ let $\mathfrak{g}_\varphi \subseteq \mathfrak{g}$ be the corresponding root space so that

$$\mathfrak{n} = \sum_{\varphi \in \Delta_+} \mathfrak{g}_\varphi . \tag{15}$$

Let $\mathfrak{n}'$ be the opposing nilpotent Lie algebra given by

$$\mathfrak{n}' = \sum_{\varphi \in \Delta_+} \mathfrak{g}_{-\varphi} \tag{16}$$

and let $N' \subseteq G$ be the corresponding subgroup.

For any $\varphi \in \Delta$ let $x_\varphi \in \mathfrak{a}$ be such that $\varphi(h) = (x_\varphi, h)$ for any $h \in \mathfrak{a}$ where the Killing form for $x, y \in \mathfrak{g}$ is denoted by (x, y). One then defines an open cone $D_{\mathbf{C}}$ in $\mathfrak{a}^*_{\mathbf{C}}$ by putting

$$D_{\mathbf{C}} = \{\mu \in \mathfrak{a}^*_{\mathbf{C}} \mid \operatorname{Re}\mu(x_\varphi) > 0 \ \text{ for all } \ \varphi \in \Delta_+\} . \tag{17}$$

The usual (restricted) open Weyl chamber D in $\mathfrak{a}^*$ is of course given by

$$D = D_{\mathbf{C}} \cap \mathfrak{a}^* . \tag{18}$$

Now let $d = \dim \mathfrak{a}$ and let $\Pi \subseteq \Delta_+$ be the set of simple positive restricted roots. Thus card $\Pi = d$ and if we write $\Pi = \{\alpha_1, \ldots, \alpha_d\}$ then the α_i are a basis of $\mathfrak{a}^*$. As in [K] let $J \subseteq \mathfrak{a}^*$ be a lattice in $\mathfrak{a}^*$ generated by Δ and let $J_+ = J \cap D^c$ where D^c is the closure of D. See page 298 in Sec. 2.6 of [K]. Let $x_i = x_{\alpha_i}$, $i = 1, \ldots, d$, so that the x_i are a basis of $\mathfrak{a}$. The following is well known but we prove it for completeness.

Lemma 1. *There exists a basis λ_j, $j = 1, \ldots, d$, of $\mathfrak{a}^*$ such that $\lambda_j \in 2J_+$.*

Proof: Let B be the $d \times d$ matrix defined by putting $B_{ik} = \alpha_k(x_i)$. From the representation of $Sl(2, \mathbf{C})$ one knows that B is an invertible matrix with rational entries. Let $C = B^{-1}$ so that there exists a positive integer r such that rC has integral entries. Thus if $\nu_j = \sum_k rC_{kj}\alpha_k$ then $\nu_j \in J$ and in fact $\nu_j \in J_+$ since $\nu_j(x_i) = r \cdot \delta_{ij}$. Clearly then, putting $\lambda_j = 2\nu_j$ establishes the lemma. QED

Henceforth λ_j, $j = 1, \ldots, d$ will be as in Lemma 1. From the proof above we note that

$$\lambda_j(x_i) = 2r \cdot \delta_{ij} \ \text{ with } \ r > 0 . \tag{19}$$

For convenience let $\chi_j = \chi_{\lambda_j}$ so that (since λ_j is real)

$$\chi_j(a) > 0 \ \text{ for all } \ a \in A . \tag{20}$$

Let C be the cone in $\mathfrak{a}$ generated by x_φ for $\varphi \in \Delta_+$. It is clearly the same as the cone generated by the x_i.

Proposition 2. *Let $a \in A$. Then $a \in \exp C$ if and only if*

$$\chi_j(a) \geq 1 \quad \text{for } j = 1, \ldots, d. \tag{21}$$

Furthermore if $e \neq a \in \exp C$ there exists j such that $\chi_j(a) > 1$.

Proof: This is immediate from (19) since if $a = \exp h$ for $h \in \mathfrak{a}$ then $\chi_j(a) = e^{\lambda_j(h)}$. QED

One also has the well known

Proposition 3. *For $j = 1, \ldots, d$, there exists a finite dimensional irreducible representation*

$$\eta_j : G \longrightarrow \operatorname{End} Z_j \tag{22}$$

of G with highest weight vector $0 \neq z_j \in Z_j$ such that for $m \in M$, $a \in A$ and $n \in N$

$$\eta_j(man)z_j = \chi_j(a)z_j\,. \tag{23}$$

Proof: Recalling (1), and since (1) is clearly faithful on A, it suffices to replace G by $G_{\mathbb{R}}$. But then recalling Lemma 1 the result is given, for example, by Theorem 2.6.1, Theorem 2.6.2 and Remark 2.6.3 (See pages 297–299) in [K]. QED

2.4. Now $\mathfrak{g}_u = \mathfrak{k} + i\mathfrak{p}$ is a compact real form of $\mathfrak{g}_{\mathbf{C}}$. But then there exists a Hilbert space structure S_j on Z_j with inner product $\{z, v\}$ for $z, v \in Z_j$ such that

$$\eta_j(x) \text{ is skew-Hermitian for any } x \in \mathfrak{g}_u\,. \tag{24}$$

By irreducibility S_j is unique up to a positive multiple. We normalize it so that $\{z_j, z_j\} = 1$.

Now for any $g \in G$ let $k(g) \in K$, $a(g) \in A$ and $n(g) \in N$ be such that

$$g = k(g)a(g)n(g)\,. \tag{25}$$

Lemma 4. *For any $g \in G$ and $j = 1, \ldots, d$, one has*

$$\{\eta_j(g)z_j, \eta_j(g)z_j\} = \chi_j(a(g))^2\,. \tag{26}$$

Proof: This is immediate from (23) since $\eta_j(k)$ is unitary for any $k \in K$. QED

Let $U_+(\mathfrak{n}')$ be the augmentation ideal $\mathfrak{n}'U(\mathfrak{n}')$ in the enveloping algebra $U(\mathfrak{n}')$ of $\mathfrak{n}'$.

LEMMA 5. *For $j = 1, \dots, d$*

$$Z_j = \mathbf{C} \cdot z_j + Z_j^+ \tag{27}$$

is a S_j orthogonal direct sum where

$$Z_j^+ = \eta_j(U_+(\mathfrak{n}'))z_j \,.$$

PROOF: By irreducibility one has $\eta_j(U(\mathfrak{g}))z_j = Z_j$. But then $\eta_j(U(\mathfrak{n}'))z_j = Z_j$ by (23). This implies the equality (27) which is necessarily a direct sum since the weights for the action of $\mathfrak{a}$ on Z_j^+ are less than λ_j. But then (27) is a S_j orthogonal direct sum since $\eta_j(h)$ is a Hermitian operator for any $h \in \mathfrak{a}$ by (24). QED

We will generally reserve the letter n' to denote elements in N'.

PROPOSITION 6. *For any $n' \in N'$ one has*

$$a(n') \in \exp C \,. \tag{28}$$

PROOF: Let $j \in \{1, \dots, d\}$. Then clearly by (26) $\eta_j(n')z_j = z_j + v$ where $v \in Z_j^+$. But then $\{\eta_j(n')z_j, \eta_j(n')z_j\} \geq 1$ by Lemma 5. This implies $\chi_j(a(n')) \geq 1$ by (26). The result then follows from Proposition 2. QED

Now let $\nu \in \mathfrak{a}_{\mathbf{C}}^*$ be arbitrary. Since the λ_j are a basis we may write

$$\nu = c_1\lambda_1 + \cdots + c_d\lambda_d \tag{29}$$

where the $c_j \in \mathbf{C}$ are arbitrary. One notes immediately from (19) that

$$\nu \in D_{\mathbf{C}} \quad \text{if and only if} \quad \operatorname{Re} c_j > 0 \quad \text{for all} \quad j \,. \tag{30}$$

Now since χ_j are positive valued χ_j^c is a well defined function on A for any $c \in \mathbf{C}$. But then by definition of χ_μ one has from (29)

$$\chi_\nu = \chi_1^{c_1} \cdots \chi_d^{c_d} \,. \tag{31}$$

This yields

PROPOSITION 7. *For any $n' \in N'$ and $\nu \in D_{\mathbf{C}}$ one has*

$$|a(n')^{-\nu}| \leq 1 \,. \tag{32}$$

PROOF: This is immediate from (21), (28), (30) and (31). QED

§ 3. The Intertwining Operators $A(s,\mu)$

3.1. Now, as usual, let $\rho \in \mathfrak{a}^*$ be defined by the relation $\rho(h) = 1/2 \operatorname{tr} \operatorname{ad} h|\mathfrak{n}$ for $h \in \mathfrak{a}$. As one knows $\rho(x_i) > 0$ for all i so that $\rho \in D \subseteq D_{\mathbf{C}}$ of M in M'. Thus if $w \in M'$ there exists a unique element in W which we write as s_w such that $w \in s_w$. Let $s \in W$ and let

$$N'_s = N' \cap w^{-1} N w \tag{33}$$

where $w \in s$. Let $d_s n'$ denote a Haar measure on N'_s. Then one knows (see [S] p. 35) that $d_s n'$ may be normalized so that

$$\int_{N'_s} a(n')^{-2\rho} d_s n' = 1\,. \tag{34}$$

We henceforth assume that $d_s n'$ is so normalized.

Now consider the translate $\rho + D_{\mathbf{C}}$ of the open cone $D_{\mathbf{C}}$ in $\mathfrak{a}^*_{\mathbf{C}}$ by ρ. Of course since $\rho \in D_{\mathbf{C}}$

$$\rho + D_{\mathbf{C}} \subseteq D_{\mathbf{C}}\,. \tag{35}$$

The following 3 propositions are, by now, classical. For proofs see [HC], [H_1], [KN–S], [KU–S], and [S].

Proposition 8. *Let $w \in M'$ and let $s = s_w$. Let $\mu \in \rho + D_{\mathbf{C}}$. Then for any $f \in C^\infty(K)$ and $k \in K$ the integral*

$$(B(w,\mu)f)(k) = \int_{N'_s} f_\mu(kwn') d_s n' \tag{36}$$

absolutely converges defining an operator $B(w,\mu)$ on $C^\infty(K)$.

Now since Ad M is compact its elements operate as measure preserving transformations of N'_s. It follows then from (11) that X^∞ is stable under $B(w,\mu)$ and that the restriction depends only on s_w. Thus for any $s \in W$ and $\mu \in \rho + D_{\mathbf{C}}$ we can define the operator

$$A(s,\mu) = B(w,\mu)|X^\infty \tag{37}$$

on X^∞ where $w \in s$.

Proposition 9. *Let $s \in W$ and $\mu \in \rho + D_{\mathbf{C}}$. Then for any $g \in G$ one has*

$$A(s,\mu)\cdot\pi_\mu(g) = \pi_{s(\mu-\rho)+\rho}(g)\cdot A(s,\mu) \tag{38}$$

on X^∞ and

$$A(s,\mu) \text{ stabilizes } X^\gamma \text{ for any } \gamma \in K^\wedge . \tag{39}$$

Now let $\gamma \in K^\wedge$, $s \in W$ and let $\mu \in \rho + D_{\mathbf{C}}$. Put

$$A^\gamma(s,\mu) = A(s,\mu)|X^\gamma \tag{40}$$

so that, since X^γ is finite dimensional, $A^\gamma(s,\mu)$ is an operator (by (39)) on a finite dimensional space. Consequently the following statement is meaningful.

Proposition 10. *Let $s \in W$ and let $\gamma \in K^\wedge$. Then the map $\mu \longrightarrow A^\gamma(s,\mu)$ is holomorphic on the open subset $\rho + D_{\mathbf{C}}$ of $\mathfrak{a}^*_{\mathbf{C}}$ and extends to a meromorphic map of $\mathfrak{a}^*_{\mathbf{C}}$ into* End X^γ.

§ 4. The P^γ Matrices

4.1. For any $u \in U = U(\mathfrak{g})$ let $p^u \in U(\mathfrak{a})$ be the component of u in $U(\mathfrak{a})$ relative to the decomposition

$$U = U(\mathfrak{a}) \oplus (\mathfrak{n}U + U\mathfrak{k}) . \tag{41}$$

We regard $U(\mathfrak{a})$ as the ring of polynomial functions on $\mathfrak{a}^*_{\mathbf{C}}$ so that $p^u(\lambda) \in \mathbf{C}$ is defined for any $\lambda \in \mathbf{C}$.

Remark 11: Recalling Proposition 1.2.2 on p. 237 in [K] where p_u was defined one easily has $p^u = (p_{u^t})^t$ where $v \longrightarrow v^t$ is the anti-involution of U such that $x^t = -x$ for $x \in \mathfrak{g}$. On page 259 in [K] a definition of p^u was given which is different from the present one. This latter definition is being replaced here by the present one since here the representation parameterized by μ is Ind χ_μ whereas in [K] it was Ind $\chi_{\kappa\mu}$. We recall that $\kappa \in W$ is the longest element. The definition of p^u in [K] is the polynomial map

$$\lambda \longrightarrow p^u(\kappa\lambda) \tag{42}$$

where p^u in (42) is our present definition.

4.2. Let $1_K \in X$ be the function on K such that $1_K(k) = 1$ for all $k \in K$. One of course has

$$(43) \qquad \mathbf{C} \cdot 1_K = X^{\gamma_0}$$

where $\gamma_0 \in K^\wedge$ is the trivial representation. The significance of p^u from our point of view arises from the easily verified relation

$$(44) \qquad \pi_\mu(u) 1_K(e) = p^u(\mu)$$

for any $u \in U$ and $\mu \in \mathfrak{a}^*_{\mathbf{C}}$.

Now let $S^* \subseteq U$ be the span of all powers x^k, $k \in \mathbf{N}$ where $x \in \mathfrak{p}_{\mathbf{C}}$. This is stable under the action of $\operatorname{Ad} K$ and let $J^* = (S^*)^K \subseteq S^*$ be the space of all $\operatorname{Ad} K$ invariants. Let $H^* \subseteq S^*$ be the span of all powers x^k, $k \in \mathbf{N}$ where $x \in \mathfrak{p}_{\mathbf{C}}$ is nilpotent. Then one knows that the map

$$(45) \qquad H^* \otimes J^* \longrightarrow H^* J^*, \quad u \otimes v \longrightarrow uv$$

is a linear isomorphism and

$$(46) \qquad U = U\mathfrak{k} \oplus H^* J^*$$

is a direct sum. See Lemma 1.4.2 and Remark 1.4.3 on p. 243 in [K].

Now for any $\gamma \in K^\wedge$ let

$$l(\gamma) = \dim V_\gamma^M = \dim (V_\gamma^*)^M .$$

Now H^* is a K-module via the adjoint representation. For $\gamma \in K^\wedge$ let $E_\gamma = \operatorname{Hom}_K(V_\gamma, H^*)$. Then one knows

$$(47) \qquad \dim E_\gamma = l(\gamma) .$$

REMARK 12: This result is stated as Theorem 19 in [K–R] for the K_θ case. However the result is true in general since, by p. 253 and p. 255 in [K], one clearly has

$$(48) \qquad K/M \cong K_{\theta,\mathbf{R}}/M_{\theta,\mathbf{R}}$$

using the notation of [K]. See also Proposition 10, p. 772 in [K–R] which implies that

$$(49) \qquad (S^*)^{K_\theta} = J^* .$$

4.3. In addition to the comments of Remark 12 there are a number of other results in [K] which are stated for K_θ but which, consequently, are easily seen to be true for K in general. We will freely use them as though they were stated for K.

Following [Wallach], we can write

$$H^* = \sum_{\gamma \in K^\wedge} V_\gamma \otimes E_\gamma \tag{50}$$

where if $v \in V_\gamma$ and $\sigma \in E_\gamma$ then $v \otimes \sigma = \sigma(v)$. As a K-module K, of course, operates only on the left factor of $V_\gamma \otimes E_\gamma$.

Now let $\gamma \in K^\wedge$. Let $v_1, \ldots, v_{l(\gamma)}$ be a basis of V_γ^M and let $\sigma_1, \ldots \sigma_{l(\gamma)}$ be a basis of E_γ. Let P^γ, as in [K], be the $l(\gamma) \times l(\gamma)$ matrix with coefficients in $U(\mathfrak{a})$ defined by putting

$$P^\gamma_{ij} = p^{\sigma_j(v_i)} . \tag{51}$$

Thus $P^\gamma(\lambda)$, for any $\lambda \in \mathfrak{a}^*_{\mathbf{C}}$, is an $l(\gamma) \times l(\gamma)$ matrix with complex coefficients.

Proposition 13. *Let $\gamma \in K^\wedge$ be arbitrary and P^γ be as above. Then for any $\mu \in \rho + D_{\mathbf{C}}$ the matrix*

$$P^\gamma(\mu) \text{ is invertible} . \tag{52}$$

Proof: See Lemma 2.10.1, p. 323 in [K]. (Recall that one has to apply κ to the statement of Lemma 2.10.1). QED

For any $\gamma \in K^\wedge$ let $(H^*)^\gamma = V_\gamma \otimes E_\gamma$ be the γ primary component of H^*.

Proposition 14. *Let $\mu \in \rho + D_{\mathbf{C}}$. Then the map*

$$\xi_\mu : H^* \longrightarrow X \tag{53}$$

and, in particular the restriction map,

$$\xi^\gamma_\mu : (H^*)^\gamma \longrightarrow X^\gamma \tag{54}$$

for any $\gamma \in K^\wedge$, is a K-isomorphism, where for $u \in H^*$

$$\xi_\mu(u) = \pi_\mu(u) 1_K . \tag{55}$$

Proof: See Theorem 2.10.3, p. 324 in [K].QED

Remark 15: Actually the references cited above in [K] assert much more subtle facts than the statements of Propsitions 13 and 14. Namely, they assert the validity of (52) and the isomorphisms of (53) and (54) for μ in the *closure* of $\rho + D_{\mathbf{C}}$. (In particular for ρ itself). However we shall not need this here.

4.4. The matrix P^γ of course depends on the choice of a basis of E_γ and $(V_\gamma)^M$. However the neat formalisms (6) and (50) introduced by Wallach in [W] suggest a basis independent definition which is directly applicable to our concerns. Indeed let $\gamma \in K^\wedge$, $\mu \in \mathfrak{a}_{\mathbf{C}}^*$ and let

$$P : E_\gamma \longrightarrow V_\gamma^* \tag{56}$$

be the linear map defined by putting, for $\sigma \in E_\gamma$ and $v \in V_\gamma$, $P(\sigma)(v) = p^{\sigma(v)}(\mu)$. It then follows from (3) in Lemma 1.7.10, p. 259 in [K] that the image of (56) is in $(V_\gamma^*)^M$. But then $P^\gamma(\mu)$ defined by (51) is just the matrix of (56) relative to the basis σ_j of E_γ and v_i' of $(V_\gamma^*)^M$ which is dual to the basis v_i of V_γ^*. Thus we can invariantly regard $P^\gamma(\mu)$ as a map

$$P^\gamma(\mu) : E_\gamma \longrightarrow (V_\gamma^*)^M \tag{57}$$

where for $\sigma \in E_\gamma$ and $v \in V_\gamma$

$$p^{\sigma(v)}(\mu) = (P^\gamma(\mu)(\sigma))(v)\,. \tag{58}$$

Let 1_γ be the identity operator on V_γ so that

$$1_\gamma \otimes P^\gamma(\mu) : V_\gamma \otimes E_\gamma \longrightarrow V_\gamma \otimes (V_\gamma^*)^M\,. \tag{59}$$

Recalling the identifications of (6) and (50) one has

Proposition 16. *Let $\gamma \in K^\wedge$ and $\mu \in \mathfrak{a}_{\mathbf{C}}^*$. Then*

$$1_\gamma \otimes P^\gamma(\mu) = \xi_\mu^\gamma \tag{60}$$

where, as in (54),

$$\xi_\mu^\gamma : (H^*)^\gamma \longrightarrow X^\gamma$$

is the map defined by

$$\xi_\mu^\gamma(u) = \pi_\mu(u) 1_K$$

for $u \in (H^*)^\gamma$.

Proof: Let $\sigma \in E_\gamma$ and $v \in V_\gamma$. Then the left side of (60) applied to $v \otimes \sigma$ is equal to $v \otimes P^\gamma(\mu)\sigma$, which as a function of $k \in K$, equals $(P^\gamma(\mu)\sigma)(k^{-1} \cdot v) = p^{\sigma(k^{-1} \cdot v)}(\mu) =$ (by (44)) $\pi_\mu(\sigma(k^{-1} \cdot v))1_K(e) = \pi_\mu(\text{Ad } k^{-1}\sigma(v))1_K(e) = \pi_\mu(\sigma(v))1_K(k)$. Thus the left hand side of (60) applied to $v \otimes \sigma = \sigma(v)$ equals $\pi_\mu(\sigma(v))1_K = \xi_\mu^\gamma(\sigma(v))$. This establishes (60). QED

§ 5. The relation between intertwining operators and the P^γ

5.1. Let $\mu \in \rho + D_{\mathbf{C}}$ and let $\gamma \in K^\wedge$ be arbitrary. Then by (52) the endomorphism (57) is invertible and hence for any $s \in W$

$$P^\gamma(s(\mu-\rho)+\rho)+\rho)\cdot(P^\gamma(\mu))^{-1} \in \mathrm{End}(V_\gamma^*)^M \tag{61}$$

Furthermore recalling the identification (6) one has

$$1_\gamma \otimes P^\gamma(s(\mu-\rho)+\rho)+\rho)\cdot(P^\gamma(\mu))^{-1} \in \mathrm{End}\, X^\gamma \tag{62}$$

Remark 17: Note that $P^\gamma(s(\mu-\rho)+\rho)\cdot(P^\gamma(\mu))^{-1}$ computed as a matrix using the definition (51) is, when regarded as an endomorphism with respect to that basis v_i' of $(V_\gamma^*)^M$ which is dual to the basis v_i of V_γ^M, just the endomorphism given by (61).

Now let $s \in W$ and let $\mu \in \rho + D_{\mathbf{C}}$. Then by Propositions 8 and 9 the integral, where $w \in s$,

$$c_s(\mu) = \int_{N_s'} (1_K)_\mu(wn')d_s n' = \int_{N_s'} a(n')^{-\mu} d_s n' \tag{63}$$

converges and one has

$$A(s,\mu)1_K = c_s(\mu)\cdot 1_K\,. \tag{64}$$

In the generality stated the following result is due to Wallach. (See [W], Theorem 3.1). Independently we knew of this result in the case where $s = \kappa$ and have used it (and will use it) in establishing Theorem 32 which was announced in the introduction of [K]. However our argument was considerably more complicated (and basis dependent) than the elementary proof due to Wallach which we give here.

Theorem 18. *Let $\mu \in \rho + D_{\mathbf{C}}$, $s \in W$, and $\gamma \in K^\wedge$. As in (40) let $A^\gamma(s,\mu)$ be the restriction of the intertwining operator $A(s,\mu)$ to X^γ. Then (see (62))*

$$A^\gamma(s,\mu) = c_s(\mu)\cdot 1_\gamma \otimes P^\gamma(s(\mu-\rho)+\rho)\cdot(P^\gamma(\mu))^{-1}. \tag{65}$$

Proof: Let $u \in (H^*)^\gamma$. Then by (38) one has

$$\begin{aligned} A^\gamma(s,\mu)\cdot\pi_\mu(u)1_K &= \pi_{s(\mu-\rho)+\rho}(u)\cdot A(s,\mu)1_K \\ &= c_s(\mu)\cdot\pi_{s(\mu-\rho)+\rho}(u)1_K \end{aligned}$$

by (64). That is, as maps $(H^*)^\gamma \longrightarrow X^\gamma$

$$A^\gamma(s,\mu)\cdot\xi^\gamma_\mu = c_s(\mu)\cdot\xi^\gamma_{s(\mu-\rho)+\rho}\,. \tag{66}$$

The result then follows from (60).QED

REMARK 19: It is clear that the meromorphic extension of the map

$$\mu \longrightarrow A^\gamma(s,\mu) \tag{67}$$

to $\mathfrak{a}^*_{\mathbf{C}}$ asserted by Proposition 10 is, by Theorem 18, explicitly given by the obvious holomorphic extension of $P^\gamma(s(\mu-\rho)+\rho)$ and meromorphic extensions of $c_s(\mu)$ and $P^\gamma(\mu)^{-1}$. Now one knows $c_s(\mu)$. See e.g. [H], formula (39) p. 446. One should note however that the parameterization here is different from that in [H]. If we let $c_s'(\lambda)$ be the function defined by formula (37) p. 446 in [H] then

$$c_s'((\mu-\rho)/i) = c_s(\mu)\,. \tag{67.5}$$

Turning to the obvious question about the poles of (67) we now note that there is an advantage to being in the K_θ case. The reason for this is that in this case one explicitly knows the determinant of the $U(\mathfrak{a})$-valued matrix P^γ as a polynomial function on $\mathfrak{a}^*_{\mathbf{C}}$ and ***knows exactly where it vanishes.*** See Theorem 2.4.6, p. 292, Theorem 2.8.8., p. 317 and Remark 2.9.2., p. 319 in [K]. In all one can make a number of statements about the poles of the map (67) and an exact statement in the split rank 1 (that is $d=1$) case.

5.2. For any $v\in W$ let $T_v\in \operatorname{Aut} X^\infty$ be the operator defined by putting

$$T_v f(k) = f(kw) \tag{68.1}$$

for any $f\in X^\infty$ and $k\in K$ where $w\in v$. Obviously the map

$$v\longrightarrow T_v$$

defines a representation of W on X^∞. Clearly this representation commutes with π (see (4)) and hence for any $v\in W$ and $\gamma\in K^\wedge$ one defines an operator $T_v^\gamma\in \operatorname{Aut} X^\gamma$ by putting

$$T_v^\gamma = T_v|X^\gamma\,. \tag{68.2}$$

On the other hand, recalling (6), since V_γ is K-irreducible this implies that there exists a unique operator $Q_v^\gamma\in \operatorname{Aut}(V_\gamma^*)^M$ such that

$$T_v^\gamma = 1_\gamma\otimes Q_v^\gamma\,. \tag{68.3}$$

From the identification (6) it is clear that the representation

$$W \longrightarrow \operatorname{End}(V_\gamma^*)^M, \quad v \longrightarrow Q_v^\gamma \tag{68.4}$$

of the Weyl group, W, is just the natural representation defined by the K-module structure on V_γ^*.

Now let $v, s \in W$ and $\mu \in \rho + D_{\mathbf{C}}$. If $\gamma \in K^\wedge$ let $(T_v A(s,\mu))^\gamma = T_v^\gamma A^\gamma(s,\mu) \in \operatorname{End} X^\gamma$ be the restriction of $T_v A(s,\mu)$ to X^γ. By (68.3) and (65) one clearly has

$$(T_v A(s\mu))^\gamma = c_s(\mu) \cdot 1_\gamma \otimes Q_v^\gamma \cdot P^\gamma(s(\mu - \rho) + \rho) \cdot (P^\gamma(\mu))^{-1}. \tag{68.5}$$

For any $\gamma \in K^\wedge$ we will write $\dim \gamma = \dim V_\gamma$. In this paper we will be concerned with the trace of $T_v^\gamma A^\gamma(s,\mu)$ for the special case where $s = \kappa$. We observe that in general one has

Corollary 20. *Let $\mu \in \rho + D_{\mathbf{C}}$, $v, s \in W$, and $\gamma \in K^\wedge$. Then*

$$\operatorname{tr} T_v^\gamma A^\gamma(s,\mu) = c_s(\mu) \cdot \dim\gamma \cdot \operatorname{tr} Q_v^\gamma \cdot P^\gamma(s(\mu-\rho)+\rho) \cdot (P^\gamma(\mu))^{-1}. \tag{69}$$

Proof: This is immediate from (68.5).QED

§ 6. The Traceability of the Intertwining Operator $A(\kappa,\mu)$

6.1. Now for any $s \in W$ choose a representative element $w(s) \in s$. We recall that the Bruhat decomposition for G asserts that

$$G = \cup_{s \in W} N' w(s) MAN \tag{70}$$

is a disjoint union and if $G^0 = N'MAN$ then G^0 is open and dense in G and the map,

$$N' \times M \times A \times N \longrightarrow G^0 \tag{71}$$

where $(n', m, a, n) \longrightarrow n'man$, is a diffeomorphism. Recalling the Iwasawa decomposition (3) it follows that if $K^0 = G^0 \cap K$ then K^0 is open and dense in K and that $k(n'm) \in K^0$ for any $n' \in N'$ and $m \in M$. Furthermore the map

$$\Psi : N'M \longrightarrow K^0 \tag{72}$$

given by

$$\begin{aligned} \Psi(n'm) &= k(n'm) \\ &= k(n')m \end{aligned} \tag{73}$$

is a diffeomorphism.

Now for any $k \in K^0$ let $b(k) \in A$ be the unique element in A such that one has

$$k = n'mb(k)n \tag{74}$$

for $n' \in N'$, $m \in M$ and $n \in N$. Since one may write $n'm = k(b(k))^{-1}n_1$ for some $n_1 \in N$ clearly

$$k(n'm) = k \tag{75}$$

and

$$\begin{aligned} a(n') &= a(n'm) \\ &= (b(k))^{-1}. \end{aligned} \tag{76}$$

Now let dm and dk be, respectively, the normalized Haar measure (i.e. having integral equal to 1) on M and K. Note that $d_\kappa n'$ is a Haar measure on N'. For convenience we will write $d_\kappa n' = dn'$. Now one knows that the representation $\pi_{2\rho}$ is equivalent to the action of G on the space of all measures on K/M which have smooth Radon–Nikodym derivative with respect to the K-invariant measure. The latter corresponds to 1_K. But for $n' \in N'$ and $m \in M$ one has $(1_K)_{2\rho}(n'm) = a(n')^{-2\rho}$. This easily recovers the following well known fact

LEMMA 21. *Let h be the function on $N'M$ defined by putting $h(n'm) = a(n')^{-2\rho}$. Let Ψ_* be the map on measures induced by the diffeomorphism (72). Then*

$$\Psi_*(h \cdot dn'dm) = dk|K^0. \tag{77}$$

Recalling (76) this clearly yields

LEMMA 22. *Let j be the function on K^0 defined by putting $j(k) = b(k)^{-2\rho}$. Then*

$$\Psi_*(dn'dm) = j \cdot dk|K^0. \tag{78}$$

Now let Q be the projection $C^\infty(K) \longrightarrow X^\infty$ defined by putting for any $f \in C^\infty(K)$ and $k \in K$

$$Qf(k) = \int_M f(km)dm. \tag{79}$$

For any $s \in W$ and $\mu \in \rho + D_{\mathbf{C}}$ we now extend the domain of definition of $A(s,\mu)$ so that $A(s,\mu)$ maps $C^\infty(K)$ into X^∞ by putting

$$A(s,\mu) = B(w,\mu)Q \tag{79.5}$$

where $w \in s$. See (36) and (37).

6.2. We now concern ourselves with the intertwining operator $A(\kappa,\mu)$. Let w be some fixed element in κ.

PROPOSITION 23. *Let $\mu \in \rho + D_{\mathbf{C}}$, $k_1 \in K$ and $f \in C^\infty(K)$. Then*

$$A(\kappa,\mu)f(k_1) = \int_{K^0} f(k_1wk)\cdot b(k)^{\mu-2\rho}dk\,. \tag{80}$$

PROOF: Since $A(\kappa,\mu) = B(w,\mu)Q$ one has by (36)

$$A(\kappa,\mu)f(k_1) = \int_{N'M} f_\mu(k_1wn'm)dn'dm \tag{81}$$

since $\mathrm{Ad}\,w$ is measure preserving on M and $\mathrm{Ad}\,m$ is measure preserving on N' for any $m \in M$. But clearly

$$\begin{aligned} f_\mu(k_1wn'm) &= a(n')^{-\mu}\cdot f(k_1wk(n'm)) \\ &= b(k)^{\mu}\cdot f(k_1wk) \end{aligned} \tag{81.5}$$

where $\Psi(n'm) = k \in K^1$. The result then follows from (78).QED

For any $\nu \in \mathfrak{a}^*_{\mathbf{C}}$ let F_ν be the function on K such that

$$F_\nu(k) = b(k)^\nu \quad \text{for} \quad k \in K^0 \tag{82}$$

and where $K - K^0$ is the complement

$$F_\nu(k) = 0 \quad \text{for} \quad k \in K - K^0\,. \tag{82.5}$$

Obviously

$$F_\nu(mk) = F_\nu(k) = F_\nu(km) \tag{83}$$

for any $m \in M$ and $k \in K$.

LEMMA 24. *If $\mu \in \rho + D_{\mathbf{C}}$ then $F_{\mu-2\rho}$ is locally integrable and hence defines a distribution on K.*

PROOF: Clearly $|F_{\mu-2\rho}| = F_{\lambda-2\rho}$ where $\lambda = \mathrm{Re}\,\mu$. But $\lambda \in \rho + D_{\mathbf{C}}$ and hence (80) converges where λ replaces μ and $f = 1_K$. This implies that $F_{\mu-2\rho}$ is locally integrable.QED

For any $f \in C^\infty(K)$ let $f^\smile \in C^\infty(K)$ be defined by putting $f^\smile(k) = f(k^{-1})$ for all $k \in K$. More generally if δ is a distribution on K let $\delta^\smile$ be that distribution on K such that $\delta^\smile(f) = \delta(f^\smile)$ for any $f \in C^\infty(K)$.

Now for any $\mu \in \rho + D_{\mathbf{C}}$ it is obvious from Lemma 24 that $F^\smile_{\mu-2\rho}$ is locally integrable and that by (83) the operation on $C^\infty(K)$ of right convolution by $F^\smile_{\mu-2\rho}$ maps $C^\infty(K)$ into X^∞. Recalling (68.1) the following is just a restatement of Proposition 23.

PROPOSITION 25. *For any $\mu \in \rho + D_{\mathbf{C}}$ the intertwining operator $A(\kappa, \mu)$ is just right convolution by $F^{\smile}_{\mu-2\rho}$ followed by T_κ so that (since $T_\kappa = T_\kappa^{-1}$) for any $f \in C^\infty(K)$ one has*

$$T_\kappa A(\kappa, \mu) f = f * F^{\smile}_{\mu-2\rho}. \tag{84}$$

6.3. Now if h is a non-negative real valued function on K and $c \in \mathbf{C}$ is such that $\operatorname{Re} c > 0$ we will understand by h^c the function on K such that if $k \in K$ then $h^c(k) = h(k)^c$ if $h(k) > 0$ and $h^c(k) = 0$ if $h(k) = 0$. If h is continuous and $\operatorname{Re} c > 0$ then h^c is clearly continuous. Furthermore one knows

LEMMA 26. *Assume $h \in C^\infty(K)$ is real and non-negative. Let $c \in \mathbf{C}$ and assume $\operatorname{Re} c > J \in \mathbf{N}$. Then h^c is at least J times differentiable. That is*

$$h^c \in C^J(K). \tag{85}$$

PROOF: The function h defines a smooth map $K \longrightarrow [0, \infty) = \mathbf{R}_+$. But clearly $t \longrightarrow t^c$ defines a function in $C^J(\mathbf{R}_+)$.QED

One notes from the proof of Lemma 26 that if ξ is a smooth vector field on K and $J > 1$ then

$$\xi h^c = c \cdot h^{c-1} \cdot \xi h \tag{86}$$

in the notation of Lemma 26.

6.4. Now let $\nu \in D_{\mathbf{C}}$ and let the notation be as in (30) so that

$$\nu = c_1 \lambda_1 + \cdots + c_d \lambda_d$$

where $\operatorname{Re} c_i > 0$. For convenience put $F_i = F_{\lambda_i}$. Since the λ_i are real F_i is non-negative on K and hence

$$F_\nu = F_1^{c_1} \ldots F_d^{c_d} \tag{87}$$

on K, recalling (31). On the other hand if $j \in \{1, \ldots, d\}$ then recalling the irreducible G-module Z_j with highest weight λ_j, highest weight vector z_j and inner product S_j (see (23) and (24)) let Ψ_j be the *analytic* function on K defined by putting

$$\Psi_j(k) = \{\eta_j(k) z_j, z_j\} \tag{88}$$

for any $k \in K$.

LEMMA 27. *The analytic function Ψ_j is non-negative valued and*

$$\Psi_j = F_j \text{ on } K^0 . \tag{89}$$

PROOF: Since K^0 is dense in K it suffices by continuity to prove (89).

Let $k \in K^0$ so that we can write $k = n'mb(k)n$ using the notation of (74). But then by (23) one has

$$\Psi_j(k) = b(k)^{\lambda_j} \cdot \{\eta_j(n')z_j, z_j\} .$$

But $b(k)^{\lambda_j} = F_j(k)$ and $\{\eta_j(n')z_j, z_j\} = 1$ by Lemma 5 since in the notation of (27) one clearly has $\eta_j(n')z_j \in z_j + Z_j^+$. This proves (89).QED

LEMMA 28. *Let $k \in K - K^0$. Then there exists $j \in \{1, \ldots, d\}$ such that*

$$\Psi_j(k) = 0 . \tag{90}$$

PROOF: Using the notation of (70) there exists $1 \neq s \in W$ such that $k \in N'w(s)MAN$. But since $s \neq 1$ there exists j such that $s\lambda_j \neq \lambda_j$. But then by (23) and Lemma 5 this clearly implies that

$$\eta_j(k)z_j \in Z_j^+ \tag{91}$$

in the notation of (27). Thus (90) follows by Lemma 5.QED

We can now establish the following smoothness result for the function F_ν.

THEOREM 29. *Let $\nu \in \mathfrak{a}_{\mathbf{C}}^*$. Using the notation of (29) write*

$$\nu = c_1\lambda_1 + \cdots + c_d\lambda_d .$$

Let $J \in \mathbf{N}$ and assume that $\operatorname{Re} c_j > J$ *for all j. Then*

$$F_\nu \in C^J(K) . \tag{92}$$

PROOF: Let

$$\Psi_\nu = \Psi_1^{c_1} \ldots \Psi_d^{c_d} .$$

This is well defined by Lemma 27. Furthermore, since the Ψ_j are analytic, $\Psi_\nu \in C^J(K)$ by Lemma 26. On the other hand $\Psi_\nu = F_\nu$ on K^0 by (89). But $\Psi_\nu = 0$ on $K - K^0$ by Lemma 28. Thus

$$\Psi_\nu = F_\nu . \tag{93}$$

QED

6.5. If F is a locally integrable function on K then the operator, R_F, of right convolution by F is the same as the integral operator with kernel $\mathcal{F}$ where

$$\mathcal{F}(k_1, k) = F(k^{-1}k_1)\,. \tag{94}$$

Therefore if $F \in L_2(K)$ it clearly follows that $\mathcal{F} \in L_2(K \times K)$ and hence R_F, regarded as an operator on $L_2(K)$, is of Hilbert-Schmidt type. Furthermore if we can write, as a finite sum,

$$F = \sum_{i=1}^{k} F_i' * F_i'' \tag{95}$$

where $F_i', F_i'' \in L_2(K)$ then F is continuous and R_F is of trace class since one knows that the product of two Hilbert-Schmidt operators is of trace class. In addition if

$$L_2(K) = \sum_{\gamma \in K^\wedge} L_2(K)^\gamma \tag{96}$$

is the Peter-Weyl decomposition of $L_2(K)$ with respect to left translation and $F^\gamma \in L_2(K)^\gamma$ is the γ-Fourier component of F then

$$F = \sum_{\gamma \in K^\wedge} F^\gamma \tag{97}$$

is point-wise absolutely and uniformly convergent and hence one easily establishes the absolutely convergent sum

$$\operatorname{tr} R_F = \sum_{\gamma \in K^\wedge} \operatorname{tr} R_{F^\gamma} \tag{98}$$

and the equation

$$\operatorname{tr} R_F = F(e)\,. \tag{99}$$

The following is well known but for convenience we will sketch a proof.

PROPOSITION 30. *Assume $F \in C^J(K)$ where $J \geq \dim K$. Then F can be written in the form (95) so that the operator, R_F, of right convolution by F on $L_2(K)$ is of trace class and one has the relations (97), (98), and (99).*

(Sketched) PROOF: It suffices to establish a relation of the form

$$\delta_e = \sum_{i=1}^{k} \partial_{u_i} F_i'' \tag{100}$$

where $u_i \in U_{\dim K}(\mathfrak{k})$, (using the notation for the standard filtration of $U(\mathfrak{k})$), ∂_{u_i} is the corresponding right invariant differential operator on K, F_i'' is a bounded measurable function on K and δ_e is the Dirac measure at e. Indeed (100) implies that

$$\delta_e = \sum_{1=1}^{k} \delta_{u_i} * F_i'' \tag{101}$$

where δ_{u_i} is the distribution with support at e corresponding to u_i. But then clearly one has (95) where $F_i' = F * \delta_{u_i}$. One notes here that F_i' is continuous since right convolution of F by δ_{u_i} is the same as applying a left invariant differential operator of degree at most $\dim K$ to F.

To establish (100) it suffices to establish it in a normal coordinate neighborhood V of e where the F_i'' have support in V and the ∂_{u_i} are replaced by constant coefficient differential operators of degree at most $\dim K$. But then by using the product structure for distributions it suffices to observe that on $\mathbf{R}$ the Dirac measure δ at $0 \in \mathbf{R}$ can be written

$$\delta = \psi + d/dt\varphi \tag{102}$$

where φ, ψ are bounded functions on $\mathbf{R}$ with support in any given neighborhood of 0. But (102) is obvious.QED

6.6. If R is an operator of trace class on a Hilbert space then clearly so is TR where T is a bounded operator on a subspace which contains the image of R. If $v \in W$ then (see (68.1)) T_v obviously extends to a unitary operator on the closure of X^∞ in $L_2(K)$.

THEOREM 31. *There exists an element $\rho_0 \in \rho + D$ such that for any $\mu \in \rho_0 + D_{\mathbf{C}}$ the intertwining operator $A(\kappa, \mu)$ on $L_2(K)$ is of trace class. Furthermore if $v \in W$ then*

$$\operatorname{tr} T_v A(\kappa, \mu) = 0 \quad \text{if} \quad v \neq \kappa \tag{103}$$

and

$$\operatorname{tr} T_\kappa A(\kappa, \mu) = 1\,. \tag{104}$$

PROOF: Let $n_0 = \dim K$. In the notation of (29) let $\lambda = n_0 \cdot (\lambda_1 + \cdots + \lambda_d)$ and put

$$\rho_0 = 2\rho + \lambda\,. \tag{105}$$

Obviously $\rho_0 \in \rho + D_{\mathbf{C}}$. But then if $\mu \in \rho_0 + D_{\mathbf{C}}$ one certainly has

$$\mu \in \rho + D_{\mathbf{C}}\,. \tag{106}$$

But also if $\nu = \mu - 2\rho$ then $\nu \in \lambda + D_{\mathbf{C}}$ and hence if c_j is defined as in (29) then clearly $\operatorname{Re} c_j > n_0$ for all j. But then $F_{\mu-2\rho} \in C^{n_0}(K)$ by Theorem 29. Hence

$$F^{\smile}_{\mu-2\rho} \in C^{n_0}(K)\,. \tag{107}$$

Thus if $F = F^{\smile}_{\mu-2\rho}$ then using (106)

$$T_\kappa A(\kappa,\mu) = R_F \tag{108}$$

by Proposition 25 and hence $T_\kappa A(\kappa,\mu)$ is of trace class by Proposition 30. But then (104) follows from (99) and the definition, (82) and (82.5), of F_ν.

Now let $1 \neq s \in W$ and put $v = s\kappa$ so that $v \neq \kappa$ and $T_v A(\kappa,\mu) = T_s R_F$. But clearly if $w(s) \in s$ as in (70) then $T_s R_F = R_{F_s}$ where $F_s(k) = F(kw(s))$ so that

$$T_v A(\kappa,\mu) = R_{F_s}\,. \tag{109}$$

But obviously $F_s \in C^{n_0}(K)$ so that not only is $T_v A(\kappa,\mu)$ of trace class – in particular $A(\kappa,\mu)$ is of trace class – but its trace equals $F_s(e)$ by (99). However $F_s(e) = F(w(s)) = F_{\mu-2\rho}(w(s)^{-1})$. But $w(s)^{-1} \in s^{-1}$ and hence $w(s)^{-1} \in K^0$ by (70). Thus

$$F_s(e) = 0 \tag{110}$$

by (82.5).QED

§ 7. The sum formula by restriction to K-types

7.1. Let Y be the closure of X^∞ in $L_2(K)$. Obviously (see (79.5)), for $v \in W$ and $\mu \in \rho_0 + D_{\mathbf{C}}$, by continuity, $T_v A(\kappa,\mu)$ maps $L_2(K)$ into Y and hence $\operatorname{tr} T_v A(\kappa,\mu) = \operatorname{tr} T_v A(\kappa,\mu)|Y$. But then by further restricting to K-types and using properties of trace class operators one has as a consequence of (103) and (104) that for any $\mu \in \rho_0 + D_{\mathbf{C}}$

$$\sum_{\gamma \in K^\wedge} \operatorname{tr} T_v^\gamma A^\gamma(\kappa,\mu) = 0 \ \text{ if } \ v \neq \kappa \tag{111}$$

and

$$\sum_{\gamma \in K^\wedge} \operatorname{tr} T_\kappa^\gamma A^\gamma(\kappa,\mu) = 1 \tag{112}$$

are absolutely convergent sums.

For any $\mu \in \mathfrak{a}^*_{\mathbf{C}}$ let $c(\mu) = c_\kappa(\mu)$ so that $\mu \longrightarrow c(\mu)$ is the c-function of Harish-Chandra which one associates to the symmetric space G/K. Now applying (69) we obtain

THEOREM 32. *For any $\gamma \in K^\wedge$, $v \in W$ and $\mu \in \mathfrak{a}_{\mathbf{C}}^*$ let $P^\gamma(\mu)$ and Q_v^γ be defined as in (57) and (68.3) respectively. Let ρ_0 be as in Theorem 31. (See (105).) Assume $\mu \in \rho_0 + D_{\mathbf{C}}$ so that $P^\gamma(\mu)$ is invertible (see (52)) and hence Q_v^γ and $P^\gamma(\kappa\mu+2\rho)\cdot(P^\gamma(\mu))^{-1}$ are operators on $(V_\gamma^*)^M$. Then*

$$c(\mu) \neq 0 \tag{113}$$

and

$$\sum_{\gamma \in K^\wedge} \dim\gamma \cdot \operatorname{tr} Q_\kappa^\gamma \cdot P^\gamma(\kappa\mu + 2\rho) \cdot (P^\gamma(\mu))^{-1} = c(\mu)^{-1} \tag{114}$$

is an absolutely convergent sum. Furthermore if $v \neq \kappa$ then

$$\sum_{\gamma \in K^\wedge} \dim\gamma \cdot \operatorname{tr} Q_v^\gamma \cdot P^\gamma(\kappa\mu + 2\rho) \cdot (P^\gamma(\mu))^{-1} = 0 \tag{115}$$

and in particular, putting $v = 1$,

$$\sum_{\gamma \in K^\wedge} \dim\gamma \cdot \operatorname{tr} P^\gamma(\kappa\mu + 2\rho) \cdot (P^\gamma(\mu))^{-1} = 0 \tag{116}$$

are also absolutely convergent sums.

PROOF: By (112) and Corollary 20 for $v = s = \kappa$

$$c(\mu) \cdot \sum_{\gamma \in K^\wedge} \dim\gamma \cdot \operatorname{tr} Q_\kappa^\gamma \cdot P^\gamma(\kappa\mu + 2\rho) \cdot (P^\gamma(\mu))^{-1} = 1 \tag{117}$$

is absolutely convergent. This implies (113) and (114). But then Corollary 20, (113) and (111) yields the absolutely convergent sum (115).QED

§ 8 THE SPLIT RANK 1 CASE AND THE GENERALIZATION OF THE KUMMER EQUATION (.4) FOR SUITABLE RESTRICTIONS OF THE VARIABLES

8.1. Now one knows that if $\mathfrak{g}$ is a simple Lie algebra where

$$d = 1 \tag{118}$$

then $\mathfrak{g}$ falls into one of 4 cases. These cases may be represented, respectively, by the Lie algebras of $SO(n,1)$, $SU(n,1)$ $Sp(n,1)$ and that real form of F_4 having Spin 9 as a maximal compact subgroup. If $d = 1$ there is only one simple root α and if (in the notation of [K])

$$s = \dim \mathfrak{g}_{2\alpha} \tag{119}$$

then the 4 cases correspond, respectively, to the values

$$s = 0, 1, 3 \text{ and } 7. \tag{120}$$

See Chapter 2, Section 1 in [K].

Henceforth assume that $d = 1$ and that we are in the K_θ case. The letter s will always subsequently be defined by (119).

REMARK 33: The fact that we are in the K_θ case makes a difference only if $s = 0$. Indeed one notes that in the proof of Lemma 2.2.8, p. 278 in [K] it is established, in the notation of [K], that $K = K_\theta$ whenever $s \neq 0$.

Now let $K_+^\wedge = \{\gamma \in K^\wedge \mid l(\gamma) \neq 0\}$. Then we have established in [K] that

$$l(\gamma) = 1 \quad \text{for all } \gamma \in K_+^\wedge \tag{121}$$

and that each $\gamma \in K_+^\wedge$ may be parameterized by a triple $\gamma = \gamma(i, j, \varepsilon)$ where $i, j \in \mathbf{Z}_+$ are arbitrary for $s = 1, 3$ and 7. For $s = 0$ then $i \in \mathbf{Z}_+$ is arbitrary but j is restricted to 0 and 1. The symbol ε may be ignored for $s = 0, 3$ and 7. For $s = 1$ it takes the values $+$ and $-$ when $j > 0$ and is ignored if $j = 0$. See Chapter 2, Sections 1 and 2 in [K] for this parameterization of $K_+^\wedge$ and the proof of (121). For a more explicit case by case description of $\gamma(i, j, \varepsilon)$ see [J] and [J-W].

Now it follows from (121) that the primary decomposition

$$H^* = \sum_{i,j,\varepsilon} (H^*)^{\gamma(i,j,\varepsilon)} \tag{122}$$

is the unique decomposition of H^* into *irreducible* K-submodules. More specifically one knows that the space H^* is naturally graded by $\mathbf{Z}_+$ (where the homogeneous subspace H_k^* corresponds under the map β, to the space of harmonic elements of degree k in H, using the notation of [K], p. 242) and that, by Theorem 2.1.28, p. 276 and Theorem 2.2.9, p. 279

$$H_k^* = \sum_{\substack{i,j,\varepsilon \\ k=2i+j}} (H^*)^{\gamma(i,j,\varepsilon)} \tag{123}$$

Now let $\gamma \in K_+^\wedge$. Choose a basal element in E_γ and one in $(V_\gamma^*)^M$ so that for any $\mu \in \mathfrak{a}_{\mathbf{C}}^*$ we can regard (with regard to these basal elements) $P^\gamma(\mu)$ as a scalar. The following was referred to in [K] as the key result.

THEOREM 34. *Let $\gamma = \gamma(i,j,\varepsilon) \in K_+^\wedge$. Let $\mu \in \mathfrak{a}_{\mathbf{C}}^*$ so that, where α is the simple root, we can write*

$$\mu = \tau\alpha \tag{124}$$

where $\tau \in \mathbf{C}$. Then there exists a non-zero $b \in \mathbf{C}$ independent of μ such that one has the "double" product

(125)

$$P^\gamma(\mu) = b \cdot [\tau(\tau+2)\ldots(\tau+2(i+j)-2)] \cdot [(\tau+1-s)(\tau+3-s)\ldots(\tau+2i-1-s)]$$

PROOF: In the notation of [K] (See p. 319) p^γ is defined to be the determinant of the $U(\mathfrak{a})$ valued matrix P^γ. The element $p_\gamma \in U(\mathfrak{a})$ is also recalled on p. 319 of [K] and on that page it is observed that p^γ is a non-zero scalar multiple of p_γ. Recalling in [K] the definition of $p_{v(i,j)}$ on p. 313 one has by Remark 2.2.14. on p. 280 and Remark 2.9.2 on p. 319 and p. 320 that up to a non-zero scalar

$$p_{\gamma(i,j.\varepsilon)} = p_{v(i,j)}\,. \tag{126}$$

But now if $w \in \mathfrak{a}$ is such that $< \alpha, w >= 1$ then by Theorem 2.8.8., p. 317 in [K], up to a non-zero scalar

(127)

$$p_{v(i,j)} = [w(w-2)\ldots(w-2(i+j)+2)] \cdot [(w-1+s)(w-3+s)\ldots.(w-2i+1+s)]$$

But now $< \mu, w >= \tau$ and

$$\kappa = -1 \textit{ in the rank } 1 \textit{ case}\ . \tag{128}$$

Hence recalling (42) one obtains (125) from (126) and (127) after factoring out all the minus signs.QED

8.2. Now we wish to apply Theorem 32 in the present (rank 1) case – the point being that all the terms in (114) and (115) can be completely determined. Put for i and j as in (126)

$$d(i,j) = \sum_\varepsilon \dim\gamma(i,j,\varepsilon)\,. \tag{129}$$

The summation in (129) has, of course, at most 2 terms. Except for the case $s = 0$ we shall not explicitly write down $d(i,j)$ although by using Weyl's dimension formula it is straightforward to do so. This is particularly true if one uses the explicit descriptions of $\gamma(i,j,\varepsilon)$ given in [J] and [J-W].

Let

$$t = \dim \mathfrak{g}_\alpha . \tag{130}$$

Let $\mu = \tau\alpha \in \mathfrak{a}^*_{\mathbf{C}}$. We wish to compute the value $c(\mu)$ of the c-function on μ and we will use the formula (9), p. 446 in [H] where however c in [H] is denoted by c' here. The element α_0 of [H] is w in this paper and, recalling (67.5) here one has that the element $i\lambda$ of [H] is $\mu - \rho$ here. But then

$$< i\lambda, \alpha_0 >= \tau - (s + 1/2 \cdot t) \tag{131}$$

since

$$< \rho, w >= s + 1/2 \cdot t . \tag{132}$$

But then by (9), p. 426 in [H]

$$c(\mu) = c_0 \cdot \frac{2^{-(\tau-(s+1/2\cdot t))} \cdot \Gamma(\tau - (s + 1/2 \cdot t))}{\Gamma(1/2 \cdot (1 + \tau - s)) \cdot \Gamma(\tau/2)} \tag{133}$$

where

$$c_0 = \Gamma(1/2 \cdot (s + t + 1)) \cdot 2^{s+1/2\cdot t} . \tag{134}$$

Recalling Lemma 1 we can put $\lambda_1 = 2\alpha$. Recall also that if, as in the proof of Theorem 31, we put $n_0 = \dim K$ then here

$$n_0 = s + t + 1 \tag{135}$$

and hence by (105) and (132) we can take

$$\rho_0 = (4s + 3t + 2)\alpha . \tag{136}$$

Now in the split rank 1 case Theorem 32 becomes Theorem 35 below, which upon the verification given in Theorem 36, can be regarded as a "2 parameter" generalization of (.4) – at least for a restriction of the variables as indicated.

THEOREM 35. *Let $\mathfrak{g}$ be a simple real Lie algebra of split rank 1. Let t be the dimension of the root space corresponding to a simple restricted root α and let s be the dimension of the root space corresponding to 2α so that s takes the possible values*

$$s = 0, 1, 3 \text{ and } 7 .$$

Then if $\tau \in \mathbf{C}$ and $\operatorname{Re}\tau > 4s + 3t + 2$ one has

$$\frac{2^{\tau-2s-t} \cdot \Gamma(1/2 \cdot (1+\tau-s)) \cdot \Gamma(\tau/2)}{\Gamma(1/2 \cdot (s+t+1)) \cdot \Gamma(\tau - (s + 1/2 \cdot t))} \tag{137}$$

$$= \sum_{i,j} (-1)^{2i+j} \cdot d(i,j) \cdot \varphi_{i,j}(\tau) \tag{138}$$

where for $\xi = -\tau + 2s + t$,

$$\varphi_{ij}(\tau) = \frac{[\xi(\xi+2)\ldots(\xi+2(i+j)-2)]\cdot[(\xi+1-s)(\xi+3-s)\ldots(\xi+2i-1-s)]}{[\tau(\tau+2)\ldots(\tau+2(i+j)-2)]\cdot[(\tau+1-s)(\tau+3-s)\ldots(\tau+2i-1-s)]}. \tag{139}$$

The sum (138) is absolutely convergent and is over all $i, j, \in \mathbb{Z}_+$ if $s \neq 0$ but j is restricted to $0, 1$ if $s = 0$. The number $d(i,j)$ is the dimension of a representation of the maximal compact subgroup of a group G having $\mathfrak{g}$ as its Lie algebra. The sum (137) is a non-zero complex number and is the inverse of the value of the c-function at $\tau\alpha$.

On the other hand if we drop the factor $(-1)^{2i+j}$ in (138) then one has the absolutely convergent sum

$$\sum_{ij} \cdot d(i,j) \cdot \varphi_{i,j}(\tau) = 0\,. \tag{140}$$

Proof: In the case at hand the Weyl group $W = \{1, \kappa\}$. Using (125), (132) and the expression (133) for the c-function the proof follows immediately from Theorem 31 as soon as we verify that if $\gamma = \gamma(i,j,\varepsilon)$ then $Q^{\gamma}_{\kappa} = (-1)^{2i+j}$ on the one dimensional subspace $(V^*_{\gamma})^M$.

Let $k \in \kappa$ and let $z \in V^M_{\gamma}$. It suffices to show that $g \cdot z = (-1)^{2i+j} \cdot z$. Now $\mathfrak{a} = \mathbf{R} \cdot w$. We will use the notation of pages 804 and 805 in [K-R] so that H' is the space of harmonic polynomial functions on $\mathfrak{p}$. By Theorem 20, p. 805 in [K-R] β_w maps $F_{\gamma} = \operatorname{Hom}_K(V^*_{\gamma}, H')$ isomorphically onto a subspace $V_{\gamma}(w) \subset V_{\gamma}$. But $V_{\gamma} \cong H_{\gamma}$. Hence by Lemma 2.2.11, p. 279 in [K] one has $V_{\gamma}(w) = V^M_{\gamma}$. Thus there exists $\sigma \in F_{\gamma}$ such that $\beta_w(\sigma) = z$. But σ is a K-map from V^*_{γ} to H'. By (123) one necessarily has

$$\sigma(V^*_{\gamma}) \subseteq H'_{2i+j}\,. \tag{141}$$

But now if $v \in V^*_{\gamma}$ then $< v, k \cdot \beta_w(\sigma) > = < k^{-1} \cdot v, \beta_w(\sigma) > = \sigma(k^{-1} \cdot v)(w) = (k^{-1} \cdot \sigma(v))(w) = \sigma(v)(k \cdot w)$. But $k \cdot w = \operatorname{Ad} k(w) = -w$ since $k \in \kappa$. But then $\sigma(v)(k \cdot w) = (-1)^{2i+j} \cdot \sigma(v)(w)$ by (141). Thus $k \cdot z = (-1)^{2i+j} \cdot z$ since $v \in V^*_{\gamma}$ is arbitrary.QED

Remark 35.5: See Remark E in § 1.4 concerning Koorwinder's "continuous parameter" extension of Theorem 35.

8.3. We now consider Theorem 35 for the special case where $s = 0$. We may take $\mathfrak{g} = \operatorname{Lie} SO(n,1)$ where $n \geq 2$ so that

$$t = n - 1\,. \tag{142}$$

Furthermore the parameterizing pair (i,j) may be replaced by $k = 2i+j$, $d(i,j)$ by $d(k)$ and we may take

$$V_{\gamma(i,j)} = H'_k \tag{143}$$

where H'_k is the space of harmonic polynomial functions on $\mathfrak{p} \cong \mathbf{R}^n$. If $S_k(\mathfrak{p})$ is the space of all polynomial functions on $\mathfrak{p}$ then in terms of binomial coefficients

$$\dim S_k(\mathfrak{p}) = \begin{bmatrix} n+k-1 \\ k \end{bmatrix}. \tag{143}$$

But if Q denotes the non-zero K-invariant in $S_2(\mathfrak{p})$ then one knows that H'_k is a complement of $Q \cdot S_{k-2}(\mathfrak{p})$ in $S_k(\mathfrak{p})$ and hence

$$d_k = \begin{bmatrix} n+k-1 \\ k \end{bmatrix} - \begin{bmatrix} n+k-3 \\ k-2 \end{bmatrix}. \tag{144}$$

Using standard relations among binomial coefficients we may rewrite (144) as

$$d_k = \begin{bmatrix} n+k-2 \\ k \end{bmatrix} + \begin{bmatrix} n+k-3 \\ k-1 \end{bmatrix}. \tag{145}$$

On the other hand we now have

$$(137) = \frac{2^{\tau-n+1} \cdot \Gamma((\tau+1)/2) \cdot \Gamma(\tau/2)}{\Gamma(n/2) \cdot \Gamma(\tau-(n-1)/2)}\,. \tag{146}$$

But, one knows

$$\Gamma((\tau+1)/2) \cdot \Gamma(\tau/2) = 2^{1-\tau} \cdot \Gamma(1/2) \cdot \Gamma(\tau)\,. \tag{147}$$

See e.g. [B], p. 5 formula (15). Thus

$$(137) = 2 \cdot \frac{2^{-n+1} \cdot \Gamma(1/2) \cdot \Gamma(\tau)}{\Gamma(n/2) \cdot \Gamma(\tau-(n-1)/2)\,.} \tag{148}$$

On the other hand now

$$(138) = \sum_{k=0}^{\infty} (-1)^k \cdot \left(\begin{bmatrix} n+k-2 \\ k \end{bmatrix} + \begin{bmatrix} n+k-3 \\ k-1 \end{bmatrix} \right) \cdot \varphi_k(\tau) \tag{149}$$

where now

$$\varphi_k(\tau) = \frac{\xi(\xi+1)\dots(\xi+k-1)}{\tau(\tau+1)\dots(\tau+k-1)} \tag{150}$$

for $\xi = -\tau + n - 1$.

Now, as in the theory of hypergeometric functions, let, for any $\beta \in \mathbf{C}$ and $k \in \mathbf{N}$,

$$(\beta)_k = \beta(\beta+1)\dots(\beta+k-1)\,. \tag{151}$$

Also one puts $(\beta)_0 = 1$. Note then that

$$\begin{bmatrix} n+k-2 \\ k \end{bmatrix} = (n-1)_k/(1)_k\,. \tag{152}$$

Now for $a, b, c \in \mathbf{C}$ where $-c \notin \mathbf{Z}_+$ then

$$F(a,b;\,c;\,z) = \sum_{k=0}^{\infty} \frac{(a)_k \cdot (b)_k}{(c)_k \cdot (1)_k} \cdot z^k \tag{153}$$

defines a hypergeometric function of z with the usual parameters for z in the interior of the unit disk. If Re $c >$ Re a + Re b then it is a result of Gauss that the series $F(a,b;\,c;\,1)$ converges and that

$$F(a,b;\,c;\,1) = \frac{\Gamma(c-a-b)\cdot\Gamma(c)}{\Gamma(c-a)\cdot\Gamma(c-b)}\,. \tag{154}$$

See e.g. [B], p. 104, equation (46). For a proof see [W-W] p. 281-2. For $z \neq 1$ on the unit circle there seems no other such general result. However there are results if there is a relation between a, b, and c. A result of Kummer asserts that

$$F(a,b;\,1+a-b;\,-1) = 2^{-a}\cdot\frac{\Gamma(1+a-b)\cdot\Gamma(1/2)}{\Gamma(1-b+a/2)\cdot\Gamma((a+1)/2)} \tag{155}$$

where $1+a-b \notin -\mathbf{Z}_+$ is the only restriction. See [B], p. 104, equation (47) or with a proof see [R], Section 42, p. 68. This result of Kummer may be obtained from (154) upon using the quadratic transformation formula of Gauss and Kummer which establishes

(156)

$$\begin{aligned} &F(a,b;\,1+a-b;\,z) \\ &\quad = (1-z)^{-a}\cdot F(a/2, -b+(a+1)/2;\,1+a-b;\,-4z/(1-z)^2) \end{aligned}$$

and putting $z = -1$. See [B], p. 64, equation (25) or with a proof see [R], top of p. 67. There is a misprint however in the [B] reference in that on the right side $1 - a + b$ is mistakenly written for $1 + a - b$.

Now we wish to rewrite (155) using different parameters. Let $\tau = 1 + a - b$, $a = m - 1$ so that $b = -\tau + m$. Then Kummer's identity (155) takes the form

$$(157) \quad F(m-1, -\tau + m; \tau; -1) = 2^{-m+1} \frac{\Gamma(\tau) \cdot \Gamma(1/2)}{\Gamma(m/2) \cdot \Gamma(\tau - (m-1)/2)}$$

where $-\tau \notin \mathbf{Z}_+$.

Now the absolutely convergent sum (see (111))

$$\sum_{\gamma \in K^\wedge} \operatorname{tr} T_\kappa^\gamma A^\gamma(\kappa, \mu) = 1$$

where $\mu \in \rho + D_{\mathbf{C}}$ for any semisimple Lie group yields the absolutely convergent expression (see (114) in Theorem 32)

$$(158) \qquad \sum_{\gamma \in K^\wedge} \dim \gamma \cdot \operatorname{tr} Q_\kappa^\gamma \cdot P^\gamma(\kappa\mu + 2\rho) \cdot (P^\gamma(\mu))^{-1} = c(\mu)^{-1}$$

for the inverse of the c-function. In the rank 1 case this becomes the very explicit sum (138) in Theorem 35. We will see that Theorem 35 (or indeed Theorem 32) can be regarded as a generalization of Kummer's identity (157), albeit for certain restrictions of the parameters, in that with these restrictions, the equality (138) for the groups $SO(n,1)$ becomes Kummer's formula.

THEOREM 36. *Let the notation be as in Theorem 35. Let* $\mathfrak{g} = \operatorname{Lie} SO(n,1)$ *so that* $s = 0$. *Then the sum (138) is just the hypergeometric series*

$$(159) \quad \sum_{i,j} (-1)^{2i+j} \cdot d(i,j) \cdot \varphi_{i,j}(\tau) = 2 \cdot F(n-1, -\tau + n; \tau; -1).$$

(That is this hypergeometric series can be regarded as a sum over K-types for the trace of a trace class operator.) On the other hand

$$(160) \qquad c(\tau\alpha)^{-1} = 2 \cdot \frac{2^{-n+1} \cdot \Gamma(1/2) \cdot \Gamma(\tau)}{\Gamma(n/2) \cdot \Gamma(\tau - (n-1)/2)}$$

so that the equality of the series (159) with (160) given by (158) (that is by Theorem 35 for $\mathfrak{g} = \operatorname{Lie} SO(n,1)$*) is Kummer's formula (157) –*

with both sides multiplied by 2 – when m is an integer $n \geq 2$ and $\operatorname{Re}\tau > 3n-1$.

PROOF: We first observe that

$$F(n-1,-\tau+n-1;\tau;z)=\sum_{k=0}^{\infty} z^k\cdot\begin{bmatrix} n+k-2 \\ k \end{bmatrix}\cdot\varphi_k(\tau)\,. \tag{161}$$

This is clear from (150) and (152). On the other hand let

$$F=F(n-1,-\tau+n;\tau;z)\,. \tag{162}$$

Using the familiar notation of (153) we adopt the classic formalism that $F(b-1)$ is the hypergeometric series (153) with $b-1$ substituted for b. A similar convention holds for a and c and also where 1 replaces -1. We note then that

$$(161)=F(b-1)\,. \tag{163}$$

We note also explicitly that if as before $\xi=-\tau+n-1$ then

$$F=\sum_{k=0}^{\infty} z^k\cdot\begin{bmatrix} n+k-2 \\ k \end{bmatrix}\cdot\frac{(\xi+1)_k}{(\tau)_k}\,. \tag{164}$$

On the other hand

$$\begin{aligned}\sum_{k=0}^{\infty} z^k\cdot\begin{bmatrix} n+k-3 \\ k-1 \end{bmatrix}\cdot\varphi_k(\tau) &= \frac{\xi}{\tau}\cdot z\cdot\sum_{k=1}^{\infty} z^{k-1}\cdot\begin{bmatrix} n+k-3 \\ k-1 \end{bmatrix}\cdot\frac{(\xi+1)_{k-1}}{(\tau+1)_{k-1}} \\ &= \frac{\xi}{\tau}\cdot z\cdot\sum_{k=0}^{\infty} z^k\cdot\begin{bmatrix} n+k-2 \\ k \end{bmatrix}\cdot\frac{(\xi+1)_k}{(\tau+1)_k} \\ &= \tfrac{a-c}{c}\cdot z\cdot F(c+1)\end{aligned} \tag{165}$$

by (164) since $a=n-1$, $c=\tau$ so that

$$\xi=a-c\,. \tag{166}$$

But now if $H=H(z)$ is the function defined by adding (161) to (165). That is, by (163),

$$H=F(b-1)+\frac{a-c}{c}\cdot z\cdot F(c+1) \tag{166}$$

then by (149)

$$(138)=H(-1)\,.$$

But

$$H = (1 - z) \cdot F \tag{167}$$

by one of the contiguity equations for hypergeometric functions. See e.g. equation (38), p. 103 in [B] with a and b interchanged. For a proof see [R], formula (20), p. 53. Thus

$$(138) = 2 \cdot F(-1)\,. \tag{168}$$

This establishes (159). But now (160) has already been proved. See (148). Thus Theorem 35 implies Kummer's identity for the values of m and τ stated in Theorem 36.QED

8.4.

REMARK 37: Theorem 36 traces back to the identity (112) for arbitrary $\mathfrak{g}$. But in this generality we have also established the (vanishing) identity (111)

$$\sum_{\gamma \in K^\wedge} \operatorname{tr} T_v^\gamma A^\gamma(\kappa, \mu) = 0 \quad \text{if} \quad v \neq \kappa$$

which more explicitly becomes (115)

$$\sum_{\gamma \in K^\wedge} \dim \gamma \cdot \operatorname{tr} Q_v^\gamma \cdot P^\gamma(\kappa\mu + 2\rho) \cdot (P^\gamma(\mu))^{-1} = 0$$

if $v \neq \kappa$. In the split rank 1 case this means that if we drop the alternation in (138) we obtain (140) or

$$\sum_{i,j} d(i,j) \frac{[\xi(\xi+2)\dots(\xi+2(i+j)-2)]\cdot[(\xi+1-s)(\xi+3-s)\dots(\xi+2i-1-s)]}{[\tau(\tau+2)\dots(\tau+2(i+j)-2]\cdot[(\tau+1-s)(\tau+3-s)\dots(\tau+2i-1-s)]} = 0\,.$$

Using hypergeometric identities we would like to verify this in the case where $G = SO(n,1)$. Indeed by (145) the equation (140) when $s = 0$ becomes

$$\sum_{k=0}^{\infty} \left(\begin{bmatrix} n+k-2 \\ k \end{bmatrix} + \begin{bmatrix} n+k-3 \\ k-1 \end{bmatrix} \right) \cdot \varphi_k(\tau) = 0. \tag{169}$$

To verify (169) note that in the notation of the proof of Theorem 36 the left side of (169) is just $H(1)$. But $H = (1-z)F$, recalling (167). Thus

$$H(1) = 0\,. \tag{170}$$

QED

REFERENCES

[B] H. Bateman, "Higher Transcendental Functions," McGraw–Hill, 1953.

[HC] Harish–Chandra, *Spherical functions on a semi-simple Lie group. I*, Amer. J. Math. **80** (1958), 241–310.

[H_1] S. Helgason, *A duality for symmetric spaces with applications to group representations*, Advances in Math. **5** (1970), 1–154.

[H_2] S. Helgason, "Groups and Geometric Analysis," Academic Press, 1984.

[J] K. Johnson, *Composition series and intertwining operators for the spherical principal series. II*, Trans. Amer. Math. Soc. **215** (1976), 269–283.

[J-W] K. Johnson and N. Wallach, *Composition series and intertwining operators for the spherical principal series. I*, Trans. Amer. Math. Soc. **229** (1977), 137–173.

[KN-S] A. Knapp, and E. Stein, *Intertwining operators for semi-simple groups*, Ann. of Math. **93** (1971), 489–578.

[Koo] T. Koornwinder, *Jacobi functions and analysis on non-compact semisimple Lie groups*, in "Special Functions: Group Theoretical Aspects and Applications," pp. 1–85; (Edited by R.A. Askey et al.) D. Reidel Publishing Company, 1984.

[K] B. Kostant, *On the existence and irreducibility of certain series of representations*, in "Lie groups and their Representations," (I.M. Gelfand ed.), Halsted, 1975, pp. 231–329.

[K-R] B. Kostant and S. Rallis, *Orbits and representations associated to symmetric spaces*, Amer. J. Math. **93** (1971), 753–809.

[KU-S] R. Kunze and E. Stein, *Uniformly bounded representations. III. Intertwining operators for the principal series on semisimple groups*, Amer. J. Math. **89** (1967), 385–442.

[R] E. Rainville, "Special Functions," Chelsea Publishing Co., 1960.

[S] G. Schiffmann, *Integrales d'entralacement et fonctions de Whittaker*, Bull. Soc. Math. France **99** (1971), 3–72.

[Sn] Sniatcyki, J., "Geometric Quantization and Quantum Mechanics, Applied Mathematical Sciences 30," Springer-Verlag.

[W] N. Wallach, *Kostant's P^γ and R^γ matrices and intertwining integrals*, in "Harmonic Analysis on Homogeneous Spaces, Proceedings of Symposia in Pure Mathematics AMS," 1973, pp. 269–273.

[W-W] E.T. Whittaker and G.N. Watson, "A Course in Modern Analysis," Cambridge University Press, 1963.

Department of Mathematics
Massachusetts Institute of Technology
Cambridge, MA 02139
USA

The Penney–Fujiwara Plancherel Formula for Symmetric Spaces

RONALD L. LIPSMAN[1]

Introduction. I am interested in the direct integral decomposition of the quasi-regular representation $\tau = \mathrm{Ind}_H^G 1$ for symmetric spaces G/H. Suppose that τ is type I and one has an equivalence

$$\tau \simeq \int_S^{\oplus} n_\tau(\pi)\pi \; d\mu_\tau(\pi),$$

where μ_τ is a Borel measure on $\hat{G}$, $n_\tau(\pi)$ is a multiplicity function and $\mathcal{S} \subset \hat{G}$ is a minimal closed μ_τ-co-null set. If we can compute explicitly the triple $(\mathcal{S}, n_\tau, \mu_\tau)$, we say that we know the direct integral decomposition of τ. Such computations have been carried out for exponential solvable symmetric spaces [1]. (In that case, the multiplicity function is identically 1.) Actually Benoist's formula is a special case of direct integral decomposition formulas for general induced representations of nilpotent homogeneous spaces [3] [7] and completely solvable homogeneous spaces [9].

There is an ingredient missing from the picture, however – namely the explicit intertwining operator which effects the equivalence between τ and the direct integral. I announced the following important observation at the conference: the intertwining operator can be computed by utilizing the distribution-theoretic version of the direct integral decomposition due to Penney [11]. Let $\mathcal{H}_\tau^\infty$ be the C^∞-vectors in the space $\mathcal{H}_\tau$ of τ and let α_τ be the canonical cyclic distribution. Suppose one can find distributions α_π (with the correct H-equivariance property – see below) such that

$$(I) \qquad \langle \tau(\varphi)\alpha_\tau, \alpha_\tau\rangle = \int_S \langle \pi(\varphi)\alpha_\pi, \alpha_\pi\rangle d\mu_\tau(\pi), \quad \varphi \in \mathcal{D}(G).$$

Then it is not difficult to see that the map

$$\tau(\varphi)\alpha_\tau \to \{\pi(\varphi)\alpha_\pi\}_{\pi \in S}$$

extends from $\mathcal{H}_\tau^\infty$ to an isometry of $\mathcal{H}_\tau$ onto $\int_S^\oplus \mathcal{H}_\pi d\mu_\tau(\pi)$ which intertwines τ and the direct integral $\int_S^\oplus \pi d\mu_\tau(\pi)$. Formula (I) (together with

[1]Supported by NSF under DMS–87–00551A01

explicit expressions for $\mathcal{S}, \mu_\tau$ and α_π) were derived for G nilpotent by Benoist in [2]. Analogous formulae were obtained by Fujuwara for some exponential solvable homogeneous spaces in [4]. I have proven analogs of (I) for arbitrary completely solvable homogeneous spaces [10]. My purpose here is to illustrate [10] with results in a very special case – a case in which the main new difficulty, that arises when passing from nilpotent to exponential solvable homogeneous spaces, is already manifest.

I assume G/H is an abelian symmetric space [6], and also that G is completely solvable, algebraic, metabelian, but not nilpotent. In other words, I will consider here semidirect products $G = VA$ where A, V are both real vector groups, V is normal and A acts semisimply on V with real (positive) eigenvalues; and $H = A$. The results of [6] and [8] say that:

$$(II) \qquad \tau = \mathrm{Ind}_A^G 1 \simeq \int_{\mathfrak{a}^\perp/A}^{\oplus} \pi_\psi d\dot{\psi} \simeq \int_{\hat{V}/A}^{\oplus} \pi_\chi d\dot{\chi},$$

where π_ψ corresponds to $\psi \in \mathfrak{a}^\perp \subset \mathfrak{g}^*$ (by Kirillov-Bernat); $\pi_\psi \simeq \pi_\chi = \mathrm{Ind}_{VA_\chi}^G(\chi \times 1)$, $\chi(\exp X) = e^{i\psi(X)}$, $X \in \mathfrak{v}$; and $d\dot{\psi} = d\dot{\chi}$ is the push-forward of the Lebesgue measure class on $\mathfrak{a}^\perp = \hat{V}$ under the action of A. We prove the following

THEOREM. *For almost all $\chi \in \hat{V}$, there exist distributions α_χ – given by the convergent integrals specified in formula (IX) below – such that*

$$(III) \qquad \langle \tau(\varphi)\alpha_\tau, \alpha_\tau \rangle = \int_{\hat{V}/A} \langle \pi_\chi(\varphi)\alpha_\chi, \alpha_\chi \rangle d\dot{\chi}, \quad \varphi \in \mathcal{D}(G).$$

The main results of this paper are the proof of the convergence of the integral (IX) and the derivation of the Penney-Fujuwara Plancherel formula (III).

Group and measure structure. Let $G = VA$ be a semidirect product of two vector groups, V normal. We assume also that A acts semisimply on V with real (therefore positive) eigenvalues. We fix Haar measures dv, da on V, A respectively. The modulus for the action of A on V is denoted $\delta(a)$; so if $\mathcal{K}(V)$ denotes the continuous functions of compact support, we have

$$\delta(a) \int_V f(ava^{-1}) dv = \int_V f(v) dv, \quad a \in A, f \in \mathcal{K}(V).$$

Then the right and left Haar measures on G are $dg = d_r g = dvda$ and $d_\ell g = \delta(a)^{-1} dvda$. The modular function is

$$\Delta_G(va) = \frac{d_r}{d_\ell}(va) = \delta(a).$$

We note for later use that $\frac{\Delta_A}{\Delta_G}(a) = \delta(a)^{-1}$. We shall write all group actions on the right: $v \cdot a = a^{-1}va$, $(\chi \cdot a)(v) = \chi(v \cdot a^{-1}) = \chi(ava^{-1})$, $(f \cdot a)(v) = f(v \cdot a^{-1}) = f(ava^{-1})$. Thus we have

$$\delta(a) \int (f \cdot a)(v) dv = \delta(a) \int f(v \cdot a^{-1}) dv = \int f(v) dv, \ f \in \mathcal{K}(V),$$

$$\delta(a)^{-1} \int (f \cdot a)(\chi) d\chi = \delta(a) \int f(\chi \cdot a^{-1}) d\chi = \int f(\chi) d\chi, \ f \in \mathcal{K}(\hat{V}),$$

where $d\chi$ is the Haar measure on $\hat{V}$, dual to dv.

We define $q(va) = \delta(a)^{-1}$. q is a continuous function on G satisfying $q(e) = 1$ and

$$q(ag) = \delta(a)^{-1} q(g) = \frac{\Delta_A}{\Delta_G}(a) q(g), \quad a \in A, \ g \in G,$$

as in [5, eqn. 3.2]. Moreover we have

$$\text{(IV)} \qquad \int_G f(g) q(g) dg = \int_V \int_A f(av) da\, dv.$$

Next we disintegrate the Haar measure $d\chi$ under the action of A. We fix once and for all a choice of pseudo-image $d\dot{\chi}$. We next recall that the modulus of A for its action on $d\chi$ is δ^{-1}. Therefore [5] almost every orbit (of A) on $\hat{V}$ has a relatively invariant measure with that modulus. Moreover these measures are uniquely determined by the choice of $d\dot{\chi}$ according to the formula

$$\text{(V)} \qquad \int_V f(\chi) d\chi = \int_{\hat{V}/A} \int_{\chi \cdot A} f(\chi \cdot a) d\mu_{\dot{\chi}} d\dot{\chi}, \ f \in \mathcal{K}(\hat{V})$$

(see [5, §2]). Note we are writing $\dot{\chi} = \chi \cdot A \in \hat{V}/A$. Now we wish to use a (slightly abusive) coset notation to describe the measure $\mu_{\dot{\chi}}$. Henceforth we discard a set of measure zero in $\hat{V}$, and consider only the orbits $\chi \cdot A$ which carry relatively invariant measures $\mu_{\dot{\chi}}$. But the action of A_χ in $A_\chi \setminus A$ is trivial – hence for these χ we have

$$A_\chi \subset \ker \delta.$$

Thus for $f \in \mathcal{K}(A; A_\chi)$ = the continuous functions on A, left-invariant under A_χ and compactly supported mod A_χ, the function $a \to f(a)\delta(a)$ also belongs to $\mathcal{K}(A; A_\chi)$. Our "abuse of notation" is to write

$$(VI) \qquad \mu_\chi : f \to \int_{A_\chi \backslash A} f(a)\delta(a)d\dot{a}.$$

In fact any choice of Haar measure da_χ on A_χ uniquely determines an A-invariant measure on $A_\chi \setminus A$ so that

$$\int_A f(a)da = \int_{A_\chi \backslash A} \int_{A_\chi} f(a_\chi a)da_\chi d\dot{a}, \quad f \in \mathcal{K}(A).$$

We make the choice so that $\delta(a)d\dot{a}$ is the relatively invariant measure of modulus δ^{-1} uniquely determined by the choice $d\dot{\chi}$. Then

$$\begin{aligned}\delta(a_1)^{-1} \int_{A_\chi \backslash A} (f \cdot a_1)(a)d\mu_\chi(\dot{a}) &= \delta(a_1)^{-1} \int_{A_\chi \backslash A} f(aa_1^{-1})d\mu_\chi(\dot{a}) \\ &= \int_{A_\chi \backslash A} f(a)d\mu_\chi(a), \quad f \in \mathcal{K}(A; A_\chi).\end{aligned}$$

The "abuse" is that the notation suggests dependence of the choice of the point χ in the A-orbit $\chi \cdot A$, but of course the (relatively) invariant measure is independent of the choice.

Before passing to the definition of the relatively invariant distributions, we note that for $k \in \mathbf{Z}$, the measure

$$\delta(a)^k d\mu_\chi(\dot{a}) = \delta(a)^{k+1} d\dot{a}$$

is relatively invariant on $A_\chi \setminus A$ with modulus $\delta^{-(k+1)}$. In particular $\delta^{-1/2}(a)d\mu_\chi(\dot{a}) = \delta^{1/2}(a)d\dot{a}$ is relatively invariant with modulus $\delta^{-1/2}$, and $\delta^{-1}(a)d\mu_\chi(\dot{a}) = d\dot{a}$ is invariant.

The distributions and the Plancherel formula. Continuing with the notation and terminology established above, we consider the quasi-regular representation

$$\tau = \mathrm{Ind}_A^G 1.$$

The representation τ acts in the space of Borel function $f : G \to \mathbf{C}$, $f(ag) = f(g)$ which satisfy $\int_{A \backslash G} |f(g)|^2 d\dot{g} < \infty$ by the group action

$$\tau(g)f(x) = f(xg)\left[\frac{q(xg)}{q(x)}\right]^{1/2} \quad \text{(see [5])}.$$

Here $d\dot{g}$ denotes the relatively invariant measure on $A \setminus G$ determined by

$$\int_G f(g)q(g)dg = \int_V \int_A f(av)da\,dv = \int_{A\setminus G} \int_A f(ag)da\,d\dot{g} \quad \text{(see (IV))}.$$

By restricting these functions to V, we may realize τ on $L^2(V)$ via the formulae

$$\tau(v)f(v_1) = f(v_1 v)$$

$$\tau(a)f(v_1) = f(v_1 \cdot a)\left[\frac{q(v_1 a)}{q(v_1)}\right]^{1/2}$$

$$= f(a^{-1}v_1 a)\delta(a)^{-1/2}, \quad f \in L^2(V).$$

The space $L^2(V)^\infty$ of C^∞-vectors of τ is contained in $C^\infty(V)$. We write $(L^2(V)^{-\infty})_{A,q^{-1/2}}$ to denote the set of anti-distributions (i.e. conjugate-linear functionals on $L^2(V)^\infty$ which are continuous in the canonical topology of the C^∞-vectors) which transform under A by the modulus $q^{-1/2}$. The canonical cyclic distribution

$$\alpha = \alpha_\tau : f \to \overline{f(e)}$$

is such a distribution. In fact

$$\langle \tau(a)\alpha, f\rangle = \langle \alpha, \tau(a)^{-1}f\rangle = \langle \alpha, f(a \cdot a^{-1})\delta(a)^{1/2}\rangle = \delta(a)^{1/2}\overline{f(e)}.$$

Let us compute the matrix coefficient of τ determined by α. Take $\varphi \in \mathcal{K}(G)$ and set

$$\varphi^*(g) = \overline{\varphi(g^{-1})}\Delta_G(g)^{-1} = \overline{\varphi(g^{-1})}q(g).$$

Then

$$\tau(\varphi^*)\psi = \int \varphi^*(g)\tau(g)\psi dg = \int \bar{\varphi}(g^{-1})q(g)\tau(g)\psi dg.$$

Therefore

$$\langle \tau(\varphi)\alpha, \psi\rangle = \langle \alpha, \tau(\varphi^*)\psi\rangle = \left[\int \bar{\varphi}(g^{-1})q(g)(\tau(g)\psi)(e)dg\right]^-$$

$$= \int \varphi(g^{-1})\bar{\psi}(g)q(g)^{1/2}q(g)dg.$$

Consequently, for $\varphi \in \mathcal{D}(G)$ = smooth functions of compact support, the distribution $\tau(\varphi)\psi$ – which we know a priori has to be a function – is given by

$$\tau(\varphi)\alpha(v) = \int_A f(v^{-1}a^{-1})q(a)^{1/2}da$$

$$= \int_A f(v^{-1}a^{-1})\delta(a)^{-1/2}da$$

(here we used (IV)). We shall employ the following notation. For $\varphi \in \mathcal{D}(G)$, we set

$$(VII) \qquad \varphi_A(g) = \int_A \varphi(ga^{-1})\delta(a)^{-1/2}da.$$

Then

$$\tau(\varphi)\alpha(v) = \varphi_A(v^{-1}), \ v \in V.$$

Finally

$$(VIII) \qquad \begin{aligned} \langle \tau(\varphi)\alpha, \alpha\rangle &= \overline{\langle \alpha, \tau(\varphi)\alpha\rangle} = \varphi_A(e) \\ &= \int_A \varphi(a^{-1})\delta(a)^{-1/2}da = \int_A \varphi(a)\delta(a)^{1/2}da. \end{aligned}$$

Next we study the spherical irreducible representations of G. These are

$$\pi_\chi = \mathrm{Ind}_{VA_\chi}^G (\chi \times 1), \ \chi \in \hat{V}.$$

In what follows we shall be concerned only with those π_χ arising from χ in general position, i.e. those for which $A_\chi \subset \ker\delta$ and $A_\chi \setminus A$ carries a relatively invariant measure of modulus δ^{-1}. Now the homogeneous space $VA_\chi \setminus VA$ has an invariant measure. Hence the representation π_χ is realized in the space of Borel functions $f : G \to \mathbf{C}$, $f(va_\chi g) = \chi(v)f(g)$ which are square integrable on $VA_\chi \setminus VA$ with respect to the invariant measure. Restricting these functions to A, we obtain a realization on left-A_χ-invariant functions on A by

$$\begin{aligned} &\pi_\chi(a)f(a_1) = f(a_1 a) \\ &\pi_\chi(v)f(a_1) = \chi(a_1 v a_1^{-1})f(a_1), \ f \in L^2(A; A_\chi; da). \end{aligned}$$

Within this realization the C^∞-vectors $L^2(A; A_\chi)^\infty$ are in the smooth functions $C^\infty(A; A_\chi)$. We define the distribution

$$(IX) \qquad \alpha_\chi : f \to \int_{A_\chi \setminus A} \bar{f}\delta^{-1/2}d\mu_\chi = \int_{A_\chi\setminus A} \bar{f}(a)\delta^{1/2}(a)d\dot{a}, \ f \in L^2(A; A_\chi)^\infty.$$

We shall prove in the next section that the integrals in (IX) are absolutely convergent for $f \in L^2(A; A_\chi)^\infty$. In the remainder of this section, we derive the Plancherel formula (III) assuming that result.

We leave to the reader the simple check that α_χ transforms under A by the modulus $q^{-1/2}$. We begin the verification of (III) with the computation of the matrix coefficients. For $\varphi \in \mathcal{D}(G)$, we have

$$\begin{aligned}
\langle \pi_\chi(\varphi)\alpha_\chi, \psi\rangle &= \langle \alpha_\chi, \pi_\chi(\varphi^*)\psi\rangle = \langle \alpha_\chi, \int_G \bar{\varphi}(g^{-1})q(g)\pi_\chi(g)\psi dg\rangle \\
&= \int \varphi(g^{-1})\langle \alpha_\chi, \pi_\chi(g)\psi\rangle q(g)dg = \iint \varphi(v^{-1}a^{-1})\langle \alpha_\chi, \pi_\chi(av)\psi\rangle da dv \\
&= \iint \varphi(v^{-1}a^{-1}) \int_{A_\chi\backslash A} \overline{(\pi_\chi(av)\psi)}(a_1)\delta^{1/2}(a_1)d\dot{a}_1 \, da dv \\
&= \iiint \varphi(v^{-1}a^{-1})\bar{\chi}(a_1ava^{-1}a_1^{-1})\bar{\psi}(a_1a)\delta^{1/2}(a_1)d\dot{a}_1 \, da dv \\
&= \iiint \varphi(v^{-1}a^{-1})\bar{\chi}(a_1va_1^{-1})\bar{\psi}(a_1)\delta^{1/2}(a_1)\delta^{-1/2}(a)d\dot{a}_1 \, da dv.
\end{aligned}$$

Therefore $\pi_\chi(\varphi)\alpha_\chi$ is the smooth function given by the formula

$$\pi_\chi(\varphi)\alpha_\chi(a_1) = \iint \varphi(v^{-1}a^{-1})\bar{\chi}(a_1va_1^{-1})\delta^{1/2}(a_1)\delta^{-1/2}(a)da dv.$$

Continuing, we evaluate the matrix coeffient

$$\begin{aligned}
&\langle \pi_\chi(\varphi)\alpha_\chi, \alpha_\chi\rangle = \overline{\langle \alpha_\chi, \pi_\chi(\varphi)\alpha_\chi\rangle} \\
&= \int_{A_\chi\backslash A} (\pi_\chi(\varphi)\alpha_\chi)(a_1)\delta^{1/2}(a_1)d\dot{a}_1 \\
&= \iiint \varphi(v^{-1}a^{-1})\bar{\chi}(a_1va_1^{-1})\delta(a_1)\delta^{-1/2}(a)da \, dv d\dot{a}_1 \\
&= \iint \varphi_A(v^{-1})\bar{\chi}(a_1va_1^{-1})\delta(a_1)dv d\dot{a}_1 \qquad \text{(see (VIII))}.
\end{aligned}$$

We can now derive the Penney-Fujiwara Plancherel formula as follows:

$$\begin{aligned}
\int \langle \pi_\chi(\varphi)\alpha_\chi, \alpha_\chi\rangle d\dot{\chi} &= \int_{\hat{V}/A}\int_{A_\chi\backslash A}\int_V \varphi_A(v^{-1})\bar{\chi}(a_1va_1^{-1})\delta(a_1)dv d\dot{a}_1 d\dot{\chi} \\
&= \int_{\hat{V}/A}\int_{A_\chi\backslash A}\int_V \varphi_A(v)\chi(a_1va_1^{-1})\delta(a_1)dv d\dot{a}_1 d\dot{\chi} \\
&= \int_{\hat{V}/A}\int_{A_\chi\backslash A} \hat{\varphi}_A(\chi\cdot a_1)\delta(a_1)d\dot{a}_1 d\dot{\chi} \\
&= \int_{\hat{V}/A}\int_{A_\chi\backslash A} \hat{\varphi}_A(\chi\cdot a_1)d\mu_\chi(\dot{a}_1)d\dot{\chi} \text{ (using (VI))} \\
&= \int_{\hat{V}} \hat{\varphi}_A(\chi)d\chi \qquad \text{(using (V))} \\
&= \varphi_A(e) \\
&= \langle \tau(\varphi)\alpha_\tau, \alpha_\tau\rangle \qquad \text{(using (VIII))}.
\end{aligned}$$

This completes the proof of the main Theorem of the paper – modulo the demonstration of the convergence of the distribution integrals (IX). Note that to this point we have used only that A is unimodular and that $\hat{V}/A$ is countably separated. It is in the proof of convergence that we will use the additional hypotheses on A. (Actually our arguments are quite general and apply to more symmetric spaces than those considered in this paper. But we leave that story to another occasion.)

Proof of Convergence. The smooth functions $f \in L^2(A; A_\chi)^\infty$ are square-integrable with respect to the invariant measure. There would seem to be no a priori reason why f should be integrable with respect to the relatively invariant measure $\delta^{-1/2} d\mu_\chi$. In the nilpotent case, the C^∞-vectors must be Schwartz functions and so rapidly decreasing. This is not so for (exponential) solvable groups. In general the exponential decay will only be in certain directions. But the extra growth conditions imposed by the operators $\pi_\chi(X)$, $X \in \mathfrak{v}$, actually guarentee sufficiently rapid decay in every direction. This is illustrated by the following example (see Fujiwara [4]).

Let $\mathfrak{g}$ be spanned by two generators T, X satisfying $[T, X] = X$. Let G be the corresponding simply connected group (the $ax + b$ – group):

$$G = \{\exp x X \exp t T : x, t \in \mathbf{R}\}.$$

Then

$$\delta(\exp x X \exp t T) = e^t.$$

There are two generic irreducible representations $\pi_\pm = \pi_{\chi_\pm}$ where $\chi_\pm(\exp xX) = e^{\pm ix}$. Then $\pi_\pm$ acts in $L^2(A)$ and

$$\alpha_{\chi_\pm} : f \to \int_A \overline{f(t)}\, e^{t/2}\, dt.$$

But we have

$$\pi_\pm(T) = \frac{d}{dt}$$
$$\pi_\pm(X) = e^t.$$

Hence $L^2(A)^\infty = \{f(t) \in C^\infty(\mathbf{R}) : e^{nt} f^{(j)} \in L^2,\ \forall n, j \geq 0\}$. Therefore for $f \in L^2(A)^\infty$, we have

$$\int_{-\infty}^{\infty} f(t) e^{t/2} dt = \int_{-\infty}^{0} f(t) e^{t/2} dt + \int_{0}^{\infty} f(t) e^{t/2} dt,$$

each of which is absolutely convergent – the latter since f is "exponentially" rapidly decreasing near ∞, and the former just because f is L^2 near $-\infty$.

We now present our generalization of this argument to the situation $G = VA$ of the main theorem.

LEMMA. *Let $G = VA$ be a semidirect product of vector groups with V normal and A acting semisimply with real eigenvalues. For $\chi \in \hat{V}$, set $\pi_\chi = \mathrm{Ind}_{VA_\chi}^G(\chi \times 1)$ and let $L^2(A; A_\chi)^\infty$ be the realization of the C^∞-vectors described previously. Then for almost all χ, the integral*

$$\alpha_\chi(f) = \int_{A_\chi \backslash A} \bar{f}\, \delta^{-1/2} d\mu_\chi = \int_{A_\chi \backslash A} \overline{f(a)}\, \delta^{1/2}(a) d\dot{a}$$

is absolutely convergent for all $f \in L^2(A; A_\chi)^\infty$. The map $f \to \alpha_\chi(f)$ defines an element of $L^2(A; A_\chi)^{-\infty}$ which transforms under A by the character $q^{-1/2}$.

PROOF: All that remains is to show that (generically in χ) the integral is absolutely convergent.

We first demonstrate that it is no loss of generality to assume $\dim A \leq \dim V$. Let $n = \dim V$, and suppose that $v_1, \ldots, v_n$ is a complete set of eigenvectors for the action of A

$$a v_j a^{-1} = v_j \cdot a^{-1} = \delta_j(a) v_j, \; a \in A, \; \delta_j \in \mathrm{Hom}\,(A, \mathbf{R}^+), \; 1 \leq j \leq n.$$

Then

$$\delta = \prod_{j=1}^n \delta_j.$$

Set

$$B = \bigcap_{j=1}^n \ker \delta_j,$$

a closed connected central subgroup of G. It is clear that

$$\tau = \mathrm{Ind}_A^G 1 \cong (\mathrm{Ind}_{A/B}^{G/B} 1) \circ p,$$

where $p : G \to G/B$ is the canonical projection. Moreover $B \subset A_\chi$ for every $\chi \in \hat{V}$. Thus it is obviously no loss of generality to assume B is trivial. But

$$\dim \bigcap_{j=1}^n \ker \delta_j \geq \dim A - n > 0$$

if $\dim A > n$. Therefore we may assume $\dim A \leq n = \dim V$.

Now the functions $\log \delta_j$ are linear functionals on A, and since $\cap_{j=1}^n \ker \log \delta_j$ is trivial, we can select a basis of $A^* = \mathrm{Hom}\,(A, \mathbf{R})$ from among them. Take $r = \dim A \leq n$ and list the eigenvalues

$$\delta_1, \ldots, \delta_r, \; \delta_{r+1}, \ldots, \delta_s, \; \delta_{s+1}, \ldots, \delta_n$$

so that $\log\delta_1,\ldots,\log\delta_r$ is a basis of A^*, $\delta_{s+1}=\cdots=\delta_n=1$, but $\delta_{r+1}\neq 1,\ldots,\delta_s\neq 1$. Note that s could be any value between r and n. Define

$$\mathcal{W}_j=\{a\in A:\log\delta_j(a)=0\},\ 1\leq j\leq s.$$

Expand $\chi\in\hat{V}$ in coordinates by

$$\chi(v)=\prod_{j=1}^{n}\chi_j(x_jv_j),\ \chi_j=\chi|V_j,\ V_j=\mathbf{R}v_j,\ v=\sum_{j=1}^{n}x_jv_j.$$

We shall consider only those χ such that $\chi_j\neq 1$, $1\leq j\leq r$. Note that for $\chi\in\hat{V}$, the equality $a\cdot\chi=\chi$ implies $a\cdot\chi_j=\chi_j$, $1\leq j\leq n$. But if $\chi_j\neq 1$, $1\leq j\leq r$, then it must also be that $a\in\ker\delta_j$, $1\leq j\leq r$. That is A_χ is trivial (so a fortiori $A_\chi\subset\ker\delta$) for all χ under consideration. Thus π_χ now acts on $L^2(A)$.

Now suppose $f\in L^2(A)^\infty$. Then $f\in C^\infty(A)$ because $\pi_\chi(Y)f\in L^2(A)$, $\forall Y\in\mathfrak{U}(\mathfrak{a})$. But we also have $\pi_\chi(W)f\in L^2(A)$, $\forall W\in\mathfrak{U}(\mathfrak{v})$. Moreover $\pi_\chi(v_j)f(a)=\chi(av_ja^{-1})f(a)=\chi(\delta_j(a)v_j)f(a)$. Since any $W\in\mathfrak{U}(\mathfrak{v})$ is a symmetric polynomial in the (log of the) v_j, the condition $\pi_\chi(W)f\in L^2$, $\forall W\in\mathfrak{U}(\mathfrak{v})$, says precisely that

$$(X)\qquad\prod_{j=1}^{n}\delta_j(a)^{m_j}f(a)\in L^2(A),\ \text{for all } m=(m_1,\ldots,m_n)\in(\mathbf{Z}^+)^n.$$

Now to prove absolute convergence of (IX), we shall prove it separately in each of the chambers inside $A\setminus\cup_{j=1}^{s}\mathcal{W}_j$. Fix a choice of signs $\varepsilon=(\varepsilon_1,\ldots,\varepsilon_s)$, $\varepsilon_j=\pm1$. This determines a chamber by

$$C=\{a\in A:\log\delta_j(a)\varepsilon_j>0,\ 1\leq j\leq s\}.$$

(If $s>r$, such a chamber could be empty – there may be some consistency relations the ε_j, $j>r$, must satisfy to insure non-emptiness. No matter – we only consider non-empty chambers below.)

Suppose $i_1,\ldots,i_k$ are the indices where ε is positive and $j_1,\ldots,j_\ell$ are the indices where ε is negative, $k+\ell=s$. Then

$$\begin{aligned}&\int f(a)\delta^{1/2}(a)da\\&=\int f(a)\delta_{i_1}(a)\ldots\delta_{i_k}(a)\prod_{x=1}^{k}\delta_{i_x}(a)^{-1/2}\prod_{y=1}^{\ell}\delta_{j_y}^{1/2}(a)\prod_{z=s+1}^{n}\delta_z^{1/2}(a).\end{aligned}$$

We observe that $f\delta_{i_1}\ldots\delta_{i_k} \in L^2$ by (X), and *in the chamber* C the function

$$\prod_{x=1}^{k} \delta_{i_x}^{-1/2} \quad \prod_{y=1}^{\ell} \delta_{j_y}^{1/2}$$

is exponentially rapidly decreasing. Hence the integral $\int f\delta^{1/2}$ is absolutely convergent in C. This concludes the proof of the lemma, so also the proof of our main theorem.

References

[1] Y. Benoist, *Multiplicite un pour les espaces symétriques exponentiels*, Mem. Soc. Math. France **15** (1984), 1-37.

[2] Y. Benoist, *Analyse harmonique sur les espaces symétriques nilpotents*, J. Funct. Analysis **59** (1984), 211-253.

[3] L. Corwin, F. Greenleaf and G. Grélaud, *Direct integral decompositions and multiplicities for induced representations of nilpotent Lie groups*, Trans. Amer. Math. Soc. **305** (1988), 601-622.

[4] H. Fujiwara & S. Yamagami, *Certaines representations monomiales d'un groupe de Lie resoluble exponentiel*; preprint.

[5] A. Kleppner & R. Lipsman, *The Plancherel formula for group extensions*, Ann. Scient. Ecole. Norm. Sup. **5** (1972), 459-516.

[6] R. Lipsman, *Harmonic analysis on non-semisimple symmetric spaces*, Israel J. Math. **54** (1986), 335-350.

[7] R. Lipsman, *Orbital parameters for induced and restricted representations*, Trans. Amer. Math. Soc. (1988); to appear.

[8] R. Lipsman, *Harmonic analysis on exponential solvable homogeneous spaces: the algebraic or symmetric cases*. Pacific J. Math., 1988, to appear.

[9] R. Lipsman, *Induced representations of completely solvable Lie groups*; preprint.

[10] R. Lipsman, *The Penney-Fujiwara Plancherel formula for multiplicity-free completely solvable homogeneous spaces*; in preparation.

[11] R. Penney, *Abstract Plancherel theorems and a Frobenius reciprocity theorem*, J. Funct. Analysis **18** (1975), 177-190.

Department of Mathematics
University of Maryland
College Park, MD 20742
USA

Embeddings of Discrete Series into Principal Series

TOSHIHIKO MATSUKI AND TOSHIO OSHIMA

§1. Introduction.

Let G be a connected real semisimple Lie group, σ an involution of G, and H an open subgroup of the group G^σ of fixed points for σ. For simplicity we assume that G has a complexification G_c. We fix a Cartan involution θ of G with $\sigma\theta = \theta\sigma$. The involutions of the Lie algebra $\mathfrak{g}$ of G induced by σ and θ are denoted by the same letters, respectively. Let $\mathfrak{g} = \mathfrak{h} + \mathfrak{q}$ and $\mathfrak{g} = \mathfrak{k} + \mathfrak{p}$ be the decompositions of $\mathfrak{g}$ into $+1$ and -1 eigenspaces for σ and θ, respectively. Let $\mathfrak{g}^d$, $\mathfrak{k}^d$ and $\mathfrak{h}^d$ be the subalgebras of the complexification $\mathfrak{g}_c$ of $\mathfrak{g}$ defined by

$$\mathfrak{g}^d = \mathfrak{k} \cap \mathfrak{h} + \sqrt{-1}(\mathfrak{k} \cap \mathfrak{q}) + \sqrt{-1}(\mathfrak{p} \cap \mathfrak{h}) + (\mathfrak{p} \cap \mathfrak{q}),$$
$$\mathfrak{k}^d = \mathfrak{k} \cap \mathfrak{h} + \sqrt{-1}(\mathfrak{p} \cap \mathfrak{h}), \quad \mathfrak{h}^d = \mathfrak{k} \cap \mathfrak{h} + \sqrt{-1}(\mathfrak{k} \cap \mathfrak{q}),$$

and let K, G^d, K^d and H^d be the analytic subgroups of G_c with the Lie algebras $\mathfrak{k}$, $\mathfrak{g}^d$, $\mathfrak{k}^d$ and $\mathfrak{h}^d$, respectively. Then the homogeneous space $X^d = G^d/K^d$ is a Riemannian symmetric space of the non-compact type and called the non-compact Riemannian form of the semisimple symmetric space $X = G/H$. The ring $D(X)$ of the invariant differential operators on X is naturally isomorphic to the ring $D(X^d)$ of invariant differential operators on X^d.

Let P^d be a minimal parabolic subgroup of G^d. In **[O1]** a K-finite eigenfunction ψ of $D(X)$ is called a spherical function on X and defines an H^d-invariant closed subset $FBI(\psi)$ of G^d/P^d. Namely, by the Flensted–Jensen isomorphism, ψ corresponds to a simultaneous eigenfunction $\tilde{\psi}$ of $D(X^d)$ and then $FBI(\psi)$ is the support of the image of $\tilde{\psi}$ under the boundary value isomorphism defined by **[KKMOOT]**. The main result in **[O1]** shows that $FBI(\psi)$ and the eigenvalue determine the leading terms in a convergent series expansion of ψ at every boundary point of X in $\tilde{X}$. Here $\tilde{X}$ is the compact G-manifold constructed in **[O2]** which contains X as an open G-orbit.

Suppose the spherical function ψ generates an irreducible Harish–Chandra module $U(\psi)$. Then by the leading terms we have embeddings of $U(\psi)$ into principal series for X, which is studied in **[O1**, Theorem 5.1]. The key lemma in **[O1]** is **[O1**, Lemma 3.2] which studies a local property of intertwining operators between class 1 principal series for G^d.

In §2 we will give another lemma for the intertwining operators. These lemmas give embeddings of $U(\psi)$ into principal series for X which are not obtained in **[O1]**. Namely, by the lemmas and the same argument as in **[O1]**, we get the embeddings which do not correspond to any leading term.

In §3 we consider the case where X is a semisimple Lie group. In this case ψ corresponds to a matrix coefficient of an irreducible Harish–Chandra module for the group, $FBI(\psi)$ coincides with the support of the $\mathcal{D}$-module realized in a complex flag manifold through Beilinson–Bernstein's correspondence (**[BB1]**, **[V]**) and we will give a simple theorem to find embeddings of any irreducible Harish–Chandra module into principal series for the group. The embeddings corresponding to $S(E)_0$ (i.e. the leading terms) in Theorem 3.2 are also studied by **[KW]** and **[BB2]**.

In §4 we consider the case where X is a classical simple Lie group and give an algorithm to express the H^d-orbit structure on G^d/P^d, which is sufficient to apply the theorem in §3. Thus we can obtain a simple combinatorial algorithm to obtain the embeddings.

The precise argument for the proof of the lemma and its application will be given elsewhere.

§2. Local properties of intertwining operators.

Retain the notation in §1. Let $G = KA_{\mathfrak{p}}N$ be an Iwasawa decomposition of G and $\mathfrak{a}_{\mathfrak{p}}$ the Lie algebra of $A_{\mathfrak{p}}$. Let Σ be the restricted root system for the pair $(\mathfrak{g}, \mathfrak{a}_{\mathfrak{p}})$, Σ^+ the positive system corresponding to N and Ψ the fundamental system of Σ. The Weyl group W of Σ is identified with the normalizer $N_K(\mathfrak{a}_{\mathfrak{p}})$ of $\mathfrak{a}_{\mathfrak{p}}$ in K modulo the centralizer M of $\mathfrak{a}_{\mathfrak{p}}$ in K and the group $P = MA_{\mathfrak{p}}N$ is a minimal parabolic subgroup of G. For any $\alpha \in \Sigma$, we denote by $w_\alpha \in W$ the reflection with respect to α.

For an open subset U of G the space $\mathcal{B}(U)$ of hyperfunctions on U is naturally a left $\mathfrak{g}$-module. Then for an element λ of the complexification $(\mathfrak{a}_{\mathfrak{p}})^*_c$ of the dual $\mathfrak{a}^*_{\mathfrak{p}}$ of $\mathfrak{a}_{\mathfrak{p}}$, the space of hyperfunction sections of class 1 principal series is defined:

$$\begin{aligned}\mathcal{B}(G/P; L_\lambda) =& \{f \in \mathcal{B}(G); f(gman) = f(g)a^{\lambda-\rho} \\ & \text{for } (g, m, a, n) \in G \times M \times A_{\mathfrak{p}} \times N\}.\end{aligned}$$

For any $\alpha \in \Psi$ there exists a function $T^\lambda_\alpha \in \mathcal{B}(G/P; L_{w\lambda})$ with the meromorphic parameter $\lambda \in (\mathfrak{a}_\mathfrak{p})^*_c$ so that the linear map

$$\begin{aligned} \mathcal{T}^\lambda_\alpha : \mathcal{B}(G/P; L_\lambda) &\to \mathcal{B}(G/P; L_{w\lambda}) \\ f(x) \mapsto (\mathcal{T}^\lambda_\alpha f)(x) &= \int_K f(k) T^\lambda_\alpha (k^{-1}x) dk \end{aligned} \tag{2.1}$$

is a G-homomorphism which satisfies

$$(\mathcal{T}^\lambda_\alpha f)(x) = \int_{\bar{N}_\alpha} f(x\bar{w}_\alpha \bar{n}_\alpha) d\bar{n}_\alpha$$

if f is continuous and $\mathrm{Re}\langle\lambda, \beta\rangle < 0$ for any $\beta \in \Psi$. Here $\bar{w}_\alpha$ is a representative of $w_\alpha, \bar{N}_\alpha = \theta(N) \cap \bar{w}_\alpha^{-1} N \bar{w}_\alpha$ and the measures dk and $d\bar{n}_\alpha$ are Haar measures on K and $\bar{N}_\alpha$, respectively.

For a subset S of G/P we define a subset $w[S]$ of G/P by $\overline{SPw^{-1}P}$. For an open subset U of G/P, which is identified with a right P invariant subset of G, we put

$$\begin{aligned} \mathcal{B}(U; L_\lambda) = \{ f \in \mathcal{B}(U);\ & f(gman) = f(g) a^{\lambda - \rho} \\ & \text{for } (g, m, a, n) \in U \times M \times A_\mathfrak{p} \times N \} \end{aligned}$$

and define

$$\mathcal{B}(S; L_\lambda) = \varinjlim_{U:\text{open} \supset S} \mathcal{B}(U; L_\lambda).$$

Then the key lemma in **[O1]** is

LEMMA 2.1 [**O1**, LEMMA 3.2]. *Fix an element α of Ψ and a point p of G/P and put $V = w_\alpha[\{p\}]$. Denoting*

$$\mathcal{B}(V, \{p\}; L_\lambda) = \{ f \in \mathcal{B}(V; L_\lambda);\ p \notin \operatorname{supp} f \},$$

the map (2.1) induces the $\mathfrak{g}$-homomorphism

$$\mathcal{T}^\lambda_\alpha : \mathcal{B}(V, \{p\}; L_\lambda) \to \mathcal{B}(\{p\}; L_{w_\alpha \lambda}) \tag{2.2}$$

for any $\lambda \in (\mathfrak{a}_\mathfrak{p})^*_c$ by analytic continuation. Moreover if

$$e_\alpha(\lambda) \neq 0 \quad \text{and} \quad -\frac{\langle\lambda, \alpha\rangle}{\langle\alpha, \alpha\rangle} \notin \{1, 2, 3, \ldots\}, \tag{2.3}$$

then (2.2) is injective.

In the above lemma, $\langle\ ,\ \rangle$ is the non-degenerate bilinear form on $(\mathfrak{a}_\mathfrak{p})^*_c$ induced from the Killing form of $\mathfrak{g}$,

$$e_\alpha(\lambda) = \Gamma\left(\frac{1}{2}\frac{\langle\lambda, \alpha\rangle}{\langle\alpha, \alpha\rangle} + \frac{1}{4}m_\alpha + \frac{1}{2}\right)^{-1} \Gamma\left(\frac{1}{2}\frac{\langle\lambda, \alpha\rangle}{\langle\alpha, \alpha\rangle} + \frac{1}{4}m_\alpha + \frac{1}{2}m_{2\alpha}\right)^{-1}$$

and m_β denotes the multiplicity of the root space for a root $\beta \in \Sigma$.

Here we give another lemma.

LEMMA 2.2. *Use the notation as in Lemma 2.1.*

1) *Suppose* λ *satisfies*

$$\frac{\langle\lambda,\alpha\rangle}{\langle\alpha,\alpha\rangle}\in\{0,1,2,\dots\}. \tag{2.4}$$

Then the function T_α^μ *has a pole of order* 1 *at* $\mu=\lambda$ *and the residue defines the* $\mathfrak{g}$*-homomorphism*

$$\operatorname{Res} T_\alpha^\lambda : \mathcal{B}(V;L_\lambda)\to\mathcal{B}(V;L_{w_\alpha\lambda}) \tag{2.5}$$

and if the support of f *in* $\mathcal{B}(V;L_\lambda)$ *is not equal to* V, *then*

$$\operatorname{supp} f=\operatorname{supp}(\operatorname{Res} T_\alpha^\lambda)f. \tag{2.6}$$

2) *If* $m_\alpha=1$ *and*

$$2\frac{\langle\lambda,\alpha\rangle}{\langle\alpha,\alpha\rangle}\in\{0,1,2,\dots\}, \tag{2.7}$$

then there exists a $\mathfrak{g}$*-homomorphism*

$$\mathcal{S}_\alpha^\lambda : \mathcal{B}(V;L_\lambda)\to\mathcal{B}(V;L_{w_\alpha\lambda}) \tag{2.8}$$

such that if the support of f *in* $\mathcal{B}(V;L_\lambda)$ *is not equal to* V, *then*

$$\operatorname{supp} f=\operatorname{supp}\mathcal{S}_\alpha^\lambda f. \tag{2.9}$$

3) *If* $e_\alpha(\lambda)e_\alpha(-\lambda)\neq 0$, *then the analytic continuation of* $\Gamma\left(\frac{\langle\lambda,\alpha\rangle}{\langle\alpha,\alpha\rangle}\right)^{-1} T_\alpha^\lambda$ *defines a bijective* $\mathfrak{g}$*-homomorphism*

$$\tilde{T}_\alpha^\lambda : \mathcal{B}(V;L_\lambda)\to\mathcal{B}(V;L_{w_\alpha\lambda}). \tag{2.10}$$

§3. Group cases.

Let G be a connected real semisimple Lie group with a simply connected complexification G_c and $G=KA_\mathfrak{p}N$ an Iwasawa decomposition of G. Let K_c be an analytic subgroup of G_c with the Lie algebra $\mathfrak{k}_c$ which is the complexification of the Lie algebra $\mathfrak{k}$ of K, B a Borel subgroup of G_c which contains $A_\mathfrak{p}N$ and $\mathfrak{j}_c$ a Cartan subalgebra of $\mathfrak{g}_c$ which satisfies $A_\mathfrak{p}\subset\exp(\mathfrak{j}_c)\subset B$. Let $\Sigma(\mathfrak{j})$ be the root system for the pair $(\mathfrak{g}_c,\mathfrak{j}_c)$ by denoting $\mathfrak{j}=\mathfrak{g}\cap\mathfrak{j}_c$, $\Sigma(\mathfrak{j})^+$ the positive system corresponding to B, $\Psi(\mathfrak{j}) = \{\alpha_1, \ldots, \alpha_\ell\}$ the fundamental system and ρ half the sum of the positive roots. The Weyl group W of $\Sigma(\mathfrak{j})$ is generated by the reflections s_j with respect to simple roots α_j $(j = 1, \ldots, \ell)$.

Let E be an irreducible Harish–Chandra module with an integral infinitesimal character $-\lambda$. Here we choose the element λ of the complex dual $\mathfrak{j}_c^*$ of $\mathfrak{j}_c$ with

$$(3.1) \qquad \langle \lambda, \alpha \rangle \geq 0 \quad \text{for any} \quad \alpha \in \Sigma(\mathfrak{j})^+.$$

Let L_λ be the holomorphic line bundle over the flag manifold $Y = G_c/B$ induced from the holomorphic character τ_λ of B which satisfies $\tau_\lambda(\exp(Z)) = \exp\langle \rho - \lambda, Z \rangle$ for $Z \in \mathfrak{j}_c$. The twisted sheaf of differential operators $\mathcal{D}_\lambda$ on Y is defined by

$$(3.2) \qquad \mathcal{D}_\lambda = L_\lambda \underset{\mathcal{O}_Y}{\otimes} \mathcal{D}_Y \underset{\mathcal{O}_Y}{\otimes} L_\lambda^{-1}.$$

Here $\mathcal{O}_Y$ (resp. $\mathcal{D}_Y$) are the sheaf of holomorphic functions (resp. that of differential operators) on Y in the Zariski topology. Let $U(\mathfrak{g})$ be the universal enveloping algebra of $\mathfrak{g}_c$. For a matrix coefficient ψ of E we put $FBI(E) = FBI(\psi)$. Then $FBI(E)$ is well-defined and a closure of a single K_c-orbit on Y and satisfies

$$(3.3) \qquad FBI(E) = \mathrm{supp}(\mathcal{D}_\lambda \underset{U(\mathfrak{g})}{\otimes} E).$$

If $\mathrm{rank}(G) = \mathrm{rank}(K)$ and E is the Harish–Chandra module belonging to the discrete series of G, then E is isomorphic to $H_V^n(Y, L_\lambda)$ with a compact K_c-orbit V on Y. Here n is the codimension of V in Y.

Let L be the centralizer of $A_\mathfrak{p}$ in G and L_0 its identity component. Then $P = LN$ is a minimal parabolic subgroup of G and $P_0 = L_0N$ is its identity component. Let π_λ be the irreducible representation of P_0 whose restriction on L_0 has the lowest weight $\rho - \lambda$ and U_λ the Harish–Chandra module of the representation of G induced from π_λ. Then U_λ is a finite direct sum of principal series of G in the category of Harish–Chandra modules.

By denoting $B_j = \overline{Bs_jB}$ for any $\alpha_j \in \Psi(\mathfrak{j})$, we have

Definition 3.1. *For any closed subset V of Y, we put*

$$S(V) = \{w \in W;\ \text{there exists a reduced expression}$$
$$w = s_{\nu(k)} \cdots s_{\nu(1)}$$
with the length k of w and a map
$$\varepsilon : \{1, \ldots, k\} \to \{0, 1\}$$
such that
$$V_k = Y \text{ and } V_{i-1} \neq V_{i-1}B_{\nu(i)} \text{ for } i = 1, \ldots, k$$
by inductively denoting
$$V_i = \begin{cases} V_{i-1}B_{\nu(i)} & \text{if } \varepsilon(i) = 1, \\ V_{i-1} & \text{if } \varepsilon(i) = 0, \end{cases} \quad \text{for } i = 1, \ldots, k.\}$$

For the irreducible Harish–Chandra module E we put $S(E) = S(FBI(E))$. Since $FBI(E) = \bar{D}$ with a K_c-orbit D of Y, each V_i in the above definition is a closure of a single K_c-orbit D_i and $\dim D_i = \dim D_{i-1} + \varepsilon(i)$. Then for a non-negative integer j, we put

$$S(E)_j = \{w \in S(E);\ \text{the length of } w \text{ equals } j + \operatorname{codim} D\}.$$

By a similar argument as in **[O1**, §4, 5] with the lemmas in §2, we have

Theorem 3.2. *Retain the above notation.*

1) *For any $w \in S(E)$, there exists an embedding of E into $U_{w\lambda}$.*
2) ([**O1**, Theorem 4.1]) *Let ψ be a matrix coefficient of E. Suppose λ is regular for simplicity. Then there exists a positive number ε, non-zero real analytic functions $a_w(g, g')$ of $G \times G$ for $w \in S(E)_0$ such that*
$$\begin{aligned}&\psi(g \cdot \exp Z \cdot g') \\ &= \sum_{w \in S(E)_0} a_w(g, g') e^{\langle w\lambda - \rho, Z\rangle} \\ &+ 0\Big(\sum_{w \in S(E)_0} \Big| e^{\langle w\lambda - (1+\varepsilon)\rho, Z\rangle}\Big|\Big)\end{aligned}$$
for $Z \in \mathfrak{a}_{\mathfrak{p}}$ and $(g, g') \in G \times G$ when $\alpha(Z) \to \infty$ for all $\alpha \in \Psi(\mathfrak{j})$ with $\alpha|_{\mathfrak{a}_{\mathfrak{p}}} \neq 0$. Here the estimate is locally uniform on $G \times G$.
3) *If E is embedded in U_μ with an element $\mu \in \mathfrak{j}_c^*$, then there exist $v \in W$ and $w \in S(E)_0$ satisfying $\mu = v\lambda$ and $v \geq w$ with respect to Bruhat ordering.*

For an element w of $S(E)_j$ we put $\partial w = \{v \in S(E)_{j-1}; v < w\}$. Then we have

CONJECTURE 3.3. *$(\partial, S(E))$ is isomorphic to a regular contractible CW-complex.*

On the other hand, we have

PROPOSITION 3.4. $\sum(-1)^j \# S(E)_j = 1$.

§4. Orbit structures on complex flag manifolds of classical type.

Let G_c be a connected complex reductive Lie group with a connected real form G. Let $\mathfrak{g} = \mathfrak{k} + \mathfrak{p}$ be a Cartan decomposition of $\mathfrak{g} = \operatorname{Lie} G$ with respect to a Cartan involution θ and K_c the analytic subgroup of G for $\mathfrak{k}_c$.

Let B be a Borel subgroup of G_c, $\mathfrak{b}$ its Lie algebra and $Y = G_c/B$ the flag manifold for G_c. Since Y is identified with the set of all Borel subalgebras in $\mathfrak{g}_c$ on which G_c acts by the adjoint action, the K_c-orbit structure on Y depends only on $\mathfrak{g}^s = [\mathfrak{g}, \mathfrak{g}]$.

Let $\tilde{K}_c$ be a subgroup of G_c such that $K_c \subset \tilde{K}_c \subset N_{G_c}(K_c)$, where $N_{G_c}(K_c)$ is the normalizer of K_c in G_c. Then all the K_c-orbits contained in a $\tilde{K}_c$-orbit are diffeomorphic to each other. Let D_1 and D_2 be two K_c-orbits on Y with $\tilde{K}_c D_1 = \tilde{K}_c D_2$. Then we can easily obtain $S(\bar{D}_1) = S(\bar{D}_2)$. Hence in order to get $S(\bar{D})$ for a K_c-orbit D in Y, we have only to study the $\tilde{K}_c$-orbit structure on Y for some $\tilde{K}_c$.

For any $gB \in Y$, the Borel subalgebra $\operatorname{Ad}(g)\mathfrak{b}$ has a split component $\mathfrak{a}$ such that $\theta\mathfrak{a} = \mathfrak{a}$ ([M1], [R]). Note that $\mathfrak{a}_c$ is a θ-stable Cartan subalgebra of $\mathfrak{g}_c$ contained in $\operatorname{Ad}(g)\mathfrak{b}$. Let Σ be the root system for the pair $(\mathfrak{g}_c, \mathfrak{a}_c)$, Σ^+ the positive system for $\operatorname{Ad}(g)\mathfrak{b}$, Ψ the set of simple roots in Σ^+ and $\mathfrak{g}(\mathfrak{a}; \alpha)$ the root space for a root $\alpha \in \Sigma$.

In this section we parametrize the $\tilde{K}_c$-orbit structure on Y when $\mathfrak{g}^s$ is a simple Lie algebra of classical type.

Suppose that $G_c = GL(n, \mathbf{C})$, $SO(2n+1, \mathbf{C})$, $\mathrm{Sp}(n, \mathbf{C})$ or $SO(2n, \mathbf{C})$. (Later we will consider the case when $\mathfrak{g}^s$ is complex simple.) Take the orthogonal basis $\{e_1, \ldots, e_n\}$ of the dual $\mathfrak{a}^*$ of $\mathfrak{a}$ such that

$$\Psi = \begin{cases} \{\alpha_1, \ldots, \alpha_{n-1}\} & \text{if } G = GL(n, \mathbf{C}), \\ \{\alpha_1, \ldots, \alpha_n\} & \text{otherwise,} \end{cases}$$

where $\alpha_1 = e_1 - e_2, \ldots, \alpha_{n-1} = e_{n-1} - e_n$ and $\alpha_n = e_n, 2e_n$ or $e_{n-1} + e_n$ if $G = SO(2n+1, \mathbf{C})$, $\mathrm{Sp}(n, \mathbf{C})$ or $SO(2n, \mathbf{C})$, respectively.

Since θ induces an involution of Σ, we have a permutation φ of $\{1,2,\ldots,n\}$ such that $\varphi^2 = \mathrm{id}$ and that

$$\theta e_i = \pm e_{\varphi(i)}$$

for every $i = 1,\ldots,n$. We can assign to the pair $(\mathfrak{a},\Psi)$ an ordered set $\{\varepsilon_1,\ldots,\varepsilon_n\}$, which we call "a clan", of n "persons" with the following structure:

Each person ε_i is an element of the set $\{+,-,\circ\}$ of three elements, the signs $+$ and $-$ and the circle $\circ$, which we call "a boy", "a girl" and "an adult", respectively. Some of the adults in a clan form pairs and no adult belongs to two different pairs. Each pair is "a young couple" or "an old couple".

A young couple and an old couple are expressed by joining the corresponding two circles with a line and an arrow, respectively. Here we ignore the direction of the arrow.

The clan has the following property:

($\pm$) If $\theta e_i = e_i$, then $\varepsilon_i = +$ or $-$. Moreover ε_i and ε_j are the same sign if and only if $\mathfrak{g}(\mathfrak{a}; e_i - e_j) \subset \mathfrak{k}_c$, that is, the root $e_i - e_j$ is a compact root.

(a) If $\theta e_i = e_j$ with $i \neq j$, then ε_i and ε_j are adults and form a young couple.

(A) If $\theta e_i = -e_j$ with $i \neq j$, then ε_i and ε_j are adults and form an old couple.

($\circ$) If $\theta e_i = -e_i$, then ε_i is an adult which does not belong to any pair, which we call "the aged".

THEOREM 4.1. *The $\tilde{K}_c$-orbits on Y and the clans with the conditions in Table 1 are in one-to-one correspondance.*

REMARK 4.2 (i) In Table 1, for example, the condition $(A,\circ)_n$ means that the clan consists of n persons and there exists no boy, no girl or no young couple.

(ii) N_+, N_- and N_A are the members of boys, girls and old couples, respectively.

(iii) For BI, $\mathfrak{g}(\mathfrak{a};\alpha_n) \subset \mathfrak{k}_c \Leftrightarrow (N_+ - N_- = p - q$ and $\varepsilon_n = +)$ or $(N_+ - N_- = p-q+1$ and $\varepsilon_n = -)$.

For CI, $\mathfrak{g}(\mathfrak{a}; 2e_i) \not\subset \mathfrak{g}_c$ if $\theta e_i = e_i$.

For CII, $\mathfrak{g}(\mathfrak{a}; 2e_i) \subset \mathfrak{k}_c$ if $\theta e_i = e_i$.

For DI, $\mathfrak{g}(\mathfrak{a};\alpha_{n-1}) \subset \mathfrak{k}_c \Leftrightarrow \mathfrak{g}(\mathfrak{a};\alpha_n) \subset \mathfrak{k}_c$.

For $DIII$, $\mathfrak{g}(\mathfrak{a};\alpha_{n-1}) \subset \mathfrak{k}_c \Leftrightarrow \mathfrak{g}(\mathfrak{a};\alpha_n) \not\subset \mathfrak{k}_c$.

(iv) For the compact orbits and open orbits, see Table 1'.

Table 1 ($p+q=n$)

Type	$\mathfrak{g}^s$	G_c	$\tilde{K}_c$	Condition for the clans
AI	$\mathfrak{sl}(n,\mathbf{R})$	$GL(n,\mathbf{C})$	$O(n,\mathbf{C})$	$(A,\circ)_n$
AII	$\mathfrak{su}^*(n)$	$GL(n,\mathbf{C})$	$Sp(n/2,\mathbf{C})$	$(A)_n$ ($n=$ even)
AIII	$\mathfrak{su}(p,q)$	$GL(n,\mathbf{C})$	$GL(p,\mathbf{C})\times GL(q,\mathbf{C})$	$(\pm,a)_n$ $N_+-N_-=p-q$
BI	$\mathfrak{so}(2p+1,2q)$	$SO(2n+1,\mathbf{C})$	$S(O(2p+1,\mathbf{C})\times O(2p,\mathbf{C}))$	$(\pm,a,A,\circ)_n$ $N_+-N_-=p-q$ or $p-q+1$
CI	$\mathfrak{sp}(n,\mathbf{R})$	$Sp(n,\mathbf{C})$	$GL(n,\mathbf{C})$	$(\pm,a,A,\circ)_n$
CII	$\mathfrak{sp}(p,q)$	$Sp(n,\mathbf{C})$	$Sp(p,\mathbf{C})\times Sp(q,\mathbf{C})$	$(\pm,a,A)_n$ $N_+-N_-=p-q$
DI	$\mathfrak{so}(2p,2q)$	$SO(2n,\mathbf{C})$	$S(O(2p,\mathbf{C})\times O(2q,\mathbf{C}))$	$(\pm,a,a,\circ)_n$ $N_+-N_-=p-q$
DI'	$\mathfrak{so}(2p+1,2q-1)$	$SO(2n,\mathbf{C})$	$S(O(2p+1,\mathbf{C})\times O(2q-1,\mathbf{C}))$	$(\pm,a,A,\circ)_n$ $N_+-N_-=p-q+1$
DIII	$\mathfrak{so}^*(2n)$	$SO(2n,\mathbf{C})$	$\mathbf{C}^\times\times PSL(n,\mathbf{C})$	$(\pm,a,A)_n$ $N_+-N_-+2N_A\equiv n \pmod 4$

Table 1'

Type	$\mathfrak{g}^s$	$\tilde{N}_c$	N_c	compact orbits	open orbits	codimension of compact orbits
AI	$\mathfrak{sl}(n,\mathbf{R})$	1	$\begin{cases} n: \text{even } 2 \\ n: \text{odd } 1 \end{cases}$	$\begin{cases} AB\cdots\cdot BA \\ AB\cdot\cdot\circ\cdot\cdot BA \end{cases}$	$(\circ)_n$	$\begin{cases} n^2/4 \\ (n^2-1)/4 \end{cases}$
AII	$\mathfrak{su}^*(n)$	1	1	$AB\cdots\cdot BA$	$AABB\cdots\cdot$	$n(n-2)/4$
AIII	$\mathfrak{su}(p,q)$	$\binom{n}{p}$	$\binom{n}{p}$	$(\pm)_n$	$ab\cdot\cdot(+)_{p-q}\cdot\cdot ba$	pq
BI	$\mathfrak{so}(2p+1,2q)$	$\binom{n}{p}$	$2\binom{n}{p}$	$(\pm)_n$	$\begin{cases} p \geq q\ \circ\cdot\cdot\circ(+)_{p-q} \\ p < q\ \circ\cdot\cdot\circ(-)_{q-p-1} \end{cases}$	$(2p+1)q$
CI	$\mathfrak{sp}(n,\mathbf{R})$	2^n	2^n	$(\pm)_n$	$(\circ)_n$	$n(n+1)/2$
CII	$\mathfrak{sp}(p,q)$	$\binom{n}{p}$	$\binom{n}{p}$	$(\pm)_n$	$AABB\cdot\cdot(+)_{p-q}$	$2pq$
DI	$\mathfrak{so}(2p,2q)$	$\binom{n}{p}$	$2\binom{n}{p}$	$(\pm)_n$	$\circ\cdot\cdot\circ(+)_{p-q}$	$2pq$
DI'	$\mathfrak{so}(2p+1,2q-1)$	$\binom{n-1}{p}$	$\binom{n-1}{p}$	$(\pm)_{n-1}\circ$	$\circ\cdot\cdot\circ(+)_{p-q+1}$	$2pq-p+q-1$
DIII	$\mathfrak{so}^*(2n)$	2^{n-1}	2^{n-1}	$(\pm)_n$	$AABB\cdots\cdot$	$n(n-1)/2$

Here $\tilde{N}_c$ is the number of the compact $\tilde{K}_c$-orbits, N_c is the number of compact K_c-orbits and $(+)_k$ denotes the row of $+$'s of length k if $k \geq 0$ and that of $-$'s of length $-k$ otherwise.

Each $\tilde{K}_c$-orbit in Table 1 can be expressed by a symbol $\varepsilon_1\varepsilon_2\ldots\varepsilon_n$ with lines and arrows, where $\{\varepsilon_1, \varepsilon_2, \ldots, \varepsilon_n\}$ is the clan corresponding to the orbit. The following are examples of $\tilde{K}_c$-orbits of Type CI with $n = 5$ in Table 1.

$$+ - + \circ - \qquad + \overset{\frown}{\circ + \circ} - \qquad + \overset{\frown}{\circ\circ\circ\circ}$$

To express the orbit more easily we give "a family name" for each pair and then we can write the above example as follows, respectively:

$$+ - + \circ - \qquad + a + a - \qquad +abaB.$$

Here each couple in a clan has a family name consisting of letter to distinguish couples in a clan. A young couple is expressed by the same small letters and an old couple is expressed by the small letter and the capital letter corresponding to the family name. In some cases, we express an old couple by the same capital letters. We remark that the following expressions also correspond the last orbit in the above example:

$$+abaB \qquad +babA \qquad +aBaB \qquad +bAbA$$

Let B_i be the parabolic subgroup of G_c for $\{-\alpha_i\} \cup \Sigma^+$. Let $Y_i = G/B_i$ and $\pi_i : Y \to Y_i$ be the canonical projection. Let D_1 and D_2 be two $\tilde{K}_c$-orbits on Y. Then we write

$$D_1 \xrightarrow{i} D_2$$

if and only if $\pi_i(D_1) = \pi_i(D_2)$ and $\dim D_1 < \dim D_2$, which implies $\dim D_2 = \dim D_1 + 1$.

Proposition 4.3. ([V, §5], [M2]). *Choose a pair $(\mathfrak{a}, \Psi)$ corresponding to an element of D_1. Then $D_1 \xrightarrow{i} D_2$ for some $\tilde{K}_c$-orbit D_2 if and only if one of the following conditions holds.*

(I) *$\theta\alpha_i = \alpha_i$ and $\mathfrak{g}(\mathfrak{a}; \alpha_i) \not\subset \mathfrak{k}_c$, that is, α_i is a non-compact simple root.*

(II)
$$\theta\alpha_i \in \Sigma^+ - \{\alpha_i\}.$$

We will give the necessary and sufficient condition for $D_1 \xrightarrow{i} D_2$ in Table 2 and examples of the $\tilde{K}_c$-orbit structure on Y of Type AI,..., DIII in Fig. 1~ Fig. 20.

Table 2

We express the orbits by rows consisting of $+, -$ and letters. Let $\delta_1 \dots \delta_n$ and $\delta'_1 \dots \delta'_n$ be the expressions corresponding to D_1 and D_2, respectively.

(i) Here $i = 1, \dots, n-1$ and the old couple (resp. young couple) is expressed by the capital letters (resp. small letters) corresponding to the family name. Then $D_1 \xrightarrow{i} D_2$ if and only if $\delta_j = \delta'_j$ for $j = 1, \dots, i-1, i+2, \dots, n$ and $\delta_i, \delta_{i+1}, \delta'_i$ and δ'_{i+1} equal to one of the following lists, where the letters p, P, q, Q correspond to suitable family names.

$\delta_i\delta_{i+1}$	$\delta'_i\delta'_{i+1}$	Condition	$\theta\alpha_i$
$+-$	pp		α_i
$-+$	pp		α_i
pp	$\circ\circ$		α_i
$p\pm$	$\pm p$	$\varphi(i) < i+1$	$e_{\varphi(i)} - e_{i+1}$
$\pm p$	$p\pm$	$i < \varphi(i+1)$	$e_i - e_{\varphi(i+1)}$
pq	qp	$\varphi(i) < \varphi(i+1)$	$e_{\varphi(i)} - e_{\varphi(i+1)}$
$\pm P$	$P\pm$		$e_i + e_{\varphi(i+1)}$
PQ	QP	$\varphi(i+1) < \varphi(i)$	$e_{\varphi(i+1)} - e_{\varphi(i)}$
pQ	Qp		$e_{\varphi(i)} + e_{\varphi(i+1)}$
$\pm\circ$	$\circ\pm$		$e_i + e_{i+1}$
$p\circ$	$\circ p$		$e_{\varphi(i)} + e_{i+1}$
$P\circ$	$\circ P$	$i+1 < \varphi(i)$	$e_{i+1} - e_{\varphi(i)}$
$\circ P$	$P\circ$	$\varphi(i+1) < i$	$e_{\varphi(i+1)} - e_i$

(ii) Here $i = n$ and the old couple (resp. young couple) is expressed by the small letter and capital letter (resp. small letter) corresponding to the family name. If δ_j and $\delta_{j'}$ are an old couple with $j < j'$, then we express δ_j by a small letter and $\delta_{j'}$ by the corresponding capital letter.

(1) In the cases BI, CI and CII, $D_1 \xrightarrow{n} D_2$ if and only if $\delta_j = \delta'_j$ for $j = 1, \dots, n-1$ and δ_n and δ'_n are one of the following:

δ_n	δ'_n	$\theta\alpha_n$
$\pm$	$\circ$	α_n
p	P	$e_{\varphi(n)}$

(2) In the case DI, $D_1 \xrightarrow{n} D_2$ if and only if $\delta_j = \delta'_j$ for $j = 1, \dots, n-2$

and δ_{n-1}, δ_n, δ'_{n-1} and δ'_n are one of the following:

$\delta_{n-1}\delta_n$	$\delta'_{n-1}\delta'_n$	Condition	$\theta\alpha_n$
$+-$	PP		α_n
$-+$	PP		α_n
pp	oo		α_n
$\pm p$	$P\pm$		$e_{\varphi(n)} + e_{n-1}$
$p\pm$	$\pm P$		$e_{\varphi(n-1)} + e_n$
pq	QP		$e_{\varphi(n-1)} + e_{\varphi(n)}$
pQ	qP	$\varphi(n-1) < \varphi(n)$	$e_{\varphi(n-1)} - e_{\varphi(n)}$
Qp	Pq	$\varphi(n) < \varphi(n-1)$	$e_{\varphi(n)} - e_{\varphi(n-1)}$
$\pm$o	o$\pm$		$e_{n-1} - e_n$
po	oP		$e_{\varphi(n-1)} - e_n$
op	Po		$e_{\varphi(n)} - e_{n-1}$

(3) In the case DIII, $D_1 \xrightarrow{n} D_2$ if and only if $\delta_j = \delta'_j$ for $j = 1, \ldots, n-2$ and δ_{n-1}, δ_n, δ'_{n-1} and δ'_n are one of the following:

$\delta_{n-1}\delta_n$	$\delta'_{n-1}\delta'_n$	Condition	$\theta\alpha_n$
$++$	PP		α_n
$--$	PP		α_n
$\pm p$	$P\mp$		$e_{\varphi(n)} + e_{n-1}$
$p\pm$	$\mp P$		$e_{\varphi(n-1)} + e_n$
pq	QP		$e_{\varphi(n-1)} + e_{\varphi(n)}$
pQ	qP	$\varphi(n-1) < \varphi(n)$	$e_{\varphi(n-1)} - e_{\varphi(n)}$
Qp	Pq	$\varphi(n) < \varphi(n-1)$	$e_{\varphi(n)} - e_{\varphi(n-1)}$

Next consider the case where $\mathfrak{g}^s$ is a classical complex simple Lie algebra. Then we may suppose $G_c = G'_c \times G'_c$ with $G'_c = GL(n, \mathbf{C})$, $SO(2n+1, \mathbf{C})$, $\mathrm{Sp}(n, \mathbf{C})$ or $SO(2n, \mathbf{C})$, $\theta(g, g') = (g', g)$ for $(g, g') \in G_c$, $K_c = \tilde{K}_c = \{(g, g); g \in G'_c\}$ and $B = B' \times B'$ with a Borel subgroup B' of G'_c. Then

$$K_c \setminus G_c / B \xrightarrow{\sim} B' \setminus G'_c / B'$$

by the map $(g, g') \mapsto g^{-1}g'$ of G_c onto G'_c. Thus the K_c-orbit structure on $Y = G_c/B$ is reduced to the structure of the Bruhat decomposition of G'_c. Then by taking the orthogonal bases $\{e_1, \ldots, e_n, e'_1, \ldots, e'_n\}$ of the dual $\mathfrak{a}^*$ of $\mathfrak{a}$ in a natural way as before, we can express the K_c-orbit structure as in Fig. 21 $\sim$ Fig. 23.

Now we give some examples of $S(E)$ in Theorem 3.2, which are easily obtained from the diagram of the $\tilde{K}_c$-orbit structure. Suppose $G = SU(2, 1)$. Then the corresponding diagram is Fig. 5. Since the closed subset V_i is a closure of one K_c-orbit on Y, we express it by the corresponding clan. Suppose E is the Harish-Chandra module belonging to the discrete series of $SU(2,1)$. Then $FBI(E) = -++$ or $+-+$ or $++-$ and we have $S(-++) = \{s_2s_1 = \binom{123}{231}\}$, $s(++-) = \{s_1s_2 = \binom{123}{312}\}$ and $S(+-+) = \{s_2s_1, s_1s_2, s_1s_2s_1 = \binom{123}{321}\}$. Here we identify the Weyl group W with the permutation group of 3 numbers 1, 2 and 3 and the elements of $S(+-+)$ are obtained as in the following table:

	V_0	$\nu(1)$	$\varepsilon(1)$	V_1	$\nu(2)$	$\varepsilon(2)$	V_2	$\nu(3)$	$\varepsilon(3)$	V_3
s_2s_1	$+-+$	1	1	$aa+$	2	1	$a+a$			
s_1s_2	$+-+$	2	1	$+aa$	1	1	$a+a$			
$s_1s_2s_1$	$+-+$	1	0	$+-+$	2	1	$+aa$	1	1	$a+a$

Suppose $G = SU(p,q)$ with $p \geq q$. For a Harish-Chandra module E belonging to a discrete series of G, we can obtain $S(E)_0$ in the following way:

Consider the ordered q pairs in the clan corresponding to the compact K_c-orbit $FBI(E)$ which satisfies the following conditions.

Let $(\varepsilon_{I(i)}, \varepsilon_{J(i)})$ denote the i-th pair with $I(i) < J(i)$.

(1) Each pair consists of a boy and a girl.
(2) There exist $p-q$ boys who do not belong to any pair.
(3) If there exist i and j with $I(i) < j < J(i)$, then there exists i' with $i' < i$ such that $I(i) < I(i') < J(i') < J(i)$ and $j \in \{I(i'), J(i')\}$.

For the ordered q pairs, we attach an element $\begin{pmatrix} 1 & \cdots & n \\ j_1 & \cdots & j_n \end{pmatrix}$ of the permutation group of n numbers $1, \ldots, n$ satisfying $j_k = J(i_k)$ and $j_{n+1-k} = I(i_k)$ for $k = 1, \ldots, q$ and $j_{q+1} < j_{q+2} < \cdots < j_{n-q}$. Identifying the permutation group with the Weyl group of $SL(n, \mathbf{C})$ and considering all the definitions of the above q pairs, we obtain all elements of $S(E)_0$.

For example, if $FBI(E)$ corresponds to $++--+$, then there exist 3 types of the ordered q pairs

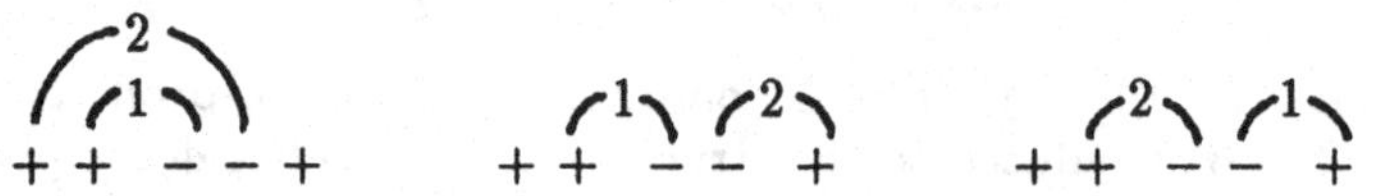

from which we have the following elements, respectively:

$$\begin{pmatrix} 12345 \\ 34512 \end{pmatrix} \quad \begin{pmatrix} 12345 \\ 35142 \end{pmatrix} \quad \begin{pmatrix} 12345 \\ 53124 \end{pmatrix}.$$

In the case where $\mathfrak{g}_c$ is classical and simple, a computer programme to calculate $S(\bar{D})$ for any K_c-orbit D on Y was written by the second named author. By the program, for example, we have

$$\#S(+-+-) = 7, \#S(+-+-+) = 35, \#S(+-+-+-) = 135, \ldots$$

in the case of type $SU(p,q)$. It takes about 2 minutes by a micro-computer (CPU:80286, Clock:10MHz) to get all $S(+-+-+-)$. The programme with source code in C is available from the second named author.

Remark 4.4. Suppose G is a classical simple Lie group whose real rank equals 1. Suppose E is a Harish-Chandra module belonging to the discrete series of G. Then comparing a result in [C], we can obtain all the embeddings of E into principal series by Theorem 3.2 except the following cases:

$FBI(E) = ++\cdots+-$ and $\mathfrak{g} = \mathfrak{sp}(p,1)$ with $p \geq 2$. In these cases there are two embeddings of E into principal series of G but Theorem 3.2 gives only one embedding corresponding to the leading term of the matrix coefficient of E.

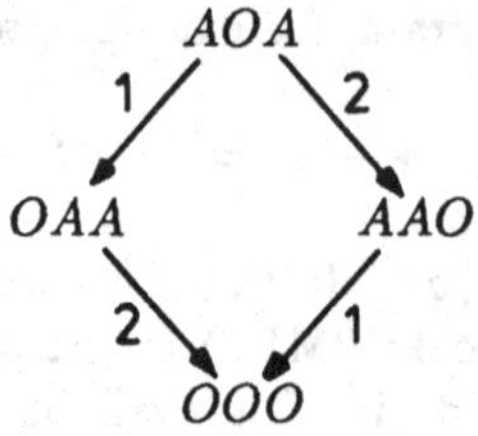

AI $\quad \mathfrak{g}^s = \mathfrak{sl}(3, \mathbf{R}) \quad O(3)\backslash GL(3, \mathbf{C})/B \quad$ o—o
$\qquad\qquad\qquad\qquad\qquad\qquad\qquad\qquad\quad$ 1 2

Fig. 1

AI $\quad \mathfrak{g}^s = \mathfrak{sl}(4, \mathbf{R}) \quad O(4)\backslash GL(4, \mathbf{C})/B \quad$ o—o—o
$\qquad\qquad\qquad\qquad\qquad\qquad\qquad\qquad\quad$ 1 2 3

Fig. 2

$$ABBA$$
$$1\big\downarrow\big\downarrow 3$$
$$ABAB$$
$$\big\downarrow 2$$
$$AABB$$

AII $\quad \mathfrak{g}^s = \mathfrak{su}^*(4) \quad Sp(2, \mathbf{C})\backslash GL(4, \mathbf{C})/B \quad$ o—o—o
$\qquad\qquad\qquad\qquad\qquad\qquad\qquad\qquad\qquad$ 1 2 3

Fig. 3

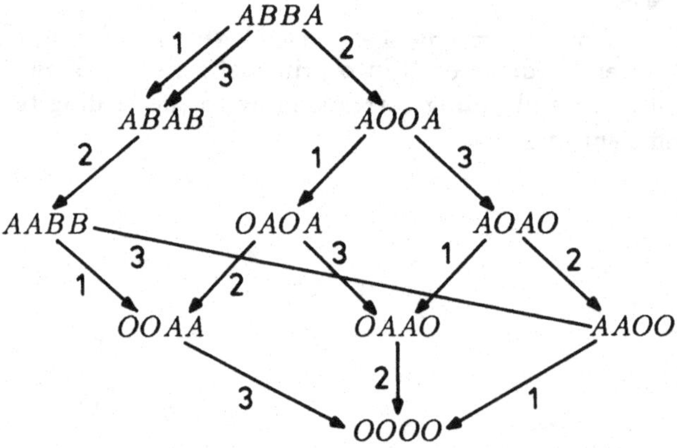

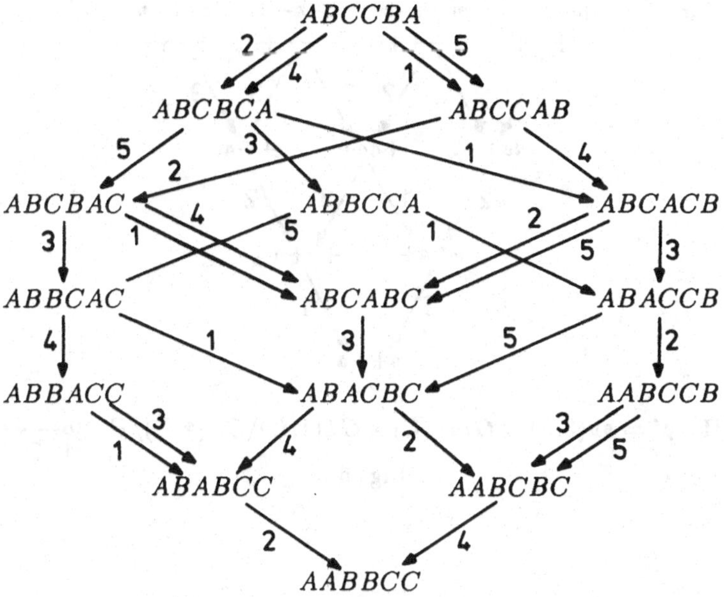

AII $\mathfrak{g}^s = \mathfrak{su}^*(6)$ $Sp(3,\mathbf{C})\backslash GL(6,\mathbf{C})/B$ 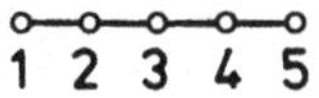

Fig. 4

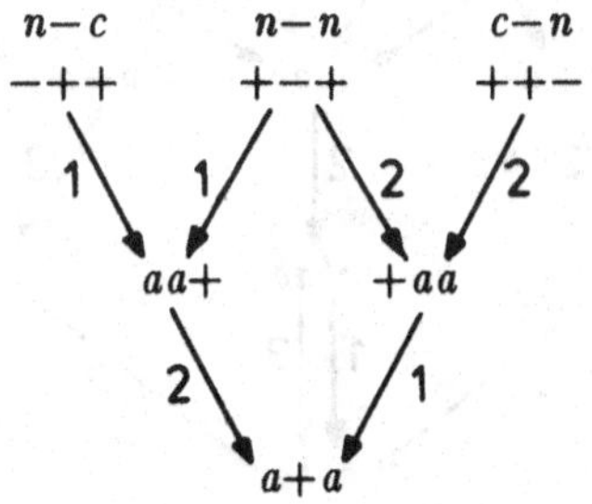

AIII $\mathfrak{g}^s = \mathfrak{su}(2,1)$ $GL(2,\mathbf{C}) \times GL(1,\mathbf{C})\backslash GL(3,\mathbf{C})/B$ 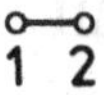

Fig. 5

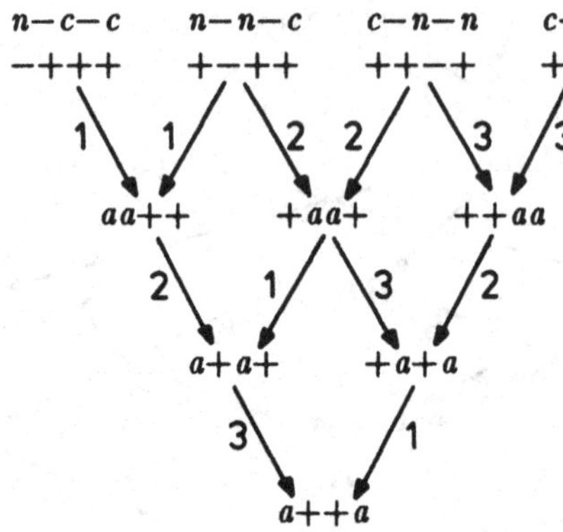

AIII $\mathfrak{g}^s = \mathfrak{su}(3,1)$ $GL(3,\mathbf{C}) \times GL(1,\mathbf{C})\backslash GL(4,\mathbf{C})/B$ 1 2 3

Fig. 6

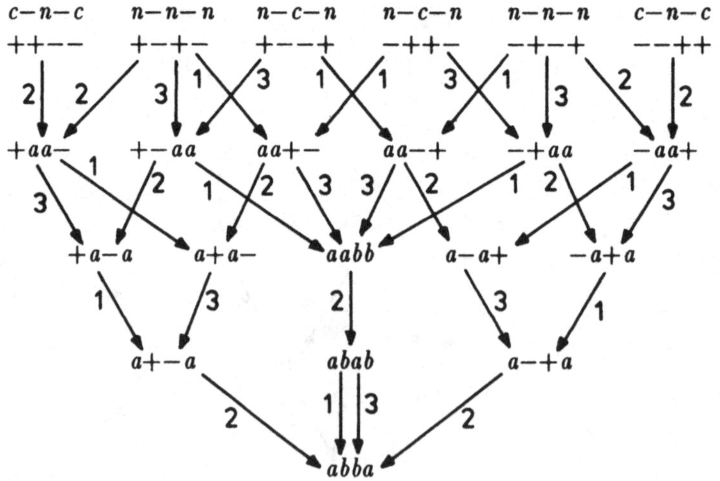

AIII $\mathfrak{g}^s = \mathfrak{su}(2,2)$ $GL(2,\mathbf{C}) \times GL(2,\mathbf{C})\backslash GL(4,\mathbf{C})/B$ 1 2 3

Fig. 7

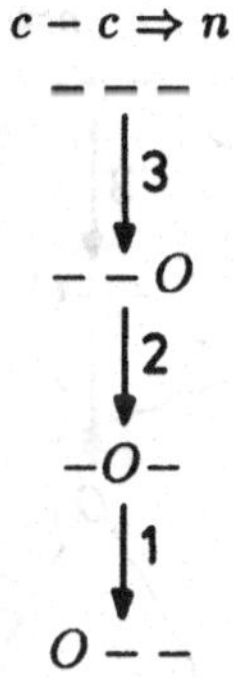

BI $\mathfrak{g}^s = \mathfrak{so}(1,6)$ $S(O(1,\mathbf{C}) \times O(6,\mathbf{C}))\backslash SO(7,\mathbf{C})/B$ o—o—o 1 2 3

Fig. 8

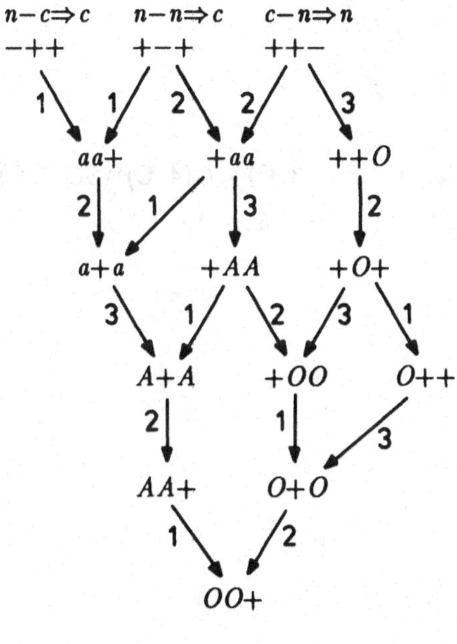

BI $\mathfrak{g}^s = \mathfrak{so}(5,2)$ $S(O(5,\mathbf{C}) \times O(2,\mathbf{C}))\backslash SO(7,\mathbf{C})/B$ o—o—o 1 2 3

Fig. 9

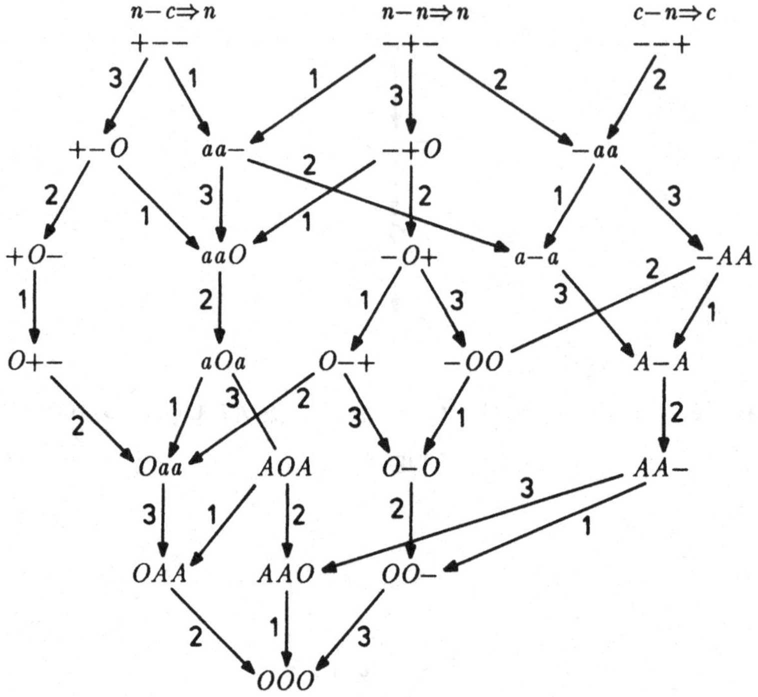

BI $\mathfrak{g}^s = \mathfrak{so}(3,4)$ $S(O(3,\mathbf{C}) \times O(4,\mathbf{C}))\backslash SO(7,\mathbf{C})/B$

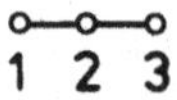

Fig. 10

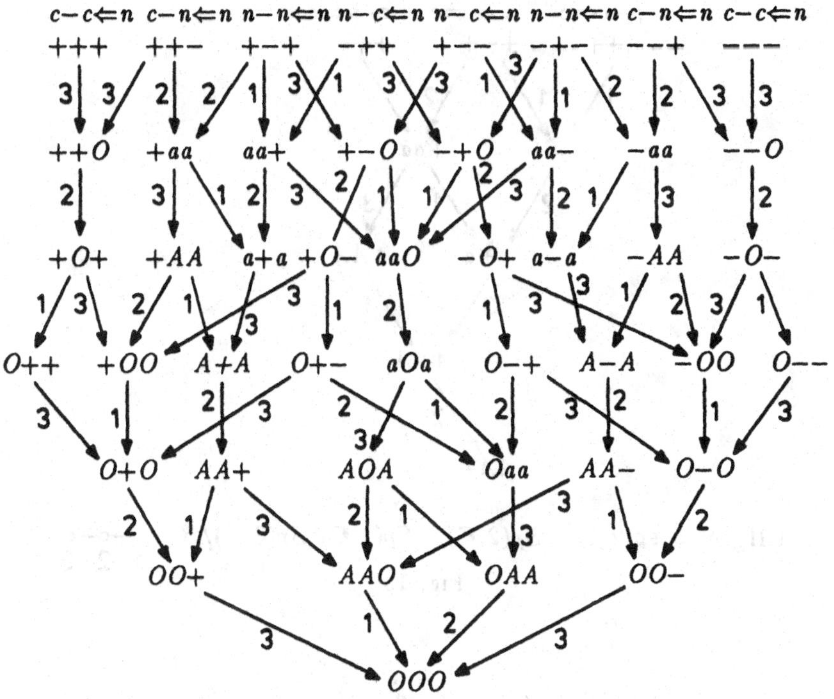

CI · $\mathfrak{g}^s = \mathfrak{sp}(3, \mathbf{R})$ $GL(3, \mathbf{C})\backslash Sp(3, \mathbf{C})/B$

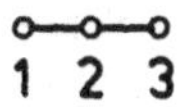

Fig. 11

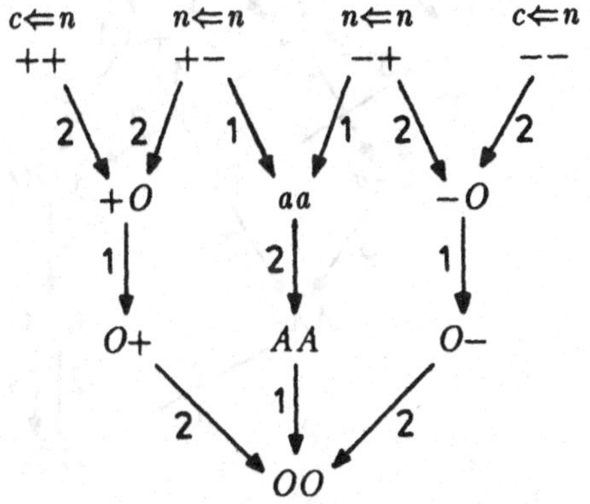

CI $\mathfrak{g}^s = \mathfrak{sp}(2, \mathbf{R})$ $GL(2, \mathbf{C})\backslash Sp(2, \mathbf{C})/B$

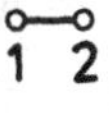

Fig. 12

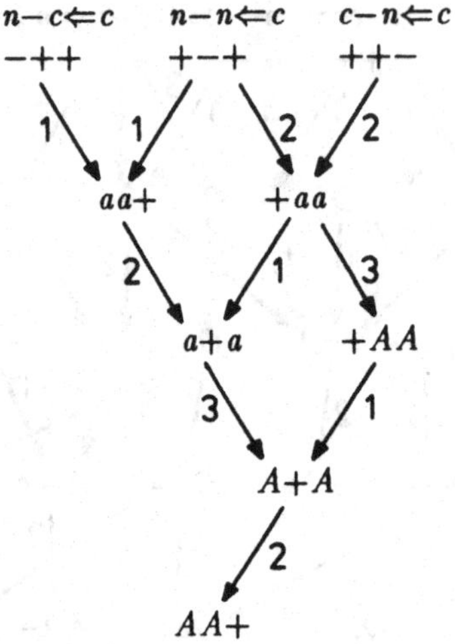

CII $\mathfrak{g}^s = \mathfrak{sp}(2,1)$ $Sp(2,\mathbf{C}) \times Sp(1,\mathbf{C})\backslash Sp(3,\mathbf{C})/B$ 1 2 3

Fig. 13

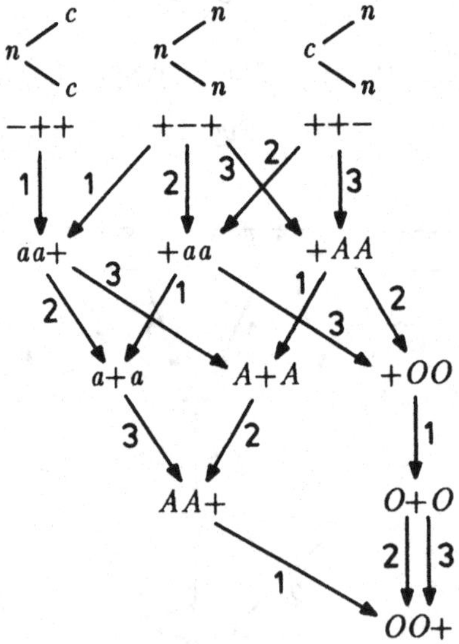

DI $\mathfrak{g}^s = \mathfrak{so}(4,2)$ $S(O(4,\mathbf{C}) \times O(2,\mathbf{C}))\backslash SO(6,\mathbf{C})/B$ 1 2 3

Fig. 14

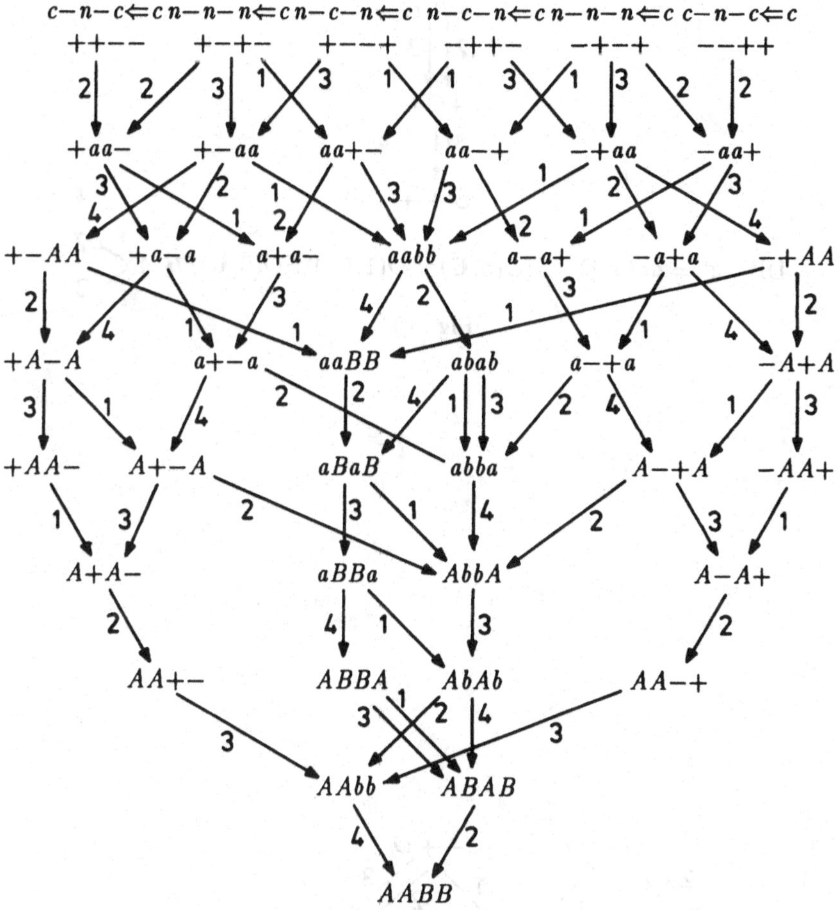

CII $\mathfrak{g}^s = \mathfrak{sp}(2,2)$ $Sp(2,\mathbf{C}) \times Sp(2,\mathbf{C}) \backslash Sp(4,\mathbf{C})/B$

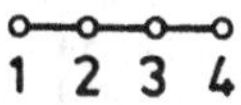

Fig. 15

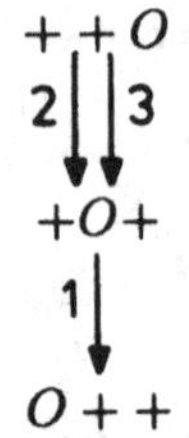

$$
\begin{array}{c}
+ + O \\
2 \big\| 3 \\
\Downarrow\Downarrow \\
+ O + \\
1 \big\downarrow \\
O + +
\end{array}
$$

DI' $\mathfrak{g}^s = \mathfrak{so}(5,1)$ $S(O(5,\mathbf{C}) \times O(1,\mathbf{C}))\backslash SO(6,\mathbf{C})/B$ $1\!\!\begin{array}{c}\circ 2\\\\\circ 3\end{array}$

Fig. 16

DI' $\mathfrak{g}^s = \mathfrak{so}(3,3)$ $S(O(3,\mathbf{C}) \times O(3,\mathbf{C}))\backslash SO(6,\mathbf{C})/B$ $1\!\!\begin{array}{c}\circ 2\\\\\circ 3\end{array}$

Fig. 17

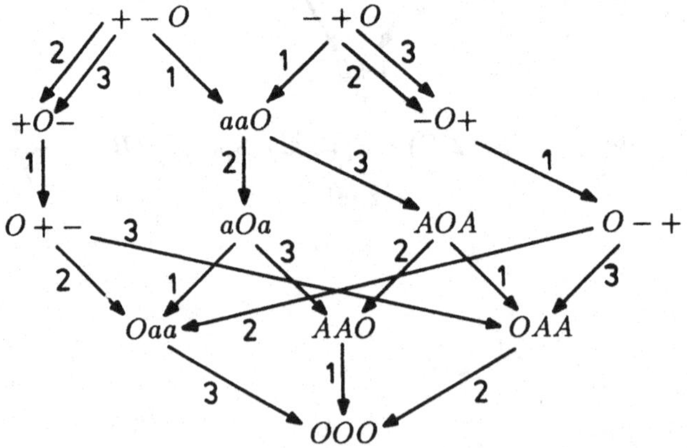

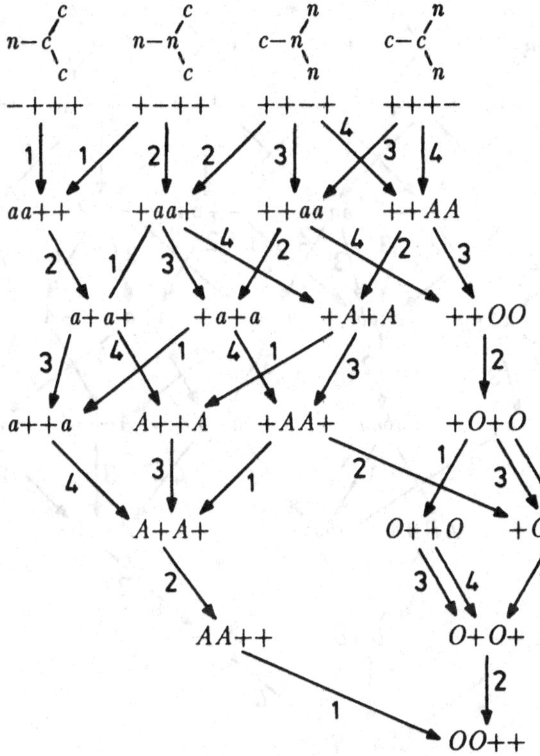

DI $\mathfrak{g}^s = \mathfrak{so}(6,2)$ $S(O(6,\mathbf{C}) \times O(2,\mathbf{C}))\backslash SO(8,\mathbf{C})/B$ 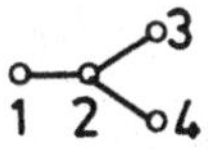

Fig. 18

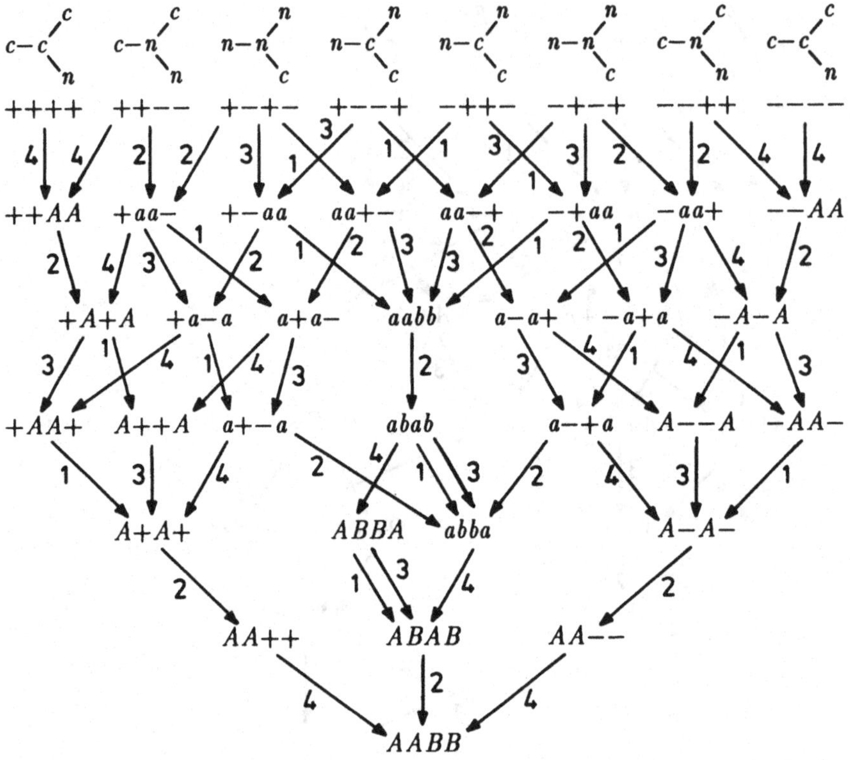

DIII $\mathfrak{g}^s = \mathfrak{so}^*(8)$ $\mathbf{C}^\times \times PSL(4, \mathbf{C})\backslash SO(8, \mathbf{C})/B$ 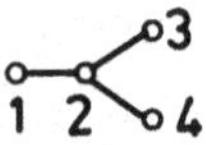

Fig. 19

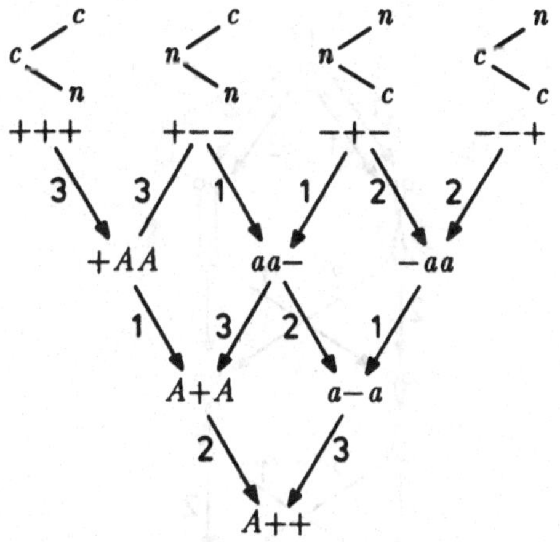

DIII $\mathfrak{g}^s = \mathfrak{so}^*(6)$ $\mathbf{C}^\times \times PSL(3,\mathbf{C})\backslash SO(6,\mathbf{C})/B$ 1, 2, 3

Fig. 20

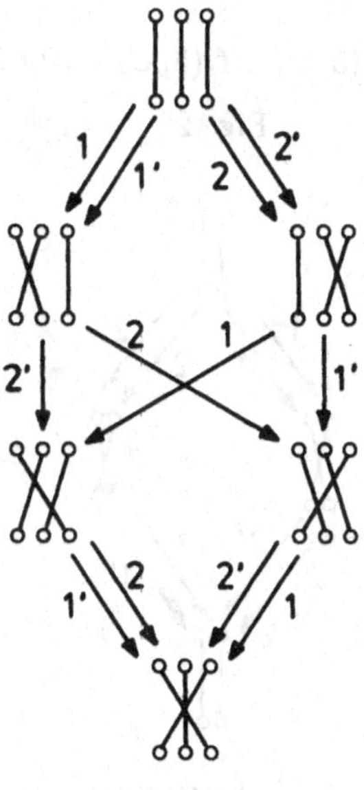

A_2 $\mathfrak{g}^s = \mathfrak{sl}(3,\mathbf{C})$ $\Delta GL(3,\mathbf{C})\backslash GL(3,\mathbf{C}) \times GL(3,\mathbf{C})/B$ 1 2 1' 2'

Fig. 21

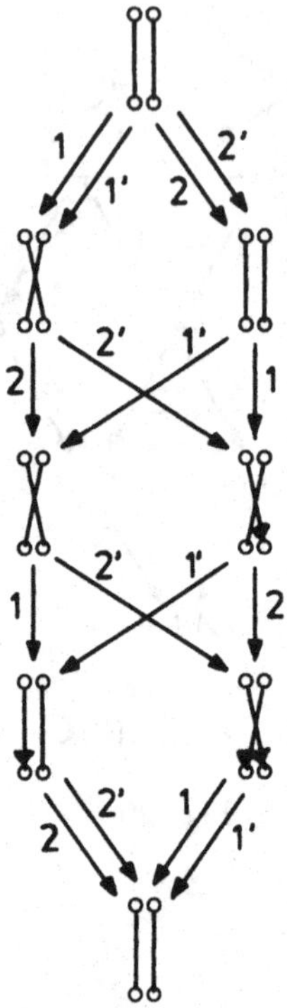

B$_2$ $\mathfrak{g}^s = \mathfrak{so}(5, \mathbb{C})$ $\Delta SO(5, \mathbb{C}) \backslash SO(5, \mathbb{C}) \times SO(5, \mathbb{C})/B$ 1 2 1' 2'

Fig. 22

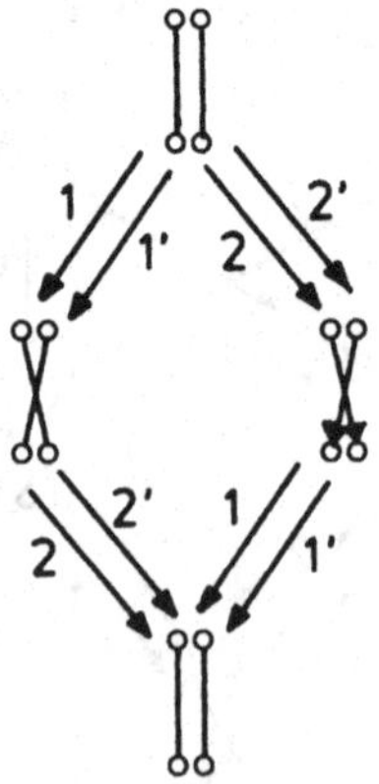

D$_2$ $\mathfrak{g}^s = \mathfrak{so}(4, \mathbb{C})$ $\Delta SO(4, \mathbb{C}) \backslash SO(4, \mathbb{C}) \times SO(4, \mathbb{C})/B$ ∘1 ∘2 ∘1' ∘2'

Fig. 23

References

[BB1] A.A. Beilinson and J. Bernstein, *Localization de $\mathfrak{g}$-modules*, C.R. Acad. Sci. Paris **292** (1981), 15-18.

[BB2] ——————, *A generalization of Casselman's submodule theorem*, in "Representation Theory of Reductive Groups;" edited by P.C. Trombi, Progress in Mathematics 40, Birkhäuser, Boston, 1983, pp. 35-52.

[C] D.H. Collingwood, *Representations of Rank One Lie Groups*. Pitman Advanced Pub. Program, Boston, 1985.

[KKMOOT] M. Kashiwara, A. Kowata, K. Minemura, K. Okamoto, T. Oshima and M. Tanaka, *Eigenfunctions of invariant differential operators on a symmetric space*, Ann. of Math. **107** (1978), 1-39.

[KW] A.W. Knapp and N.R. Wallach, *Szegő kernels associated with discrete series*. Invent. Math. **34** (1976), 163-200, **62** (1980), 341-346.

[M1] T. Matsuki, *The orbits of affine symmetric spaces under the action of minimal parabolic subgroups*, J. Math. Soc. Japan **31** (1979), 331-357.

[M2] ————, *Closure relations for orbits on affine symmetric spaces under the action of minimal parabolic subgroups*, Advanced Studies in Pure Math. **14** (1988), 541-559.

[O1] T. Oshima, *Asymptotic behavior of spherical functions on semisimple symmetric spaces*, Advances Studies in Pure Math. **14** (1988), 561-601.

[O2] ————, *A realization of semisimple symmetric spaces and construction of boundary value maps*, Advanced Studies in Pure Math. **14** (1988), 603-650.

[R] W. Rossman, *The structure of semisimple symmetric spaces*, Canad. J. Math. **31** (1979), 157-180.

[V] D. Vogan, *Irreducible characters of semisimple Lie groups III*, Invent. Math. **71** (1983), 381-417.

Toshihiko Matsuki
Faculty of General Education
Tottori University
Tottori 680, Japan

Toshio Oshima
Department of Mathematics
Faculty of Science
University of Tokyo
Tokyo 113, Japan

Is There an Orbit Method for Affine Symmetric Spaces?

GESTUR 'OLAFSSON AND BENT ØRSTED*

Abstract. For affine symmetric spaces of Hermitian type we realize certain discrete representations in a natural way on orbits in the tangent space of the origin. The orbits are with respect to the isotropy group of the origin, and their structure is related to that of bounded symmetric complex domains.

0. INTRODUCTION

Far from giving an answer to the question raised above, our aim is to present some evidence for the semi-simple case that encourages further investigations. The case of nilpotent symmetric spaces has been settled very satisfactorily by Y. Benoist, both geometrically and algebraically. But for semi-simple spaces the situation has been a little mysterious. Roughly speaking, for $X = G/H$ and $\mathfrak{q}_0$ the tangent space at the origin, one would like a correspondence between H-orbits in $\mathfrak{q}_0$ (admissible in a certain sense) and unitary representations of G with an H-fixed distribution vector. When the orbit is of maximal dimension, the representation should contribute to the Plancherel formula for X. Although this philosophy seems to be right, no direct geometric construction of representations or characters has been found in analogy with the symplectic geometry of the Kirillov–Kostant geometric quantization. For example, naive guesses at Kirillov character formulas via Fourier transforms of measures produce spherical functions with too few regular singular points, as is already seen for $SL(2,\mathbf{R})$ modulo the diagonal matrices.

Our positive evidence is for X of Hermitian type and orbits through a certain central element generating a proper convex cone in $\mathfrak{q}_0$. We show that some holomorphic discrete series representations on X (these we construct in Chapter 1) in a natural way live on L^2-spaces on such orbits. The intertwining operators are Fourier–Laplace transformations composed with our previously found intertwining operators into the discrete part of $L^2(X)$.

Thus the situation is very special indeed as far as a general orbit method is concerned; still, we hope that the explicit formulas here will shed some light on the general problem.

---

*Lecture by Bent Ørsted

1. Spaces of Hermitian type and their holomorphic discrete series

Among the discrete series representations of a semi-simple Lie group, the first to be constructed directly were the holomorphic series in case the Riemannian symmetric space had a complex structure. For this Harish-Chandra found very explicit formulas for both the lowest K-type and the intertwining operator into $L^2(G)$. It is this construction that we wish to generalize to construct what we call the holomorphic discrete series of $L^2(X)$.

We follow the notation of [5] and [6], and we also refer to these for proofs and more detailed results. Let G be a non-compact connected semi-simple Lie group contained in a simply connected complex Lie group $G_{\mathbf{C}}$ with Lie algebras $\mathfrak{g}_0$ and $\mathfrak{g}$ respectively. Similar notation is used for other Lie groups and for complexifications of vector spaces. Fix a non-trivial involution τ of G and a Cartan involution θ commuting with τ. H will be an open subgroup of the fixpoint group G^τ, and $K = G^\theta$. The affine symmetric space we shall study is $X = G/H$ with tangent space at the origin $\mathfrak{q}_0$, the -1 eigenspace of τ in $\mathfrak{g}_0$.

Assume that X is of Hermitian type, i.e.

(1.1.i) There is no non-trivial ideal of $\mathfrak{g}$ contained in $\mathfrak{h}$.

(1.1.ii) Let $\mathfrak{c}$ be the centralizer of $\mathfrak{q}_{\mathfrak{k}} = \mathfrak{q} \cap \mathfrak{k}$ in $\mathfrak{q}_{\mathfrak{k}}$, then $\mathfrak{z}_{\mathfrak{q}}(\mathfrak{c}) = \mathfrak{q}_{\mathfrak{k}}$.

In more geometric terms, one may also characterize a Hermitian type involution τ by operating as a conjugation on G/K; or, alternatively, that there exists a proper (i.e. open and containing no straight lines) H-invariant convex cone $C \subseteq \mathfrak{q}_0$ such that $\mathfrak{q}_{\mathfrak{k}} \cap C^0 \neq \emptyset$. Note that this last condition could be used as a definition in the non-semi-simple case. But for purposes of dealing with root systems etc., (1.1) is the most convenient definition of X (or τ) to be of Hermitian type.

In case $X = G_1$, a semi-simple Lie group, (1.1) says that G_1 has a Hermitian symmetric space. Just as in this group case, (1.1) in general means that there is an H-invariant convex cone in $\mathfrak{q}_0$ containing no straight lines. It is a consequence of (1.1) that G/K is a Hermitian symmetric space, and that there exist complex coordinates here, so that τ operates as complex conjugation (see [4]). Furthermore, we can construct root systems analogous to Harish-Chandra's in the group case:

Let $\mathfrak{a}$ be a maximal abelian subalgebra of $\mathfrak{q}$ containing $\mathfrak{c}$; then $\mathfrak{c} \subseteq \mathfrak{a} \subseteq \mathfrak{q}_{\mathfrak{k}}$, and we may consider the roots Δ of $\mathfrak{a}$ in $\mathfrak{g}$ as the disjoint union of the compact and the non-compact roots, Δ_c and Δ_n respectively. Choose an ordering so that $\mathfrak{c}^*$ comes first and let Δ^+ be the positive roots. Then the subalgebras

$$\mathfrak{p}^+ = \bigoplus_{\alpha \in \Delta_n^+} \mathfrak{g}_\alpha \quad \text{and} \quad \mathfrak{p}^- = \bigoplus_{\alpha \in \Delta_n^+} \mathfrak{g}_{-\alpha}$$

spanned by the non-compact positive, respectively negative, root spaces, are abelian. In fact, these are exactly those of Harish-Chandra. Inside $G_{\mathbf{C}}$ we have the corresponding subgroups P^+ and P^-, as well as $K_{\mathbf{C}} = G_{\mathbf{C}}^\theta$ and $H_{\mathbf{C}} = G_{\mathbf{C}}^\tau$. From [5] we have

Theorem 1.1. *With notation as above, $G \subseteq H_{\mathbf{C}} K_{\mathbf{C}} P^-$, and we may define the analytic function $k_H : G \to K_{\mathbf{C}}/K_{\mathbf{C}} \cap H_{\mathbf{C}}$ by*

$$x \in H_{\mathbf{C}} k_H(x) P^- \qquad (x \in G). \tag{1.2}$$

This also defines an embedding $G/K \to H_{\mathbf{C}}/H_{\mathbf{C}} \cap K_{\mathbf{C}}$ as well as (by (1.2)) an embedding $X \to K_{\mathbf{C}}/K_{\mathbf{C}} \cap H_{\mathbf{C}}$.

We can make the function k_H in (1.2) very explicit by choosing a maximal set of strongly orthogonal roots $\alpha_1, \ldots, \alpha_s \in \Delta_n^+$. Let $t_i \in \mathfrak{a}$, $x_i \in \mathfrak{p}_{\alpha_i}$ and $y_i = x_i - \tau x_i$ so that

$$\begin{aligned} \tau \bar{x}_i &= -x_i \quad \text{(conjugation relative to } \mathfrak{g}_0) \\ [x_i, \bar{x}_i] &= t_i \\ \alpha_i(t_i) &= 2 \end{aligned}$$

for $i = 1, \ldots, s$. Then the y_i's span a maximal abelian subalgebra $\mathfrak{b}_0$ in $\mathfrak{q}_{0\mathfrak{p}}$ ([5], Lemma 2.3), and we have the formula

$$k_H(b^{-1}) = \prod_{i=1}^{s} \exp(\tfrac{1}{2}\log(\cosh 2s_i) t_i), \quad b = \prod_{i=1}^{s} \exp s_i y_i \tag{1.3}$$

on the corresponding analytic subgroup B.

Now we can construct the holomorphic discrete series of X, writing down their lowest K-types in terms of the polar coordinates KB on X. By (1.3) the formulas below are quite explicit.

Suppose δ is an irreducible unitary representation of K, μ its highest weight, and ν^0 a non-zero, K^τ-invariant vector in the contragradient representation $\delta^\vee$. Then we have ([6], Theorem 5.2).

Theorem 1.2. *For every vector v in the representation space of δ, define the function*

$$\Phi_v(x) = \langle v, \delta^\vee(k_H(x^{-1})^{-1})\nu^0 \rangle. \tag{1.4}$$

This is in $L^2(X)$ whenever $\langle \mu+\rho, \alpha\rangle < 0$ for all $\alpha \in \Delta_n^+$, where ρ is half the sum of positive roots in Δ. In this case Φ_v generates an irreducible submodule of $L^2(X)$ with K-types having highest weights of the form

$$\nu = \mu - \Sigma n_\alpha \alpha$$

with $\alpha \in \Delta_n^+$ and $n_\alpha \in \mathbf{Z}^+$.

The representations π_μ described in this theorem constitute by definition the holomorphic discrete series of X; note that one may combine (1.3) and (1.4) to get very explicit formulas for their lowest K-types. This may also be viewed as a concrete way of calculating the integrals in the formulas of Flensted–Jensen for these K-types; see [**1**]. these representations were considered by different methods (using the parametrizations and comparing to discrete series representations of the group) in [**3**].

In the simplest example, we have $G = SU(1,1)$ and the symmetric subgroup

$$H = \left\{ \pm \begin{pmatrix} \cosh t & \sinh t \\ \sinh t & \cosh t \end{pmatrix} | t \in \mathbf{R} \right\}$$

with symmetric space the hyperboloid of one sheet

$$X = \left\{ \exp\theta \begin{pmatrix} i & 0 \\ 0 & -i \end{pmatrix} \exp t \begin{pmatrix} 0 & i \\ -i & 0 \end{pmatrix} \cdot \begin{pmatrix} 0 & 1 \\ 1 & 0 \end{pmatrix} | 0 \leq \theta \leq 2\pi, t \in \mathbf{R} \right\}$$

realized in the Lie algebra.

For the lowest K-type we have the formula in terms of the polar coordinates θ and t,

$$h_n = e^{in\theta}(\cosh 2t)^{-n/2}, \quad n = 2, 4, 6, \ldots$$

and this generates the irreducible representation π_n inside $L^2(X)$; π_n is equivalent to the usual holomorphic discrete series representation with parameter n. Also, we have the formula

$$\phi_n = (1 - z^2)^{-n/2}$$

for the H-invariant distribution vector in the usual Hilbert space of holomorphic functions in the unit disc $\{z \in \mathbf{C} \mid |z| < 1\}$. Note the (not accidental) resemblance with the reproducing kernel.

A nice feature of the above construction is that the functions in (1.4) essentially via convolution define the intertwining operators between the standard models on G/K of these representations and their realization inside $L^2(X)$, see [**6**].

Let us briefly recall how this goes, since it is important in establishing our connection between H-orbits in $\mathfrak{q}_0$ and the holomorphic discrete series of X, see (2.1) below.

Every discrete series representation π of G has a realization in a reproducing kernel Hilbert space of sections of a vector bundle over G/K; what we find is that any embedding of π into $L^2(X)$ is determined by an analytic function ϕ on X with values in the space of the lowest K-type V of π such that (assuming the center of G is finite, and π is integrable)

$$f(z) \longrightarrow \int_{G/K} (f(z), \phi(z^{-1}x))_V \, dz \tag{1.5}$$

gives the intertwining operator from sections over G/K to functions on X. The inverse is

$$h(x) \longrightarrow \int_X h(x)\phi(y^{-1}x)dx. \tag{1.6}$$

Note that both are convolution operators, and we may formally write them as $f * \phi$ and $h * \phi$ respectively. Still on a formal level (but still instructive) we have the following formula

$$\Theta = \phi * \phi \tag{1.7}$$

for the H-invariant distribution on X, where convolution by Θ defines the orthogonal projection onto the irreducible invariant subspace of $L^2(X)$ where π lives.

Thanks to our concrete analysis of the holomorphic discrete series of X, we may in that case give the explicit formula for $\phi = \phi_\mu$ with μ the parameter in Theorem 1.2:

$$\phi_\mu(x) = \delta^\vee(k_H(x^{-1})^{-1})\nu^0. \tag{1.8}$$

Actually, the formula (1.7) may be given a meaning as an identity for holomorphic functions on a certain subdomain of $X_{\mathbf{C}}$; we return to this domain in Chapter 3.

2. The associated orbit structures

To simplify the exposition, we assume in this chapter that the Hermitian symmetric space G/K is of tube type. Our plan is to consider the orbit $H/H \cap K$ inside $\mathfrak{q}_0$ and associate to the parameter μ in Theorem 1.2 a measure dm_μ on $H/H \cap K$. For μ the parameter of a holomorphic discrete series representation of the group we may consider the standard model (following [7] and [8]) inside $HL^2(G/K, d\nu_\mu)$, holomorphic sections square integrable with respect to a certain measure $d\nu_\mu$. What we want is to construct natural integral operators

$$(2.1) \qquad L^2(H/H \cap K, dm_\mu) \longrightarrow HL^2(G/K, d\nu_\mu) \longrightarrow L^2(X)$$

with image π_μ. The first operator in (2.1) is a Fourier-Laplace operator, and the second is our intertwining operator mentioned at the end of Chapter 1, namely (1.5).

The main technical point in establishing (2.1) is to study the fine structure of G/K relative to the involution τ; in fact, $H/H \cap K$ is exactly the dual to the cone in the description of G/K as a Siegel domain. To see this, we need to see how τ operates on the data describing the structure of bounded Hermitian domains.

Extend $\mathfrak{a}$ to a compact Cartan subalgebra $\mathfrak{t}$, choosing order so that $\mathfrak{a}$ comes first. Let $\{\gamma_1, \ldots, \gamma_r\}$ be a maximal set of positive, non-compact, and strongly orthogonal roots of $\mathfrak{t}$ in $\mathfrak{g}$. With τ of Hermitian type as above, one may partition the γ_j's into two classes: those that are fixed under $-\tau$, and those that are pairwise interchanged under $-\tau$, i.e. $-\tau\gamma_j = \gamma_{j-1}$. This means that the usual constructions associated with a Hermitian symmetric space can be done in a τ-equivariant way since $-\tau$ is just permuting the strongly orthogonal roots in a specific way.

Of particular interest is the H-orbit $H/H \cap K$ through the origin in G/K. This orbit may also be realized inside $\mathfrak{q}_0$ as the H-orbit through a point in the center of $\mathfrak{q}_\mathfrak{k}$ (recall that this is also in the center of $\mathfrak{k}$; in fact the complex structure of G/K is defined via such an element in case G is simple).

Note that when $\tau = Ad(c^2)$, where c is the Cayley transform, we are in the situation of [8], where

$$(2.2) \qquad H/H \cap K = \Omega^*$$

is the dual cone to the cone Ω in the Siegel realization of G/K:

$$(2.3) \qquad G/K = V + i\Omega.$$

Here Ω is a proper convex cone in the real vector space V. The point is that modulo a natural diffeomorphism, (2.2) is true for a general τ of Hermitian type:

PROPOSITION 2.1 ([5]). *Let G and τ be as above; then via a linear change in exponential coordinates, $H/H \cap K$ is diffeomorphic to Ω^*, where G/K is the tube domain over Ω.*

Now we are in a position to implement our objective as in (2.1). Recall from [7] that the holomorphic discrete series for G naturally lives in an

L^2-space on Ω^* ([8] deals with the scalar series only, and we only formulate our result in that case; the general case requires little more than direct extension to vector–valued functions). Also, via a Fourier-Laplace transformation, one arrives at the standard model of these representations on G/K:

$$\widehat{\psi}(z) = \int_{\Omega^*} \psi(y) e^{i(z,y)}\, dm_\mu(y) \tag{2.4}$$

with $z \in V_{\mathbf{C}}$ and dm_μ a measure on Ω^* corresponding to the parameter μ of the representation π_μ. (2.4) represents the first arrow in (2.1), and the second arrow comes from the convolution operator into $L^2(X)$ given in [4]. Summarizing, we have

THEOREM 2.2. *Suppose G/K is of tube type, τ an involution of Hermitian type commuting with the Cartan involution, and μ the parameter of a holomorphic discrete series representation π_μ (for X) which is also a scalar holomorphic discrete series representation for G. Then π_μ is in a natural way realized on the Hilbert space $L^2(H/H \cap K, dm_\mu)$, where $H/H \cap K$ is an H-orbit in $\mathfrak{q}_0$ generating an H-invariant proper convex cone, and dm_μ a measure on this orbit associated to μ.*

It would be desirable to find a direct geometric meaning of the measure dm_μ, so that μ in a natural way determines the (admissible) orbit and its measure.

3. FURTHER REMARKS

As we have seen above, there is at least in some special cases a correspondence between the holomorphic discrete series of X and certain H-orbits in $\mathfrak{q}_0$ generating a proper convex cone C. The geometry of X and its tangent bundle $G \times_H \mathfrak{q}_0$ is in several ways influenced by this cone. In this section we shall briefly describe how X carries certain Hardy spaces of functions; we believe these spaces arise in a similar way as natural integral transforms of L^2-spaces on cones C as above — thus they too fit into our version of an orbit philosophy for X. The results in this section have been obtained jointly with J. Hilgert [2]. From the construction of the representations π_μ and their distributions Θ_μ, one finds a natural domain in $X_{\mathbf{C}}$ into which the K-finite functions and also Θ_μ can be holomorphically continued. This was conjectured in [5]; the basic results can be stated as follows:

Let C_1 be the minimal proper convex G-invariant cone in $\mathfrak{g}_0$ and consider the subsemigroup

$$\Gamma = G \exp iC_1 \subseteq G_{\mathbf{C}}$$

and its image Ξ under the projection

$$G_{\mathbf{C}} \longrightarrow X_{\mathbf{C}} : x \longrightarrow x^{-1}H_{\mathbf{C}}.$$

We also consider the corresponding cone in $\mathfrak{q}_0$:

$$C = C_1 \cap \mathfrak{q}.$$

Definition 3.1. *By* $\mathbf{H}$ *we denote the space of all holomorphic functions* $f : \Xi^0 \to \mathbf{C}$ *with*

$$\|f\| = \sup_{\gamma \in \Gamma^0} \|\gamma \cdot f\|_{L^2(X)} \tag{3.1}$$

where $\gamma \cdot f$ *denotes the left-regular action.*

This is the definition of the Hardy space associated to C_1 and C.

Theorem 3.2. $\mathbf{H}$ *is a Hilbert space with norm (3.1), and there is a* G*-equivariant isometric embedding* $I : \mathbf{H} \to L^2(X)$ *given by* $If = \lim \gamma_j f$ *for any sequence in* Γ^0 *converging to 1. Furthermore,* $I(\mathbf{H})$ *is the biggest invariant subspace where all elements of* iC_1 *have non-positive spectrum, and we have a direct sum decomposition*

$$I(\mathbf{H}) = \oplus_\mu \pi_\mu$$

(sum over all holomorphic discrete series of X*).*

As a final result tying a connection to the geometry of the tangent bundle we have

Theorem 3.3. *The domain* Ξ^0 *is biholomorphic to* $G \times_H (-iC^0)$ *in case* H *is the full isotropy group:* $H = G^\tau$.

References

[1] M. Flensted-Jensen, *Analysis on non-Riemannian symmetric spaces*, CBMS Reg. Conf. **61** (1986).
[2] J. Hilgert, G. 'Olafsson and B. Ørsted, *Hardy spaces on affine symmetric spaces*, Preprint 1989.
[3] S. Matsumoto, *Discrete series for an affine symmetric space*, Hiroshima Math. J. **11** (1981), 53–79.
[4] G. 'Olafsson, *Symmetric spaces of Hermitian type*, Preprint 1989.
[5] G. 'Olafsson and B. Ørsted, *The holomorphic discrete series for affine symmetric spaces, I*, J. Funct. Anal. **81** (1988), 126–159.
[6] G. 'Olafsson and B. Ørsted, *The holomorphic discrete series of an affine symmetric space and representations with reproducing kernels*, Preprint, Odense 1988.
[7] H. Rossi and M. Vergne, *Representations of certain solvable Lie groups on Hilbert spaces of holomorphic functions and the application to the holomorphic discrete series of a semisimple Lie group*, J. Funct. Anal. **13** (1973), 324–389.
[8] H. Rossi and M. Vergne, *Analytic continuation of the holomorphic discrete series of a semi-simple Lie group*, Acta Math. **136** (1976), 1–59.

Mathematisches Institut der Georg-August-Universität
Bunsenstrasse 3-5
Göttingen
BDR

Matematisk Institut
Odense Universitet
Campusvej 55
5230 Odense M

On a Property of the Quantization Map for the Coadjoint Orbits of Connected Lie Groups

L. PUKANSZKY[1]

Introduction. Let M be a differentiable manifold, $L \xrightarrow{p} M$ a complex line bundle and ∇ a linear connection on L. Its curvature form gives rise to a closed 2-form on M. Conversely, $\omega \in Z^2(M)$ arises in the said fashion, if it is of an integral de Rham class. Below we assume, that ω is also real and nondegenerate. We consider the totality $\mathcal{E}(M)$ of all smooth functions on M as a Lie algebra with respect to the Poisson bracket $\{\ ,\ \}$. Following B. Kostant (cf.[2]), we can canonically associate with the said data a morphism δ of $\mathcal{E}(M)$, called the prequantization map, in the Lie algebra of endomorphisms of all smooth sections $\mathcal{E}(L)$ by setting

$$\delta(\varphi) = X_\varphi - 2\pi i\varphi \qquad (\varphi \in \mathcal{E}(M))$$

where the Hamiltonian vector field X_φ is defined by $d\varphi = -\iota(X_\varphi)\omega$. – Given an integrable Lagrangean distribution $\mathcal{H}$ on M, we write $\mathcal{V}_{\mathcal{H}}(M)$ for the Lie algebra of vector fields with values in $\mathcal{H}$. We call a section polarized, if it is annihilated by all ∇_X $(X \in \mathcal{V}_{\mathcal{H}}(M))$, and shall write $\mathcal{E}^0(L)$ for their totality. Let $\mathcal{E}^0$ be the collection of all elements of $\mathcal{E}(M)$, which satisfy $X\varphi \equiv 0$ $(\varphi \in \mathcal{E}(M), X \in \mathcal{V}_{\mathcal{H}}(M))$. $\mathcal{E}^0$ is a maximal abelian subalgebra of $\mathcal{E}(M)$; we denote its normalizer by $\mathcal{E}^1$. Then the image of $\mathcal{E}^1$ via δ leaves $\mathcal{E}^0(L)$ invariant. The restriction of δ to $\mathcal{E}^1$ and $\mathcal{E}^0(L)$ is the quantization map.

As it is well known, the above construction plays an important role in the unitary representation theory, in particular, of solvable Lie groups.

Let G be such and also connected and simply connected. Any coadjoint orbit O of G gives rise canonically to a symplectic G-space (O, ω_O). If ω_O is integral and (L, ∇) corresponds to it as at the start, L is the associated bundle of a principal bundle with a total space equal to that of G. Hence $\mathcal{E}(L)$ and, along with it $\mathcal{E}^0(L)$, can be canonically identified to certain spaces of functions on G.

In a recent paper (cf.[3]), N. V. Pedersen showed, that for any (not necessarily solvable) G, if O admits a real G-invariant Lagrangean distribution, then the quantization map defines an isomorphism of Lie algebras between $\mathcal{E}^1$ and the collection $B_0^1(L)$ of all differential operators or order ≤ 1 on $\mathcal{E}^0(L)$. He then proceeds to use this to establish the existence

[1]This work was supported by a grant of the National Science Foundation

of global Darboux coordinates for any coadjoint orbit of an exponential solvable Lie group.

The objective of the present paper is to establish the existence, in the solvable case, for any minimal integral covering of a coadjoint orbit, the existence of a polarization which, via the standard procedure described in [4] leads to a factor representation, and for which, at the same time the analogue of Pedersen's result holds.[2]– In particular, this is the case for any polarization in the nilpotent case. We then proceed, following Pedersen to apply this to show the existence of global Darboux coordinates for any simply connected coadjoint orbit.

The organization of our paper is as follows. Section 1 discusses the requisite polarizations. Here we found useful and appropriate to develop ab ovo results, which are known in a more general context. Sections 2 and 3 aim at rendering the corresponding Lagrangean distributions more transparent. In Section 4 we discuss the local structure of $\mathcal{E}^0$ and $\mathcal{E}^1$. Section 5 brings the proof for the generalization of Pedersen's isomorphism. Finally, Section 6 discusses the question of the global Darboux coordinates. – As to quantization only the basic rules, listed in Section 5, are assumed.

Our whole approach is substantially inspired by ideas of Pedersen's paper.

1. On certain polarizations. Below $\mathfrak{g}$ is a finite dimensional real solvable Lie algebra, and G a corresponding connected and simply connected Lie group. We fix an element g of $\mathfrak{g}^*$, and define the bilinear form B on $\mathfrak{g} \times \mathfrak{g}$ by $B(x,y) = ([x,y],g)$ $(x, y \in \mathfrak{g})$. We write also $\mathfrak{d} = [\mathfrak{g}, \mathfrak{g}]$, and $f = g|\mathfrak{d}$. – Let $\mathfrak{h}_1$ be a fixed maximal selforthogonal subspace, with respect to B, of $\mathfrak{g}_f$.

Lemma 1. *We have* $\mathfrak{h}_1 \supset \mathfrak{g}_g + \mathfrak{d}_f$.

Proof: Since $\mathfrak{h}_1$ is, by construction, maximal self-orthogonal with respect to B in $\mathfrak{g}_f$, our claim will follow by proving, that the radical of $B|(\mathfrak{g}_f \times \mathfrak{g}_f)$ is equal to $\mathfrak{g}_g + \mathfrak{d}_f$. Since $\mathfrak{g}_f = \mathfrak{d}^{\perp}_{B}$ and thus $(\mathfrak{g}_f)^{\perp}_{B} = \mathfrak{d} + \mathfrak{g}_g$, we have

$$\begin{aligned}\mathrm{rad}(B|\mathfrak{g}_f \times \mathfrak{g}_f) &= \mathfrak{g}_f \cap (\mathfrak{g}_f)^{\perp}_{B} = \mathfrak{g}_f \cap (\mathfrak{d} + \mathfrak{g}_g)\\ &= (\mathfrak{g}_f \cap \mathfrak{d}) + \mathfrak{g}_g = \mathfrak{d}_f + \mathfrak{g}_g\end{aligned}$$

and hence also $\mathfrak{h}_1 \supset \mathfrak{d}_f + \mathfrak{g}_g$.

[2]This introduction can provide only an approximate picture of the whole; for the precise statement of our results we refer to the text below.

Lemma 2. *$\mathfrak{h}_1$ is a subalgebra of $\mathfrak{g}$.*

Proof: This is clear from

$$\mathfrak{h}_1^2 \subset \mathfrak{d} \cap \mathfrak{g}_f = \mathfrak{d}_f$$

and that, by Lemma 1, $\mathfrak{d}_f \subset \mathfrak{h}_1$.

Lemma 3. *We have $(\mathfrak{h}_1)_B^{\perp} = \mathfrak{d} + \mathfrak{h}_1$.*

Proof: The right-hand-side is contained in the left-hand side since both $\mathfrak{h}_1 \subseteq \mathfrak{g}_f$ and $\mathfrak{d}$ are orthogonal, with respect to B, to $\mathfrak{h}_1$. In this manner the desired conclusion follows by showing, that

$$\dim((\mathfrak{h}_1)_B^{\perp}) = \dim(\mathfrak{d} + \mathfrak{h}_1).$$

We have, by $\mathfrak{g}_g \subseteq \mathfrak{h}_1$, on the left: $\dim(\mathfrak{g}) - \dim(\mathfrak{h}_1) + \dim(\mathfrak{g}_g)$. On the right, by $\mathfrak{d} \cap \mathfrak{h}_1 = \mathfrak{d}_f$, we find $\dim(\mathfrak{d}+\mathfrak{h}_1) = \dim(\mathfrak{d}) + \dim(\mathfrak{h}_1) - \dim(\mathfrak{d}_f)$. In this fashion our claim is correct if and only if

$$2\dim(\mathfrak{h}_1) = \dim(\mathfrak{g}) - \dim(\mathfrak{d}) + \dim(\mathfrak{d}_f) + \dim(\mathfrak{g}_g).$$

But the right-hand-side can easily be seen to be equal to

$$\dim(\mathfrak{g}_f) + \dim(\mathfrak{d}_f + \mathfrak{g}_g).$$

The resulting equation, however, is valid, since $\mathfrak{h}_1$ is maximal self-orthogonal with respect to $B|(\mathfrak{g}_f \times \mathfrak{g}_f)$.

We observe that, if $\mathfrak{a}$ is a subspace of $\mathfrak{g}$, we have clearly $(\mathfrak{a}g)^{\perp} = \mathfrak{a}_B^{\perp}$. Below we shall write $\mathfrak{c} = \mathfrak{d} + \mathfrak{h}_1$. Hence we get

Lemma 4. *We have $\mathfrak{c}^{\perp} = \mathfrak{h}_1 g$.*

We recall, that a complex subalgebra $\mathfrak{h}$ of $\mathfrak{g}_{\mathbf{C}}$ is called a polarization with respect to g if 1) It is maximal self-orthogonal with respect to B, 2) $\mathfrak{h} + \overline{\mathfrak{h}}$ is a subalgebra. – We recall also, that there is a polarization $\mathfrak{h}'$ in $\mathfrak{d}_{\mathbf{C}}$ with respect to f, which is invariant by G_f (cf. [1], Theorem II.3.2, p.304). We make a fixed choice of $\mathfrak{h}'$ and set $\mathfrak{h} = \mathfrak{h}' + (\mathfrak{h}_1)_{\mathbf{C}}$.

Lemma 5. *$\mathfrak{h}$, as just defined, is a polarization with respect to g.*

Proof: Let us prove first, that $\mathfrak{h}$ is a subalgebra. We have $\mathfrak{h}^2 = (\mathfrak{h}')^2 + [\mathfrak{h}', \mathfrak{h}_1] + (\mathfrak{h}_1^2)_{\mathbf{C}}$. Since $\mathfrak{h}'$ is a subalgebra of $\mathfrak{d}_{\mathbf{C}}$, the first summand is in $\mathfrak{h}'$. Since $\mathfrak{h}'$ is left invariant by G_f, $\mathfrak{h}'$ is normalized by $\mathfrak{g}_f$ and thus $[\mathfrak{h}', \mathfrak{h}_1] \subset \mathfrak{h}'$. Finally, we have $(\mathfrak{h}_1^2)_{\mathbf{C}} \subseteq (\mathfrak{d}_f)_{\mathbf{C}} \subseteq \mathfrak{h}'$ by the proof of Lemma 2. – We note next, that $\mathfrak{h} + \overline{\mathfrak{h}}$ is a subalgebra of $\mathfrak{g}_{\mathbf{C}}$. We have, in fact $\mathfrak{h} + \overline{\mathfrak{h}} = (\mathfrak{h}' + \overline{\mathfrak{h}}') + (\mathfrak{h}_1)_{\mathbf{C}}$ and the subalgebra $\mathfrak{h}' + \overline{\mathfrak{h}}'$ is normalized by $\mathfrak{h}_1$. – To show, that $\mathfrak{h}$ is self-orthogonal with respect to g it is enough to note again that

$$\mathfrak{h}^2 = (\mathfrak{h}_1)^2_{\mathbf{C}} + (\mathfrak{h}')^2 + [\mathfrak{h}_1, \mathfrak{h}']$$

and that the first two terms are orthogonal to g by construction and the third by $[\mathfrak{h}_1, \mathfrak{h}'] \subseteq [\mathfrak{g}_f, \mathfrak{d}_{\mathbf{C}}]$. – Suppose, that l is B-orthogonal to $\mathfrak{h}$. Then, in particular, so is it to $\mathfrak{h}_1$ and thus, by Lemma 3, we can write $l = l_1 + l_2$ where $l_1 \in \mathfrak{d}_{\mathbf{C}}$ and $l_2 \in (\mathfrak{h}_1)_{\mathbf{C}}$. Since l_1 and l_2 are B-orthogonal to $\mathfrak{h}'$, so is l_1 and thus $l_1 \in \mathfrak{h}'$, since $\mathfrak{h}'$ is maximal self-orthogonal in $\mathfrak{d}_{\mathbf{C}}$. Hence so is $\mathfrak{h}$ in $\mathfrak{g}_{\mathbf{C}}$ proving, that $\mathfrak{h}$ is a polarization with respect to g.

REMARK 1. Let us set $d = \mathfrak{h} \cap \mathfrak{g}$, $d' = \mathfrak{h}' \cap \mathfrak{g}$, $e = (\mathfrak{h} + \overline{\mathfrak{h}}) \cap \mathfrak{g}$, $e' = (\mathfrak{h}' + \overline{\mathfrak{h}}') \cap \mathfrak{g}$. For later use we note the following easy implications of the definitions.

1) a) $d = d' + \mathfrak{h}_1$, b) $d' = d \cap \mathfrak{d}$, 2) a) $e = e' + \mathfrak{h}_1$, b) $e' = e \cap \mathfrak{d}$.

Next we assume also, that $\mathfrak{h}'$ is a positive polarization with respect to f. We recall, that this means, that if $x + iy \in \mathfrak{h}'$, then $([x,y], f) \geq 0$ and equality holds if and only if $x, y \in d'$. An implication, to be used often below, is that $(e')^2 \subseteq d'$ (cf. [1], Theorem I.4.10, p.278).

REMARK 2. Let E be the connected subgroup of G belonging to e. We claim, that d' is E-variant. In fact, let E' and H_1 be the connected subgroups belonging to e' and $\mathfrak{h}_1$ resp. Then E is semi-direct product of E' and of H_1. Hence it is enough to note that a) Since $\mathfrak{h}'$ is H_1-invariant, the same holds for $d' = h' \cap \mathfrak{d}$, b) d' is E'-invariant by $(e')^2 \subseteq d'$. Below we write D and D' for the connected subgroups belonging to d and d' resp.

LEMMA 6. *With the previous notation, Dg is closed in $\mathfrak{g}^*$.*

PROOF: D is semi-direct product of D' and H_1. Since $\mathfrak{h}_1 \subseteq \mathfrak{g}_f$, we have $H_1 g = g + \mathfrak{h}_1 g$ and thus, by Lemma 4, $H_1 g = g + \mathfrak{c}^{\perp}$. From this we conclude, that $Dg = D' H_1 g = D'(g + \mathfrak{c}^{\perp}) = D'g + \mathfrak{c}^{\perp}$. The orbit $D'(g|\mathfrak{c})$, being that of a connected unipotent group, is closed in $\mathfrak{c}^* = g^*/\mathfrak{c}^{\perp}$. Thus Dg is closed in $\mathfrak{g}^*$.

REMARK 3. We recall that by Lemma 6 we have $Dg = g + e^{\perp}$. In fact, if $l \in d$, $k \in e$ and $j \geq 0$, we have $(k, l^j g) = (-l)^{j-1}([(\operatorname{ad} l)^{j-1} k, l], g) = 0$ and thus $Dg \subseteq g + e^{\perp}$. Hence it will be enough to show, that $\dim(e) + \dim(d) = \dim(\mathfrak{g}) + \dim(\mathfrak{g}_g)$. To this end we note first, that if $x + iy \in \mathfrak{h}$, then $([x,y], g) \geq 0$ and the equality holds only if $x + iy \in d_{\mathbf{C}}$. In fact, we can write $x + iy = x' + iy' + (u + iv)$, where $x' + iy' \in \mathfrak{h}'$ and $u, v \in \mathfrak{h}_1$. Thus, since $\mathfrak{h}_1$ is self-orthogonal in $\mathfrak{g}_f$, we have

$$([x,y], g) = ([x', y'], f) \geq 0,$$

$\mathfrak{h}'$ being a positive polarization. If equality holds we have $x', y' \in d'$ and thus $x, y \in d' + \mathfrak{h}_1$. – In this manner we conclude first, that $\dim(e/d) = 2\dim(\mathfrak{h}/d_{\mathbf{C}})$ and thus also $\dim(e) + \dim(d) = \dim(\mathfrak{g}) + \dim(\mathfrak{g}_g)$.

2. Parametrizations. Below we write y_0 for a fixed element of $\mathfrak{g}^*$. We assume to be given a fixed polarization as in Lemma 5, and we continue with the same notation.

Our assumptions on $\mathfrak{h}'$ imply the existence of elements $(a_i)_{1\le i\le t}$, $(b_j)_{1\le j\le t}$ in e' such that we have

$$([a_i, a_j], y_0) \equiv 0 \equiv ([b_i, b_j], y_0) \quad \text{and}$$
$$([a_i, b_j], y_0) \equiv \delta_{ij} \ (1 \le i, j \le t).$$

Setting $V = (\oplus_{j=1}^t \mathbf{R}a_j) \oplus (\oplus_{k=1}^t \mathbf{R}b_k)$ we get $e' = d' \oplus V$.

LEMMA 7. *We have*

$$Ey_0 = \{y;\, y = y_0 + ly_0 + \tfrac{1}{2}l^2 y_0 + h \ (l \in V, h \in e^{\perp})\}.$$

PROOF: We write $\Lambda = \exp(V)$. Since $(e')^2 \subseteq d'$, we have $E' = \Lambda \cdot D'$ setwise. Hence we get

$$E = E' \cdot H_1 = \Lambda D' \cdot H_1 = \Lambda D \text{ or } E = \Lambda \cdot D.$$

Thus, by Remark 3 we obtain

$$Ey_0 = \Lambda D y_0 = \Lambda(y_0 + e^{\perp}) = \Lambda y_0 + e^{\perp}.$$

In this manner it is enough to note, that if $l \in V, k \in e$ and $j \ge 3$, we have $(k, l^j y_0) = 0$. In fact, this can be written as $([l, [l, \overline{k}]], y_0)$ with $\overline{k} \in e \cap \mathfrak{d} = e'$, and thus it is equal to zero.

We set $r = \operatorname{codim}(e)$ and $s = r + 2t$. – Let $(e_j^*)_{1\le j\le r}$ be a basis in e and

$$l = \sum_{k=1}^{t} (y_{r+k} a_k + y_{r+t+k} b_k).$$

LEMMA 8. *The map*

$$(y_1, \ldots, y_s) \mapsto y_0 + ly_0 + \tfrac{1}{2}l^2 y_0 + \sum_{k=1}^{r} y_k e_k^*$$

is a smooth bijection $\mathbf{R}^s \to Ey_0$.

PROOF: Our map is surjective by Lemma 7. That it is also injective follows from

$$(a_i, y - y_0) = (a_i, ly_0) = ([a_i, l], y_0) = y_{r+t+i}$$
$$(b_i, y - y_0) = (b_i, ly_0) = ([b_i, l], y_0) = -y_{r+i}.$$

We write $f = y_0|\mathfrak{d}$, and denote by ω and η the pullback of the canonical 2-forms on Gy_0 and Lf to Ey_0 and Ef resp. – Let μ be the restriction map $g^* \to d^*$.

Lemma 9. *With notation as above we have* $\omega = \mu^* \eta$.

Proof: (i) Suppose, that $y \in Ey_0$ and that $t, \bar{t} \in T_y(Ey_0)$. There are $l, \bar{l} \in e$ such that $t = ly$, $\bar{t} = \bar{l}y$, and thus $\omega(t \wedge \bar{t}) = ([l, \bar{l}], y)$. If $y = \sigma y_0$ $(\sigma \in E)$ this is the same as $([\sigma^{-1}l, \sigma^{-1}\bar{l}], f)$. Writing $\sigma^{-1}l = l' + l_1$, $\sigma^{-1}\bar{l} = \bar{l}' + \bar{l}_1$ $(l' \in e', l_1 \in \mathfrak{h}_1$ etc.), we can conclude by $\mathfrak{h}_1 \subseteq \mathfrak{g}_f = \mathfrak{d}_B^{\perp}$ that $\omega(t \wedge \bar{t}) = ([l', \bar{l}'], f)$. – We have at the same time $T_y(\mu)(t) = l\mu(y) = l(\sigma f) = \sigma(\sigma^{-1}l)f = \sigma(l'f)$ and $T_y(\mu)(\bar{t}) = \bar{l}(\sigma f)$ and hence

$$(\mu^* \eta)(t \wedge \bar{t}) = ([l', \bar{l}'], f) = \omega(t \wedge \bar{t}).$$

We have thus proved, that $\mu^* \eta = \omega$.

Given j $(1 \leq j \leq s)$ we write ∂_j for the vector field on Ey_0, the value of which at y is derived according to y_j (cf. Lemma 8). – Given $t \in T_y(Ey_0)$, we shall, if conducive to clarity, denote by $\tau_y(t)$ the element of g^* such that $t = (d/d\tau)(y + \tau \cdot \tau_y(t))|_{\tau=0}$. However, if convenient, we write just t.

Lemma 10. *With the previous notation we have*

$$\omega(\partial_i|_y \wedge \partial_j|_y) = \begin{cases} 1 & \textit{if } r < i \leq r+t,\ j = i+t \\ 0 & \textit{otherwise.} \end{cases}$$

Proof: We assume again $y = \sigma y_0$ $(\sigma \in E)$. – (i) We have for $1 \leq j \leq r : \tau_y(\partial_j|_y) = e_j^*$. Also, since $d_B^{\perp} = e$, $e_j^* \in e^{\perp} = \sigma(e^{\perp}) = \sigma(dy_0)$. Hence there is $l_j \in d$ such that $\tau_y(\partial_j|_y) = \sigma l_j \cdot y$. From this we conclude, that

$$\omega(\partial_i|_y \wedge \partial_j|_y) = ([\sigma l_i, \sigma l_j], y) = ([l_i, l_j], y_0) = 0$$

since d is self-orthogonal with respect to B. – (ii) Let us write for $1 \leq j \leq t$: $t_j = \mu_{*y}(\partial_{r+j}|_y)$, $\bar{t}_j = \mu_{*y}(\partial_{r+j+t}|_y)$. We have $\tau_{\mu(y)}(t_j) = \Sigma_1 + \Sigma_2$, where $\Sigma_1 = a_j f$, $\Sigma_2 = \frac{1}{2}(la_j + a_j l)f$. Let us consider these separately. Ad Σ_1) We have $\Sigma_1 = a_j f \in e'f = (d')^{\perp}$. By Remark 2 above, this is the same as $\sigma(d')^{\perp} = \sigma(e'f)$. Hence we can find $k_j \in e'$ such that $a_j f = \sigma(k_j f)$. – Ad Σ_2) We have clearly $\Sigma_2 \in (e')^{\perp} = \sigma(e')^{\perp} = \sigma(d'f)$, and thus there is u_j in d' such that $\Sigma_2 = \sigma(u_j f)$. – Summing up, we can write $\tau_{\mu(y)}(t_j) = \sigma((k_j + u_j)f)$, where $k_j \in e'$ is such that $a_j f = \sigma(k_j f)$ and $u_j \in d'$. Similarly, we have $\tau_{\mu(y)}(\bar{t}_j) = \sigma((\bar{k}_j + v_j)f)$, where $\bar{k}_j \in e'$ satisfies $b_j f = \sigma(\bar{k}_j f)$ and $v_j \in d'$. – (iii) Using $\omega = \mu^* \eta$ (cf. Lemma 9) we conclude, that a) If $1 \leq i \leq r$, $1 \leq j \leq t$ we have

$$\omega(\partial_i|_y \wedge \partial_{r+j}|_y) = ([l_i, (k_j + u_j)], f)$$

which is zero, since e and d are orthogonal with respect to B. b) If $1 \leq i, j \leq t$, we get by (ii):

$$\omega(\partial_{i+r}|_y \wedge \partial_{r+j}|_y) = ([k_i + u_i, k_j + u_j], f) = ([k_i, k_j], f).$$

We have on the other hand

$$\begin{aligned}([k_i, k_j], f) &= (k_i, k_j f) = (\sigma k_i, \sigma(k_j f)) \\ &= (\sigma k_i, a_j f) = ([\sigma k_i, a_j], f).\end{aligned}$$

Since $\sigma f - f$ is orthogonal to d', we can conclude, that

$$\begin{aligned}([k_i, k_j], f) &= ([\sigma k_i, a_j], \sigma f) = -(a_j, \sigma(k_i f)) \\ &= -(a_i, a_j f) = ([a_i, a_j], f) = 0\end{aligned}$$

by construction and thus $\omega(\partial_{r+i}|_y \wedge \partial_{r+j}|_y) = 0$. Similarly, we can show, that $\omega(\partial_{r+t+i}|_y \wedge \partial_{r+t+j}|_y) = 0$ $(1 \leq i, j \leq t)$. – c) Proceeding as above, we have finally:

$$\omega(\partial_{r+i}|_y \wedge \partial_{r+t+j}|_y) = ([k_i, \overline{k}_j], f) = ([a_i, b_j], f) = \delta_{ij}.$$

Summing up, we have shown, that if $1 \leq i < j \leq s$,

$$\omega(\partial_i|_y \wedge \partial_j|_y) = \begin{cases} 1 & \text{if } r < i \leq r+t,\ j = i+t \\ 0 & \text{otherwise.} \end{cases}$$

We shall show below (cf. Lemma 24), that any polarization, as in Lemma 5, with respect to y_0, is invariant by G_{y_0}. Hence we can form the corresponding G-invariant complex distribution $\mathcal{H}$. Let us denote its value at $y \in O$ by H_y. We have thus, if $y = \sigma y_0$ $(\sigma \in G)$: $\tau_y(H_y) = \sigma(hy_0)$.

Lemma 11. *With notation as above we have for all $y \in Ey_0$:*

$$H_y = (\oplus_{j=1}^r \mathbf{C}(\partial_j|_y)) \oplus (\oplus_{j=1}^t \mathbf{C}(\partial_{r+j}|_y + i(\partial_{r+j+t}|_y))).$$

Proof: We have, as in the proof of the previous lemma

$$\begin{aligned}\tau_y(\partial_{r+k}|_y) &= a_k y_0 + \tfrac{1}{2}(l a_k + a_k l) y_0 \\ \tau_y(\partial_{r+t+k}|_y) &= b_k y_0 + \tfrac{1}{2}(l b_k + b_k l) y_0.\end{aligned}$$

We set $t_k = \partial_{r+k}|_y + i(\partial_{r+t+k}|_y)$ and thus $\tau_y(t_k) = (a_k + ib_k)y_0 + \frac{1}{2}(l(a_k + ib_k) + (a_k + ib_k)l)y_0$. Assume $y = \sigma y_0$ $(\sigma \in E)$. (i) We claim, that $(a_k + ib_k)y_0 \in \sigma(\mathfrak{h} y_0)$. To this end we note that, as in Lemma 7, we can write $E = E' \cdot H_1$. E' leaves $\mathfrak{h}'$ invariant by $(e')^2 \subseteq d' \subseteq \mathfrak{h}'$. Invariance with respect to H_1 is implied by the G_f-invariance of $\mathfrak{h}'$. We remark next, that $\mathfrak{h}'f = \sigma(\mathfrak{h}'f)$ since, by what we have just seen, both are equal to $(\mathfrak{h}')^\perp$. In this manner we can conclude, that $(a_k + ib_k)f \in \sigma(\mathfrak{h}'f)$ and thus there is $t \in \ker(T_y(\mu))$ and $h \in \mathfrak{h}'$ such that $(a_k + ib_k)y_0 = \sigma(\mathfrak{h}y_0) + \tau_y(t)$. – We observe now, that $\tau_y(\ker(T_y(\mu))) = \sigma(\mathfrak{h}_1 y_0)$. In fact, if t is given in $T_y(Ey_0)$ there is $l \in e$ such that $\tau_y(t) = ly$ and then $\tau_{\mu(y)}(T_y(\mu)t) = l\mu(y)$. Hence $t \in \ker(T_y(\mu))$ if and only if $l \in e \cap \mathfrak{g}_{\mu(y)} = \sigma(e \cap \mathfrak{g}_f)$. But, since $e = e' + \mathfrak{h}_1$ (cf. Remark 1) we can conclude that $e \cap \mathfrak{g}_f = \mathfrak{h}_1 + (e \cap \mathfrak{g}_f) = \mathfrak{h}_1 + \mathfrak{d}_f$. Hence we have $t \in \ker(T_y(\mu))$ if and only if, for some $l \in \mathfrak{h}_1$, $\tau_y(t) = \sigma(ly_0)$. – In this fashion we can infer, that there is $l \in \mathfrak{h}_1$ such that $(a_k + ib_k)y_0 = \sigma((h + l)y_0)$. By $\mathfrak{h} = \mathfrak{h}' + (\mathfrak{h}_1)_{\mathbf{C}}$ this implies $(a_k + ib_k)y_0 \in \sigma(\mathfrak{h}y_0)$. (ii) We claim next, that for some $l' \in d$ we have $(a_k l + la_k)y_0 = \sigma(l'y_0)$. In fact, since $a_k, l \in e'$ and $(e')^2 \subseteq d'$, $(a_k l + la_k)f$ belongs to $(e')^\perp = \sigma(e')^\perp = \sigma(d'f)$. As in (i) above, there is $l_1 \in \mathfrak{h}_1$ such that $(a_k l + la_k)y_0 = \sigma((l' + l_1)y_0)$ and thus it is enough to recall, that $d = d' + \mathfrak{h}_1$. (iii) Since $\tau_y(t_k) = (a_k + ib_k)y_0 + \frac{1}{2}((a_k + ib_k)l + l(a_k + ib_k))$ we derive form (i) and (ii), that $\tau_y(t_k) \in \sigma(\mathfrak{h}y_0)$ and hence $t_k \in H_y$. – (iv) We know from (i) in the proof of Lemma 10, that, for $1 \le j \le r$, $\tau_y(\partial_j|_y) \in \sigma(dy_0)$ and thus $\partial_j|_y \in H_y$. – (v) We conclude by (iii)–(iv) that

$$(\oplus_{j=1}^r \mathbf{C}(\partial_j|_y)) \oplus (\oplus_{j=1}^t \mathbf{C}(\partial_{r+j}|_y + i(\partial_{r+j+t}|_y)) \subseteq H_y.$$

But, by comparing the dimensions on both sides, we have the assertion of our lemma.

3. Some coordinates. For our later purposes, it will be convenient to have another system of coordinates on Ey_0. We introduce these in the following steps.

1) We recall, that $(e_j^*)_{1 \le j \le r}$ is a basis in $e^\perp$. We set $e_{r+k}^* = a_k y_0$, $e_{r+t+k}^* = b_k y_0 (1 \le k \le r)$; the system $(e_j^*)_{1 \le j \le s}$ is a basis in $d^\perp$.

2) Let $(e_i)_{1 \le i \le r} \subset \mathfrak{g}$ be such, that $(e_i, e_j^*) = \delta_{ij}$ $(1 \le i, j \le r)$. We can clearly assume, that $(e_i, e_{r+k}^*) \equiv 0$ $(1 \le k \le 2t)$. Setting $e_{r+j} = -b_j$, $e_{r+t+j} = a_j$ $(1 \le j \le t)$ we obtain $(e_i, e_j^*) = \delta_{ij}$ $(1 \le i, j \le s)$.

3) We note that a) If $1 \le j \le t$ we have $y_{r+j} = -(b_j, y - y_0) = (e_{r+j}, y) - (e_{r+j}, y_0)$ and $y_{r+t+j} = (a_j, y - y_0) = (e_{r+t+j}, y) - (e_{r+t+j}, y_0)$.

4) We now define new coordinates $(y_j')_{1 \le j \le s}$ on Ey_0 as follows: $y_j' = (e_j, y) - (e_j, y_0)$ $(1 \le j \le r)$ and $y_i' = y_i$ if $i > r$. In fact, we have for $1 \le i \le r : y_i' = y_i + f_i(y_{r+1}, \ldots, y_s)$ $(1 \le i \le r)$, where f_i is a

polynomial, homogeneous of degree 2. In this fashion, by 3), we have for all k such that $1 \leq k \leq s$: $y_j = (e_j, y) - (e_j, y_0)$ $(1 \leq j \leq s)$.

5) Let ∂'_j be the operator of differentiation according to y'_j $(1 \leq j \leq s)$. We claim, that

$$\omega(\partial'_i|_y \wedge \partial'_j|_y) = \begin{cases} 1 & \text{if } r < i \leq r+t,\ j = i+t \\ 0 & \text{otherwise.} \end{cases}$$

In fact, we have by Lemma 10: $\omega = \sum_{j=1}^{t} dy_{r+j} \wedge dy_{r+t+j}$ and hence it is enough to recall, that $y'_j = y_j (j > r)$.

6) We note finally, that for any $y \in Ey_0$ we have

$$H_y = (\oplus_{j=1}^{r} \mathbf{C}(\partial'_j|_y)) \oplus (\oplus_{k=1}^{t} \mathbf{C}(\partial'_{r+k}|_y + i(\partial_{r+t+k}|_y)).$$

To see this we observe that, by Lemma 11, the analogous relations, if the primes are dropped, are valid and we have clearly $\partial_i|_y = \partial'_i|_y$ and $\tau_y(\partial_i|_y) = \tau_y(\partial'_i|_y) + h_i(y)$ for $r < i \leq s$, where $h_i(y) \in e^{\perp}$. Hence it suffices to observe, that

$$\tau_y(\oplus_{j=1}^{r} \mathbf{R}(\partial_j|_y)) = e^{\perp}.$$

We set $\tilde{E} = EG_{y_0}$. Since Ey_0 is closed in $\mathfrak{g}^*$, $\tilde{E}$ is a closed subgroup of G.

Let us consider now the principal $\tilde{E}$-bundle $G \xrightarrow{\pi} G/\tilde{E}$. Let V be an open subset in $G/\tilde{E}$, such that 1) There is a diffeomorphism τ of V with an open subset of $\mathbf{R}^r$, 2) There is a local section $\gamma \in \Gamma_V(G)$. We set $\mathcal{O} = \pi^{-1}(V)$ and propose to introduce coordinates $(y_j)_{1 \leq j \leq d}$ on $\mathcal{O}$ with the following properties:

H0) For $1 \leq i < j \leq s$ and $y \in \mathcal{O}$:

$$\omega(\partial_i|_y \wedge \partial_j|_y) = \begin{cases} 1 & \text{if } r < i \leq r+t,\ j = i+t \\ 0 & \text{otherwise.} \end{cases}$$

H1) For any $y \in \mathcal{O}$:

$$H_y = (\oplus_{j=1}^{r} \mathbf{C}(\partial_j|_y)) \oplus (\oplus_{k=1}^{t} \mathbf{C}(\partial_{r+k}|_y + i(\partial_{r+t+k}|_y)).$$

H2) Writing $\mathcal{E}^0$ for the totality of elements of $\mathcal{E}(\mathcal{O})$ annihilated by $\mathcal{H}$, we have $X_{y_j} E^0 \subseteq E^0$ $(1 \leq j \leq s)$. (We recall, that given $\varphi \in \mathcal{E}(\mathcal{O})$, X_φ is defined by $d\varphi = -\iota(X_\varphi)\omega_{\mathcal{O}}$.)

Below by coordinates on Ey_0 we mean those introduced in 4) above. – Our definition of coordinates on $\mathcal{O}$ is as follows: If $1 \leq j \leq s$, we set $y_j = ((\gamma(\pi(y)))^{-1}y)_j$ and for $j > s : y_j = (\tau(\pi(y)))_{j-s}$. By what we saw in 4) above we have in this manner: $y_j = (\gamma(\pi(y))e_j, y) - (e_j, y_0)$ $(1 \leq j \leq s)$. – Below ω_O will stand for the canonical 2-form on $O = Gy_0$, and we continue to write ω for its pullback to Ey_0.

LEMMA 12. *In the coordinates $(y_j)_{1\leq j\leq d}$ as above we have the following relations (cf. H0) above. If $1 \leq i \leq j \leq s$:*

$$\omega_O(\partial_i|_y \wedge \partial_j|_y) = \begin{cases} 1 & \text{if } r < i \leq r+t,\ j = i+t \\ 0 & \text{otherwise.} \end{cases}$$

PROOF: Let us put for $y \in \mathcal{O} : y' = (\gamma(\pi(y)))^{-1}y \in Ey_0$; we have then: $\partial_j|_y = T_y(\gamma(\pi(y)))(\partial_j|_{y'})$ $(1 \leq j \leq s)$. The desired conclusion is implied by (i) The G-invariance of ω_O, (ii) The validity of the analogous relations on Ey_0 (cf. 5) above).

LEMMA 13. *With notation as above, we have for all $y \in \mathcal{O}$ (cf. H1) above)*

$$H_y = (\oplus_{j=1}^r \mathbf{C}(\partial_j|_y)) \oplus (\oplus_{k=1}^t \mathbf{C}(\partial_{r+k}|_y + i(\partial_{r+t+k}|_y)).$$

PROOF: To this end it is enough to note, that (i) If $y = \gamma(\pi(y))y'$ $(y' \in Ey_0)$ we have: $H_y = \gamma(\pi(y))H_{y'}$, (ii) By 6) above, the relation to be established is valid on Ey_0.

LEMMA 14. *We have for $1 \leq j \leq s$: $X_{y_j}E^0 \subseteq E^0$ (cf. H2) above).*

PROOF: (i) Given $1 \in \mathfrak{g}$ fix, we write ψ_l for the function $y \mapsto (l, y)$ on O, as well as for its restriction to $\mathcal{O}$. This being so we prove first, that any y_j (j fix $l \leq j \leq s$) is of the form constant + linear combination, with coefficients in $\pi^*(E_V(G/\tilde{E}))$, of $(\psi_l)_{l\in g}$. In fact, as noted earlier, we have $y_j \equiv (\gamma(\pi(y))e_j, y) - (e_j, y_0)$. Let $(l_k)_{1\leq k\leq m}$ be a basis in $\mathfrak{g}$. Then there is a system $(f_k)_{1\leq k\leq m} \subset E(V)$ such that $\gamma(v)e_j = \sum_{k=1}^m f_k(v)l_k$ $(v \in V)$. But then we have also:

$$y_j = (\gamma(\pi(y))e_j, y) - (e_j, y_0) = \sum_{k=1}^m (\pi^* f_k)(y)\psi_{l_k}(y) - (e_j, y_0)$$

proving our statement.

(ii) Given $l \in \mathfrak{g}$ fix, we write L for the vector field on O defined by $L_y = (d/dt)\exp(-tl)y|_{t=0}$. We have clearly $L\mathcal{E}^0 \subseteq \mathcal{E}^0$ $(l \in \mathfrak{g})$. As it is well-known and easy to verify, we have $X_{\psi_l} = L$. Hence also $X_{\psi_l}\mathcal{E}^0 \subseteq \mathcal{E}^0$ $(l \in \mathfrak{g})$. – Let $\varphi \in \mathcal{E}^0$ be fixed. By (i) it is enough to show, that $X_{\varphi\psi_l}\mathcal{E}^0 \subseteq \mathcal{E}^0$. But we have $X_{\varphi\psi_l} = \varphi X_{\psi_l} + \psi_l X_\varphi$, and shall show below (cf. Lemma 14; note, that it does not depend on H2)), that $\{\mathcal{E}^0, \mathcal{E}^0\} = 0$, which completes the proof of Lemma 14. –

4. Some properties of symplectic structures. Here our objective is to study certain aspects of the following situation, motivated by the end of the previous section. Let U be an open subset of $\mathbf{R}^r$ and let us write $\mathcal{O}$ for $\mathbf{R}^s \times U$ ($s = r + 2t$). We assume to be given a real, closed, nondegenerate 2-form ω on O and a complex distribution $\mathcal{H}$ with the following properties:

H0) For $1 \leq i, j \leq s$ we have:

$$\omega(\partial_i|_y \wedge \partial_j|_y) = \begin{cases} 1 & \text{if } r < i \leq r+t,\ j = i+t \\ 0 & \text{otherwise.} \end{cases}$$

H1) We assume that $\mathcal{H}$ is Lagrangean and that for all $y \in \mathcal{O}$:

$$H_y = (\oplus_{j=1}^r \mathbf{C}(\partial_j|_y)) + (\oplus_{k=1}^t \mathbf{C}(\partial_{r+j}|_y + i(\partial_{r+t+j}|_y))).$$

Let us write $z_k = x_{r+k} + iy_{r+t+k}$ and

$$\partial_{\bar{z}_k} = \frac{1}{2}(\partial_{r+k} + i\partial_{r+t+k})\ (1 \leq k \leq t).$$

Then H1) implies, that

$$H_y = (\oplus_{j=1}^r \mathbf{C}(\partial_j|_y)) \oplus (\oplus_{k=1}^t \mathbf{C}\partial_{\bar{z}_k}|_y).$$

We denote by $\mathcal{E}^0$ the collection of functions in $\mathcal{E}(\mathcal{O})$, which are annihilated by all $(H_y)_{y\in\mathcal{O}}$. We observe, incidentally, that we have $\varphi \in \mathcal{E}^0$ if and only if there is $f \in \mathcal{E}(\mathcal{O})$ such that $\varphi(y) \equiv f(y_{r+1}, \ldots, y_d)$ and f is analytic in $(z_k)_{1\leq k\leq t}$.

H2) We assume:

$$X_{y_j}\mathcal{E}^0 \subseteq \mathcal{E}^0\ (1 \leq j \leq s).$$

Let $V_{\mathcal{H}}(\mathcal{O})$ be the collection of all vector fields, with values in $\mathcal{H}$, on $\mathcal{O}$. We set

$$\mathcal{E}^1 = \{\varphi\,;\ \varphi \in \mathcal{E}(O)\,,\ [X_\varphi, V_{\mathcal{H}}(O)] \subseteqq V_H(O)\}.$$

Our objective below is to show, that $\mathcal{E}^1$ is freely generated, over $\mathcal{E}^0$, by $(y_j)_{1\leq j\leq r}$ and $(\bar{z}_k)_{1\leq k\leq t}$.

Lemma 15. *Assume, that φ is in $\mathcal{E}(\mathcal{O})$. Then X_φ belongs to $V_{\mathcal{H}}(\mathcal{O})$ if and only if φ lies in $\mathcal{E}^0$.*

Proof: (i) We have for any $\varphi \in \mathcal{E}(\mathcal{O})$

$$X_\varphi = \sum_{j=1}^r \varphi_{y_j} X_{y_j} + \sum_{k=1}^t (\varphi_{z_k} X_{z_k} + \varphi_{\bar{z}_k} X_{\bar{z}_k}) + \sum_{j>s} \varphi_{y_j} X_{y_j}.$$

(ii) We observe, that the system $(X_{y_j}|_y)_{j>s}, (X_{z_k}|_y)_{1\leq k\leq t}$ is a basis in H_y. In fact a) It is linearly independent by $\omega(X_{y_i} \wedge \partial_k) = \delta_{ik}$ $(1 \leq i, k \leq d)$. b) It is orthogonal, with respect to ω_y, to H_y. Since the latter, by H1), is maximal self-orthogonal, it does, indeed, form a basis in H_y. – (iii) From here we can conclude, that $X_\varphi \in \mathcal{V}_\mathcal{H}(\mathcal{O})$, if and only if $\varphi_{y_j} \equiv 0$ $(1 \leq j \leq r)$ and $\varphi_{\bar{z}_k} \equiv 0$ $(1 \leq k \leq t)$ which is equivalent to $\varphi \in \mathcal{E}^0$.

LEMMA 16. *$\mathcal{E}^0$ is maximal abelian with respect to $\{\ ,\ \}$.*

PROOF: Let φ and ψ be elements in $\mathcal{E}^0$. Since $\{\varphi, \psi\} \equiv \omega(X_\varphi \wedge X_\psi)$ and X_φ, X_ψ, by the previous lemma, are in $\mathcal{V}_\mathcal{H}(\mathcal{O})$, the fact that $\mathcal{E}^0$ is abelian is implied by the selforthogonality of H_y with respect to ω_y for all $y \in \mathcal{O}$. Suppose then, that $f \in \mathcal{E}(\mathcal{O})$ is such, that $\{f, \varphi\} \equiv 0$ for all φ in $\mathcal{E}^0$; we have to show, that $f \in \mathcal{E}^0$. In fact, proceeding as above and using the maximal selforthogonality of H_y (cf. H1)) we conclude first, that $X_f \in \mathcal{V}_\mathcal{H}(\mathcal{O})$. From here, the desired result is implied by Lemma 15.

LEMMA 17. *$\mathcal{E}^1$ is the normalizer of $\mathcal{E}^0$ in $\mathcal{E}(\mathcal{O})$.*

PROOF: Let ψ be a smooth function on $\mathcal{O}$. It belongs to $\mathcal{E}^1$ if and only if, for any $\varphi \in \mathcal{E}^0, [X_\varphi, X_\psi] = X_{\{\varphi,\psi\}}$ is in $\mathcal{V}_\mathcal{H}(\mathcal{O})$. By Lemma 15, however, this is equivalent to $\{\varphi, \psi\} \in \mathcal{E}^0$. Hence $\mathcal{E}^1$ is equal to the normalizer of $\mathcal{E}^0$ in $\mathcal{E}(\mathcal{O})$.

LEMMA 18. *$\mathcal{E}^1$ is an $\mathcal{E}^0$-module.*

PROOF: Let $\varphi \in \mathcal{E}^0$ and $\psi \in \mathcal{E}^1$ be fixed elements. By Lemma 16 and 17 it is enough to show, that $\{\varphi\psi, f\}$ is in $\mathcal{E}^0$ for all $f \in \mathcal{E}^0$. We have $\{\varphi\psi, f\} = \psi\{\varphi, f\} + \varphi\{\psi, f\} = \varphi\{\psi, f\}$ by Lemma 16. But Lemma 17 implies, that $\{\psi, f\}$ is in $\mathcal{E}^0$, proving our lemma.

We recall, that H2) stipulates, that $X_{y_j}\mathcal{E}^0 \subseteq \mathcal{E}^0 (1 \leq j \leq s)$. *We have not used this so far.*

LEMMA 19. *$(y_j)_{1\leq j\leq r}, (\bar{z}_k)_{1\leq k\leq t}$ are in $\mathcal{E}^1$.*

PROOF: This is clear from H2) and Lemma 17.

LEMMA 20. *We have $\{z_k, \bar{z}_j\} = \frac{2}{i}\delta_{kj} (1 \leq k, j \leq t)$.*

PROOF: (i) Setting $\partial_{\bar{z}_k} = \frac{1}{2}(\partial_{r+k} + i\partial_{r+t+i})$ and $\partial_{z_k} = \frac{1}{2}(\partial_{r+k} - i\partial_{r+t+k})$ $(1 \leq k \leq t)$ we conclude by H0), that

$$\omega(\partial_{z_j} \wedge \partial_{z_k}) \equiv 0 \equiv \omega(\partial_{\bar{z}_j} \wedge \partial_{\bar{z}_k})$$

and that

$$\omega(\partial_{z_j} \wedge \partial_{\bar{z}_k})) = 2i\delta_{jk},\ (1 \leq j, k \leq t).$$

In this manner we have also

$$\omega = \frac{i}{2}\sum_{j=1}^{t}(dz_j \wedge \overline{dz_j}) + \rho$$

where ρ is in the ideal of differential forms generated by $(dy_j)_{j>s}$.

(ii) We conclude from here, that

$$-dz_j = \iota(X_{z_j})\omega = -\frac{i}{2}(\sum_{k=1}^{t}(\iota(X_{z_j})\overline{dz}_k)dz_k) + \iota(X_{z_j})\rho.$$

However, by $(y_j)_{j>s} \subset \mathcal{E}^0$ we have $\iota(X_{z_j})\rho = 0$. (iii) We obtain in this fashion

$$dz_j = \frac{i}{2}(\sum_{k=1}^{t}\{z_j, \overline{z}_k\}dz_k)$$

or $\{z_j, \overline{z}_k\} = \frac{2}{i}\delta_{jk}$, completing the proof of our lemma.

Lemma 21. *The operators $(X_{z_i})_{1\leq i\leq t}, (X_{y_j})_{j>s}$ are linear combinations, with coefficients in $\mathcal{E}^0$, of $(\partial_j)_{1\leq j\leq r}, (\partial_{\overline{z}_k})_{1\leq k\leq t}$.*

Proof: Ad X_{y_j}) ($j > s$): We have by Lemma 15:

$$X_{y_j} = \sum_{k=1}^{r}\{y_j, y_k\}\partial_k + \sum_{u=1}^{t}\{y_j, \overline{z}_u\}\partial_{\overline{z}_u}$$

form where our conclusion follows by noting, that a) We have $\{y_j, \overline{z}_u\} \equiv \{y_j, z_u\} \equiv 0$, b) By H2): $\{y_j, y_k\} = -X_{y_k}y_j \in \mathcal{E}^0$. Ad X_{z_k}) ($1 \leq k \leq t$): We have by Lemma 15 and 20 that

$$X_{z_k} = \sum_{u=1}^{r}\{z_k, y_u\}\partial_u + \frac{2}{i}\partial_{\overline{z}_k}.$$

But, again by H2), $\{z_k, y_u\} \in \mathcal{E}^0$ ($1 \leq u \leq r$), completing the proof of our lemma.

Lemma 22. *The operators $(\partial_k)_{1\leq k\leq r}, (\partial\overline{z}_v)_{1\leq v\leq t}$ are linear combinations, with coefficients in $\mathcal{E}^0$, of $(X_{y_j})_{j>s}$ and $(X_{z_k})_{1\leq k\leq t}$.*

Proof: This is implied by the previous lemma in conjunction with the observation, that either $(X_{y_j})_{j>s}$ and $(X_{z_k})_{1\leq k\leq t}$ or $(\partial_k)_{1\leq k\leq r}$ and $(\partial\overline{z}_v)_{1\leq v\leq t}$ is a basis in H_y ($y \in \mathcal{O}$).

LEMMA 23. *Given ψ in $\mathcal{E}^1$, there is $(\varphi_j)_{0\leq j\leq r}$ and $(\omega_k)_{1\leq k\leq t}$ in $\mathcal{E}^0$ such that*

$$\psi = \varphi_0 + \sum_{j=1}^{r} y_j\varphi_j + \sum_{k=1}^{t} \overline{z}_k\omega_k.$$

PROOF: This is clear, since $(\partial_k\psi)_{1\leq k\leq r}$, $(\partial_{\overline{z}_v}\psi)_{1\leq v\leq t}$ are in $\mathcal{E}^0$ by Lemma 17 and 22.

5. The isomorphism. The principal objective of this section is the proof of Pedersen's isomorphism (cf. Theorem 1). Important for this will be Lemma 31 and 32. We resume with the notation of Section 2. Let $\mathfrak{h}$ be as in Lemma 5, and H the connected subgroup, determined by $\mathfrak{h}$, of $G_{\mathbf{C}}$.

LEMMA 24. *GH is a closed submanifold of $G_{\mathbf{C}}$.*

PROOF: (i) We prove first, that $E_{\mathbf{C}} = EH$. In fact, we recall that, by definition $\mathfrak{h} = \mathfrak{h}' + (\mathfrak{h}_1)_{\mathbf{C}}$. Let H' be the connected subgroup of $L_{\mathbf{C}}$ corresponding to $\mathfrak{h}'$. We claim, that $E'_{\mathbf{C}} = E'\cdot H'$. To this end we note, that trivially $e' + \mathfrak{h}' = \overline{\mathfrak{h}}' + \mathfrak{h}' = e_{\mathbf{C}}$. In addition e' normalizes $\mathfrak{h}'$, since $[e',\mathfrak{h}'] \subset d'_{\mathbf{C}} \subset \mathfrak{h}'$. Hence indeed $E'_{\mathbf{C}} = E'\cdot H'$. – We have the semi-direct product decompositions $E = E'\cdot H_1$, $H = H'(H_1)_{\mathbf{C}}$, whence also $E_{\mathbf{C}} = E'_{\mathbf{C}}\cdot(H_1)_{\mathbf{C}}$. Thus, by our previous remark we can conclude, that $E_{\mathbf{C}} = E'H'(H_1)_{\mathbf{C}} = E'H = E'H_1\cdot H = E\cdot H$, proving our claim, that $E_{\mathbf{C}} = E\cdot H$. – (ii) If we consider $G_{\mathbf{C}}$ as a principal right $E_{\mathbf{C}}$-bundle, since spacewise the latter is diffeomorphic to $\mathbf{R}^n$, there is always a global section $G_{\mathbf{C}}/E_{\mathbf{C}} \to G_{\mathbf{C}}$. Below we shall need a slightly more precise statement, and therefore start from the beginning. We shall show next, that there is a global section ϑ such that $\vartheta(G/E) \subset G$. In fact, let $e' = \mathfrak{a}_0 \subset \mathfrak{a}_1 \subset \cdots \subset \mathfrak{a}_\rho = d$ be a sequence of subalgebras, such that $\dim(\mathfrak{a}_j/\mathfrak{a}_{j-1}) \equiv 1$. We choose $l_j \in \mathfrak{a}_j - \mathfrak{a}_{j-1}$ and set $g'_j(t) \equiv \exp(tl_j)$ $(1\leq j\leq\rho)$. If $T = (t_1,\ldots,t_\rho) \in \mathbf{C}^\rho$ we write $g'(T) = g_1(T_1)\ldots g_\rho(t_\rho)$. Then $(a,T) \to ag'(T)$ is an analytic isomorphism $E'_{\mathbf{C}} \times \mathbf{C}^\rho \to L_{\mathbf{C}}$. By $E = E'H_1$, we can even claim, that $(b,T) \mapsto bg'(T)$ is of the same kind from $E_{\mathbf{C}} \times \mathbf{C}^\rho$ onto $L_{\mathbf{C}}\cdot E_{\mathbf{C}}$, and the restriction to $E\times\mathbf{R}^\rho$ is a bijection with LE. – We write again $\mathfrak{c} = \mathfrak{d} + e$. Let $(k_u)_{1\leq u\leq\sigma}$ be a basis in $\mathfrak{g}$, supplementary to $\mathfrak{c}$. We set $g''_j(t) = \exp(tk_j)$ and, if $T = (t_1,\ldots,t_\sigma) \in \mathbf{C}^\sigma$, $g''(T) = g''_1(t_1)\ldots g''_\sigma(t_\sigma)$. Finally, if $r = \rho+\sigma$, $T = (T',T'') \in \mathbf{C}^r$, we write $g(T) = g'(T')g''(T'')$. Then $(a,T) \to ag(T)$ is an analytic isomorphism from $E_{\mathbf{C}} \times \mathbf{C}^r$ onto $G_{\mathbf{C}}$, and its restriction to $E\times\mathbf{R}^r$ is a bijection with G. – The map $T \mapsto g(T)$ modulo $E_{\mathbf{C}}$ being an analytic isomorphism $\mathbf{C}^r \to G_{\mathbf{C}}/E_{\mathbf{C}}$ we can form its inverse η. Then finally, $\vartheta = (g\circ\eta)^{-1}$ will be as required. – (iii) The map $(v,a) \to \vartheta(v)a$ is an analytic isomorphism $G_{\mathbf{C}}/E_{\mathbf{C}} \times E_{\mathbf{C}} \to G_{\mathbf{C}}$. To conclude the proof of our lemma, it is now enough to note, that the image of $G/E \times E_{\mathbf{C}}$ is equal to $GE_{\mathbf{C}}$ which, by what we saw earlier, is the same as $GEH = GH$.

Lemma 25. *G_f leaves invariant $\mathfrak{h}, d$ and e.*

Proof: We have $\mathfrak{h} = \mathfrak{h}' + (\mathfrak{h}_1)_{\mathbf{C}}$ and $\mathfrak{h}'$, by construction, is invariant under G_f. We show the same for $\mathfrak{h}_1$. – Since G is solvable, given $a \in G, l \in \mathfrak{g}$ we have $al - l \in \mathfrak{d}$. If, in addition, $a \in G_f$ and $l \in \mathfrak{g}_f$, then also $al - l \in \mathfrak{d}_f \subset \mathfrak{h}_1$ by Lemma 1, proving our statement. Hence $\mathfrak{h} = \mathfrak{h}' + (\mathfrak{h}_1)_{\mathbf{C}}$ is invariant under G_f. – The statements relative to e and d follow from $d = \mathfrak{h} \cap \mathfrak{g}$ and $e = (\mathfrak{h} + \overline{\mathfrak{h}}) \cap \mathfrak{g}$.

Let $\overline{G}_g$ be an open subgroup of G_g. By the previous lemma we can form the subgroups $\overline{D} = D\overline{G}_g$, $\overline{E} = E\overline{G}_g$ and $\overline{H} = H \cdot \overline{G}_g$.

Lemma 26. *The subgroups $\overline{D}$, $\overline{E}$ and $\overline{H}$ are closed.*

Proof: Below we consider only $\overline{H}$. Since $\overline{D} = G \cap \overline{H}$, this settles the case of $\overline{D}$. That of $\overline{E}$ is similar to $\overline{H}$. – (i) We show first, that HG_g is closed. To this end we start by noting, that Hg is closed. In fact, by $H = H'(H_1)_{\mathbf{C}}$, proceeding as in the proof of Lemma 6 we conclude, that $Hg = H'(H_1)_{\mathbf{C}}g = H'(g + \mathfrak{c}_{\mathbf{C}}) = H'g + \mathfrak{c}_{\mathbf{C}}$. Hence it is enough to note, that $H'(g|\mathfrak{c}_{\mathbf{C}})$ is closed. – From this we derive first, that $H(G_{\mathbf{C}})_g$ is closed. Hence, by Lemma 14, $HG \cap H(G_{\mathbf{C}})_g = HG_g$ is also closed. – (ii) From what we have just seen we conclude, that Hg is also simply connected, and therefore $H \cap (G_{\mathbf{C}})_g = ((G_{\mathbf{C}})_g)_0$. Thus we obtain:

$$H \cap G_g = (H \cap (G_{\mathbf{C}})_g) \cap G = (G_g)_0.$$

Hence the connected components of HG_g are in bijection with those of G_g, proving our lemma.

Lemma 27. *There is a closed, analytic submanifold $H' \subset H$ such that we have $E_{\mathbf{C}} = H' \cdot H$ setwise.*

Proof: We consider $E_{\mathbf{C}}$ as a principal (right) H-bundle. Since H is diffeomorphic to $\mathbf{R}^n$, there is a global section $\vartheta : E_{\mathbf{C}}/H \to E_{\mathbf{C}}$. It suffices to define $H' = \vartheta(E_{\mathbf{C}}/H)$.

Before proceeding we recall the following. Given a Lie group G and closed subgroups $G \supset T \supset K$, we have canonically a fibration $G/K \to G/T$ with the fiber T/K. By Lemma 26, we have the sequence of closed subgroups $G \supset \overline{E} \supset \overline{D} \supset \overline{G}_g$. We set $\overline{O} = G/\overline{G}_g$, $\overline{O}/\overline{D} = G/\overline{D}$ and

$\overline{O}/\overline{E} = G/\overline{E}$. We obtain thus the following diagram:

$$\begin{array}{ccccc} G & & & & \\ & \searrow^{q} & & & \\ & & \overline{O} & \xrightarrow{p} & \overline{O}/\overline{E} \\ & \nearrow_{\rho} & & & \\ \overline{O} & & & & \end{array}$$

We put also $\delta = p \circ \rho$, $\pi = p \circ q$. The fiber of the right half is $\overline{E}/\overline{D} \simeq E/D \simeq E'/D'$, and thus has a canonical complex structure. We denote by $A(\overline{O}/\overline{D})$ the totality of all those smooth functions on $\overline{O}/\overline{D}$, the restriction of which to any of the fibers is analytic.

Let V be an open subset of $G/\overline{E}$, such that there is a local section ϑ to G. We write $\tilde{V}' = \vartheta(V) \cdot H'$, where H' is as in Lemma 27. $\tilde{V}'$ is a submanifold of GH and is setwise equal to $\vartheta(V) \times H_1$; whence a partially holomorphic structure on V'. – Let j be the canonical morphism $G_{\mathbf{C}} \to G_{\mathbf{C}}/\overline{H}$ and observe, that $j : GH = G\overline{H} \to G\overline{H}/\overline{H} = \overline{O}/\overline{D}$.

LEMMA 28. *With the above notation let us set $W = p^{-1}(V)$. Then $j|\tilde{V}'$ is a diffeomorphism $\tilde{V}' \to W$ such that*

$$\begin{array}{ccc} \tilde{V}' & \xrightarrow{\quad j \quad} & W \\ \delta \circ \mathrm{pr}_1 \searrow & & \swarrow p \\ & V & \end{array}$$

and j is analytic on the fibers of $\tilde{V}' \to V$.

PROOF: (i) We start by noting that, if $\overline{E}_{\mathbf{C}} = E_{\mathbf{C}}\overline{G}_g$, we have a) By Lemma 27, that $\overline{E}_{\mathbf{C}} = H' \cdot \overline{H}_1$ b) By (i) in the proof of Lemma 24: $\overline{E}_{\mathbf{C}} = E \cdot \overline{H}$. – (ii) Let us prove first, that $j|\tilde{V}'$ is bijective. a) We have by (i):

$$j(\tilde{V}') = j(\vartheta(V) \cdot H') = j(\vartheta(V) \cdot \overline{E}_{\mathbf{C}}) = j(\vartheta(V) \cdot \overline{E}) = q(\delta^{-1}(V)) = W;$$

hence $j|\tilde{V}'$ is surjective. b) Suppose, that $x, x_1 \in V$, $h', h_1' \in H'$ are such, that $j(\vartheta(x)h') = j(\vartheta(x_1)h_1')$. Then there is $\overline{h}$ in $\overline{H}$ with $\vartheta(x_1)h_1' = \vartheta(x)h'\overline{h}$. Since $G \cap \overline{E}_{\mathbf{C}} = \overline{E}$ we conclude from this, that $x = x_1$ and $h_1' = h' \cdot h$. But then also $\overline{h} \in H$ and thus $h_1' = h'$. Hence $j|\tilde{V}'$ is bijective. – (iii) To establish the diagram, we have to prove, that for any $x \in V$, $h' \in H'$ we have $p(j(\vartheta(x)h')) = x$. We can write: $h' = ah$ $(a \in E, h \in \overline{H})$ and thus the left-hand-side is equal to $p(j(\vartheta(x)a)) = (p \circ q)(\vartheta(x)a) =$

$\delta(\vartheta(x)a) = x$. – (iv) Since $j|\tilde{V}'$ is a submersive bijection $\tilde{V}' \to W$, it is also a diffeomorphism. (v) $j : \tilde{V}' \to W$ is holomorphic on the fibers by $\overline{E}_{\mathbf{C}}/\overline{H} = \overline{E}\overline{H}/\overline{H} = \overline{E}/\overline{D}$ proving our lemma.

Let again g be a fixed element of $\mathfrak{g}^*$. Below by $\overline{G}_g$ we shall mean the reduced stabilizer of g (cf. [4], Definition 4.1, p. 492). We recall, that $\overline{G}_g = G_g$ occurs if and only if the de Rham class of the canonical 2-form ω_O on the coadjoint orbit O is integral. – There is a character φ of $\overline{G}_g$, such that $\varphi_* = 2\pi i(g|\mathfrak{g}_g)$, and also a morphism χ of $\overline{H}$ into $\mathbf{C}^*$ such that $\chi_* = 2\pi i(g|\mathfrak{h})$ and $\chi|\overline{G}_g = \varphi$. We write χ_0 for the restriction of χ to $\overline{D}$. This being so we set

$$\mathcal{E}(G,\chi_0) = \{f; f \in \mathcal{E}(G) \text{ s.t. } f(x\delta) \equiv (\chi_0(\delta))^{-1} f(x)\ (\delta \in \overline{D}, x \in G)\}.$$

We recall, incidentally, that the motivation for introducing $\mathcal{E}(G,\chi_0)$ is that there is a canonical bijection between its elements and the totality of all smooth sections in the complex line bundle $G \times^{\overline{D}} \mathbf{C}$. – Since $\overline{D} = G \cap \overline{H}$, given $f \in E(G,\chi_0)$, there is a function F on GH well-determined by the condition, that $F(xh) = (\chi(h))^{-1} f(x)$ $(x \in G, h \in H)$. We write

$$\begin{aligned}\mathcal{E}(G,\chi) = \{f\,; f \in \mathcal{E}(G,\chi_0) \text{ s.t. for any } a_0 \text{ fix in } G\\ a \mapsto F(a_0 a) \text{ is holomorphic on } E_{\mathbf{C}}\}.\end{aligned}$$

We recall also, that $E(G,\chi)$ corresponds to the collection of all polarized sections. – For our purposes below, it is convenient to introduce "localized" versions of these spaces as follows. Let V be as in Lemma 28. We set $\tilde{V} = \delta^{-1}(V)$ and define

$$\mathcal{E}_{\tilde{V}}(G,\chi_0) = \{f; f \in \mathcal{E}(\tilde{V}) \text{ s.t. } f(x\delta) \equiv \chi_0(\delta)^{-1} f(x)\ (x \in G,\ \delta \in \overline{D})\}.$$

Similarly as earlier, given $f \in \mathcal{E}_{\tilde{V}}(G,\chi_0)$ there is F well-determined on $\tilde{V}H = \tilde{V}'\overline{H}$ by $F(xh) \equiv \chi(h)^{-1} f(x)$ $(x \in V, h \in \overline{H})$. Next we set

$$\begin{aligned}\mathcal{E}_{\tilde{V}}(G,\chi) = \{f\,; f \in \mathcal{E}_{\tilde{V}}(G,\chi_0) \text{ s.t. for all } v \in V,\\ a \mapsto F(\vartheta(v)a) \text{ is holomorphic on } E_{\mathbf{C}}\}.\end{aligned}$$

We write also $A_W(\overline{O}/\overline{D})$ for the totality of all smooth functions on W, the restrictions of which to the fibers of $\overline{O}/\overline{D} \to \overline{O}/\overline{E}$ are analytic.

By Lemma 28 we can form the map $\sigma = (j^{-1}|\tilde{V}') : W \to \tilde{V}'$. For $s \in VH$ we set $s = \sigma(j(s))h(s)$. Note, that the smooth map $h : \tilde{V}H \to \overline{H}$ is analytic on the $E_{\mathbf{C}}$-fibers. – This being so 1) Given $f \in \mathcal{E}_{\tilde{V}}(G,\chi)$ put $(Af)(x) \equiv f(\sigma(x))$. Note, that $Af \in A_W(\overline{O}/\overline{D})$, 2) For g in $A_W(\overline{O}/\overline{D})$ we set $(Bg)(s) \equiv \chi(h(s)^{)}-1g(j(s))$ $(s \in \tilde{V}H)$. Then

$Bg \in E_{\tilde{V}}(G,\chi)$. – We leave to the reader the easy verification of the fact, that $A : \mathcal{E}_{\tilde{V}}(G,\chi) \to A_W(\overline{O}/\overline{D})$ is a linear isomorphism, the inverse of which is equal to B. – We write γ and γ' for the left regular representation of G on $\overline{O}$ and $\overline{O}/\overline{D}$ resp. d_γ and $d\gamma'$ act on $\mathcal{E}_{\tilde{V}}(G,\chi)$ and $A_W(\overline{O}/\overline{D})$. In fact, if $l \in \mathfrak{g}$ and $s \in \tilde{V}H$, then $\exp(tl)s$, for t close enough to zero, lies again in $\tilde{V}H$.

LEMMA 29. *Given $z \in \mathfrak{g}$, there is $\alpha_z \in A_W(\overline{O}/\overline{D})$ such that $Ad\gamma(z)A^{-1} = d\gamma'(z) + \alpha_z$.*

PROOF: If $s \in \tilde{V}H$, we write below: $j(s) = \dot{s}$. For $g \in A_W(\overline{O}/\overline{D})$ we have for all $s \in \tilde{V}H$:

$$\gamma(\exp(tz))(A^{-1}g)(s) \equiv \chi(h(\exp(-tz)s))^{-1} \cdot g(\exp(-tz)\dot{s}).$$

We set $\alpha'_z(s) = (d/dt)(\chi(h(\exp(-tz)s)))^{-1}|_{t=0}$ and note, that if $a_0 \in \vartheta(V)$ is fix, the map $a \to \alpha'_z(a_0 \cdot a)$ is analytic on $E_{\mathbf{C}}$. We obtain:

$$d\gamma(z)(A^{-1}g)(s) \equiv \alpha'_z(s)g(\dot{s}) + (\chi(h(s)))^{-1} \cdot (d\gamma'(z)g)(\dot{s}).$$

Setting $\alpha_z(x) \equiv \alpha'_z(\sigma(x))$, we have $\alpha_z \in A_W(\overline{O}/\overline{D})$. Observing, that $h(\sigma(x)) =$ unity, we get finally

$$(Ad\gamma(z)A^{-1}g)(x) \equiv \alpha_z(x)g(x) + (d\gamma(z)g)(x)$$

completing the proof of our lemma.

Below we fix a basis $(l_j)_{1\le j\le m}$ such that 1) $(l_j)_{1\le j\le \dim(\mathfrak{d})}$ is a basis in $\mathfrak{d}$, 2) $(l_j)_{j>\dim(\mathfrak{c})}$ is a basis, supplementary to $\mathfrak{c}$. – Let now y_0 be as g above, and $\overline{y}_0 \in \overline{O}$ over y_0. Given $y \in \overline{O}$, we define $d_y = \sigma(d)$, if $y = \sigma\overline{y}_0$ ($\sigma \in G$);the possibility of this we infer from Lemma 25. Given a subset f of s elements of the integers between l and m, we set

$$O_f = \{y\,;\, y \in \overline{O} \text{ s.t. } g = d_y \oplus (\oplus_{j\in f}\mathbf{R}l_j)\}.$$

We write F for the collection of f's, for which O_f is non empty.

LEMMA 30. *O_f is open and saturated with respect to the fibration $\overline{O} \xrightarrow{\pi} \overline{O}/\overline{E}$.*

PROOF: It is clear, that O_f is open. – Given $g \in \overline{O}$, we set $d'_g = \sigma(d')$ if $g = \sigma\overline{y}_0$ ($\sigma \in G$). We write also $f' = f \cap [l, \dim(\mathfrak{d})]$. Then $g \in O_f$ if and only if $d = d'_g \oplus (\oplus_{j\in f'}\mathbf{R}l_j)$. To prove, that O_f is saturated with respect to the fibration $\overline{O} \to \overline{O}/\overline{E}$ we can assume $g = \overline{y}_0$, and complete our proof by showing that $\overline{E}\overline{y}_0 \subset O_f$. But if $a \in \overline{E} = EG_g$, we have $d'_{a\overline{y}_0} = a(d') = d'$ by Remark 2 in Section 1, and the G_f-invariance of $\mathfrak{h}'$.

Below, given $f \in F$, we set $U_f = \pi(O_f)$; then $(U_f)_{f\in F}$ is a finite open covering of $\overline{O}/\overline{E}$. – We denote by $B^1(G,\chi)$ the set of all linear differential operators, of order less or equal to one, on $\mathcal{E}(G,\chi)$.

LEMMA 31. *Let $(l_j)_{1\le j\le m}$ be a basis in $\mathfrak{g}$. Given D in $B^1(G,\chi)$ there is $(\beta_j)_{0\le j\le m} \subset A(\overline{O}/\overline{D})$ such that we have:*

$$D = q^*(\beta_0) + \sum_{k=1}^{m} q^*(\beta_k) d\gamma(l_k).$$

PROOF: We can obviously assume, that our basis is as in our previous lemma. – (i) Let $f \in F$, $f = \{0 < j_1 < \cdots < j_s \le m\}$ be fixed. We choose an open set $V \subset U_f$, admitting a local section into G. We set $\tilde{V} = \delta^{-1}(V)$ and write D' for the restriction of D to $\tilde{V}$. With the notation of Lemma 28, there is $(\alpha_k)_{0\le k\le s} \subset A_W(\overline{O}/\overline{D})$ such that

$$AD'A^{-1} = \alpha_0 + \sum_{k=1}^{s} \alpha_k \cdot d\gamma'(l_{j_k}).$$

By Lemma 29 we can write:

$$d\gamma'(l_{j_k}) = Ad\gamma(l_{j_k})_k A^{-1} - \eta_k$$

where $(\eta_k)_{0\le k\le s} \subset A_W(\overline{O}/\overline{D})$. In this fashion

$$\alpha_k \cdot d\gamma'(l_{j_k}) = Aq^*(\alpha_k) d\gamma(l_{j_k}) A^{-1} - \eta_k.$$

Putting $\beta_k = \alpha_k$ $(1 \le k \le s)$ and $\beta_0 = \alpha_0 - \sum_{k=1}^{s} \eta_k$ we thus obtain

$$D' = q^*(\beta_0) + \sum_{k=1}^{s} q^*(\beta_k) \cdot d\gamma(l_{j_k}).$$

(ii) Let $\mathcal{V} = (V_j)_{j\in J}$ be a locally finite open covering of $\overline{O}/\overline{E}$, where each V_j is as V in (i). By what we have just seen, for each $j \in J$, if $\tilde{V}_j = \delta^{-1}(V_j)$, we can write

$$D|V_j = q^*(\beta_0^{(j)}) + \sum_{k=1}^{m} q^*(\beta_k^{(j)}) \cdot d\gamma(l_k).$$

Let $(\omega_j)_{j\in J}$ be a partition of one on $\overline{O}/\overline{E}$, subordinated to $\mathcal{V}$. Then $\delta^*(\omega_j)$ is in $q^*(A(\overline{O}/\overline{D}))$. To obtain the representation of D claimed in our lemma, it is enough to set

$$\beta_k = \sum_{j\in J} \delta^*(\omega_j)\beta_k^{(j)} \quad (0 \le k \le m).$$

By Lemma 25 we can form a complex distribution $\overline{\mathcal{H}}$ on $\overline{O}$, determined by $\tau_y(\overline{H}_y) = \sigma(\mathfrak{h}\overline{y}_0)$ if $y = \sigma\overline{y}_0$ $(\sigma \in G)$. We write $\lambda : \overline{O} \to O$ for the canonical projection. If $\mathcal{H}$ corresponds to $\mathfrak{h}$ on O (cf. Section 2), we have $\lambda_*(\overline{\mathcal{H}}) = \mathcal{H}$. Below $\mathcal{E}^1, \mathcal{E}^0$ on $\overline{O}$ are to be understood with respect to $\overline{\mathcal{H}}$. We note, incidentally, that $\mathcal{E}^0 = \rho^*(A(\overline{O}/\overline{D}))$. – Given $l \in \mathfrak{g}$, we set $\overline{\psi}_l = \lambda^*(\psi_l)$.

LEMMA 32. *Given $\psi \in \mathcal{E}^1$ on $\overline{O}$, there is $(\varphi_j)_{0\le j\le m} \subset \mathcal{E}^0$ such that*

$$\psi = \varphi_0 + \sum_{j=1}^{m} \varphi_j \overline{\psi}_{l_j}.$$

PROOF: We assume, that V, as in Section 3, is connected and simply connected. Then the same is valid for $\mathcal{O}$ as loc.cit. We define $\overline{\mathcal{O}} = \lambda^{-1}(\mathcal{O})$, and denote by $\overline{\mathcal{O}}_c$ a connected component of $\overline{\mathcal{O}}$. Then $(\overline{\mathcal{O}}_c, \lambda|\overline{\mathcal{O}}_c)$ is a connected covering of $\mathcal{O}$ and thus $\lambda|\overline{\mathcal{O}}_c$ is a diffeomorphism onto $\mathcal{O}$. We set $\psi' = \psi|\overline{\mathcal{O}}_c$ and write Ψ for the corresponding function, belonging to $\mathcal{E}^1(\mathcal{O})$, on $\mathcal{O}$. By Lemma 23 there is $(\varphi_j)_{0\le j\le s} \subset \mathcal{E}^0(\mathcal{O})$ such that

$$\psi = \varphi_0 + \sum_{k=1}^{s} y_k \varphi_k.$$

By the proof of Lemma 14, each y_k is of the form $\omega_0 + \sum_{j=1}^{m} \omega_j \psi_{l_j}$, where $(\omega_j)_{0\le j\le m} \subset \mathcal{E}^0(\mathcal{O})$. We conclude from all this, that we have a representation

$$\psi' = \varphi_0' + \sum_{j=1}^{m} \varphi_j' \overline{\psi}_{l_j},$$

where $(\varphi_j')_{0\le j\le m}$ is in $\mathcal{E}^0(\overline{\mathcal{O}}_c)$. Let finally $\mathcal{V} = (V_j)_{j\in J}$ be a locally finite covering of $O/\overline{E}$ with sets as V above. We can select from the set of all connected components of the corresponding $\overline{\mathcal{O}}$'s a locally finite covering of $\overline{O}$. Then it suffices to take a partition of one, with elements in $\mathcal{E}^0$, subordinated to it, and complete the proof as at the end of the previous lemma.

Before proceeding, we recall the following rules of computation governing the quantization map δ_χ (cf. [3], 1.4): 1) Given $\varphi \in \mathcal{E}^0(\overline{O})$, we denote by φ' the element of $A(\overline{O}/\overline{D})$ well-determined by $\varphi = \rho^*(\varphi')$. Then $\delta_\chi(\varphi) = cq^*(\varphi')$, where we put $c = 2\pi i$, 2) We have $\delta_\chi(\overline{\psi}_l) = d\gamma(l)$ ($l \in \mathfrak{g}$). The following rule is implied by 1)-2) above and by Lemma 32: 3) If $\varphi \in \mathcal{E}^0(\overline{O})$ and $\psi \in \mathcal{E}^1(\overline{O})$ we have: $\delta_\chi(\varphi\psi) = q^*(\varphi') \cdot \delta_\chi(\psi)$. 4) δ_χ is a Lie algebra morphism $\mathcal{E}^1(\overline{O}) \to B^1(G,\chi)$. Given a coadjoint orbit O, below by a standard covering of O we shall mean that corresponding to the reduced stabilizer (cf. [5], Definition 14, p.838).

THEOREM 1. *Suppose, that $\overline{O}$ is a standard covering corresponding to a coadjoint orbit O. Then there is a polarization, such that the quantization map is a Lie algebra isomorphism from $\mathcal{E}^1(\overline{O})$ onto $B^1(G,\chi)$.*

PROOF: Surjectivity: Suppose $D \in B^1(G,\chi)$ be given. By Lemma 31 there is $(\beta_j)_{0 \le j \le m} \subset A(\overline{O}/\overline{D})$ such that

$$D = q^*(\beta_0) + \sum_{j=1}^{m} q^*(\beta_j) d\gamma(l_j).$$

We define $\psi = \rho^*(\beta_0)/c + \sum_{j=1}^{m} (\rho^*(\beta_j)/c)\psi_{l_j}$. Then our rules of computation imply: $\delta_\chi(\psi) = D$.

Injectivity: Below $\mathcal{O}$ will stand for a set $\overline{\mathcal{O}}_c$ in the notation of the proof of Lemma 32. We recall, that the restriction of the canonical projection $\lambda : \overline{O} \to O$ to $\mathcal{O}$ is a diffeomorphism with its image. Hence we can and shall assume on O the same situation, as we have established on the similarly denoted object in Section 4. We can also suppose, that $\mathcal{O} \subset O_f$ for some $f \in F$, $f = \{0 < j_1 < \cdots < j_s \le m\}$, say. – Using $\delta_\chi(\varphi\psi) = q^*(\varphi')\delta_\chi(\varphi)$ $(\varphi \in \mathcal{E}^0(\overline{O}),\ \psi \in \mathcal{E}^1(\overline{O}))$, it is enough to show, that if $\psi \in \mathcal{E}^1(\overline{O})$ has its support in O and $\delta_\chi(\psi) = 0$, then $\psi = 0$. –

(i) We set $v_t = l_{j_t}$ $(1 \le t \le s)$. By virtue of Lemma 23 we can find $\{a_{tk}; 1 \le t \le s,\ 0 \le k \le s\} \subset \mathcal{E}^0(O)$ such that

$$\overline{\psi}_{v_t} = a_{to} + \sum_{k=1}^{s} a_{tk} y_k \quad (1 \le t \le s).$$

We set $F \equiv \det(a_{tk}\,;\, 1 \le t, k \le s)$ and claim, that the set of zeros N os F has no interior. In fact, we recall, that we can write $F \equiv f(z_1, \ldots, z_t, y_{s+1}, \ldots, y_d)$, where f is smooth and anlytic in $z_1, \ldots, z_t$. Hence, if the interior of N is non empty, by a suitable change of $\mathcal{O}$ we could arrange, that $F \equiv 0$. If so, however, we can find a system $(\omega_t)_{1 \le t \le s} \subset \mathcal{E}^0(\overline{O})$, not identically zero and with support in $\mathcal{O}$, such that

$$\sum_{t=1}^{s} \omega_t \overline{\psi}_{v_t} \equiv \omega_0 \in \mathcal{E}^0(\overline{O}).$$

Applying δ_χ to both sides we conclude that $\sum_{t=1}^{s} q^*(\omega_t') d\gamma(v_t) = q^*(\omega_0')$ which, by aid of the operator A as in the proof of Lemma 31, easily implies $\omega_t' \equiv 0$ $(1 \le t \le s)$ and thus a contradiction. We conclude from this, that there exists a system $\{b_{kt}\,;\, 0 \le t \le s;\ 1 \le k \le s\} \subset \mathcal{E}^0(O)$ such that

$$F y_k = b_{ko} + \sum_{t=1}^{s} b_{kt} \psi_{v_t}$$

and $\det(b_{kt}; 1 \le k, t \le s) \ne 0$ if $F \ne 0$.

(ii) Using Lemma 23 we can write

$$\psi = \varphi_0 + \sum_{k=1}^{s} y_k \varphi_k$$

where $(\varphi_k)_{0 \leq k \leq s} \subset \mathcal{E}^0(\overline{O})$, having their support in $\mathcal{O}$. We have then

$$\begin{aligned} F\psi &= F\varphi_0 + \sum_{k=1}^{s} \varphi_k (b_{ko} + \sum_{t=1}^{s} b_{kt} \psi_{v_t}) \\ &= (F\varphi_0 + \sum_{k=1}^{s} \varphi_k b_{ko}) + \sum_{t=1}^{s} (\sum_{k=1}^{s} b_{kt} \varphi_k) \psi_{v_t}. \end{aligned}$$

Hence, writing ϕ_0 for the first term, we can conclude, that

$$0 = \delta_\chi(\psi F) = cq^*(\phi_0') + \sum_{t=1}^{s} (\sum_{k=1}^{s} cq^*(b'_{kt}) q^*(\varphi_k)) d\gamma(v_t)$$

and thus, as above, also

$$\sum_{k=1}^{s} b_{kt} \varphi_k \equiv 0 \quad (1 \leq t \leq s)$$

which implies, that $\varphi_k \equiv 0 (1 \leq k \leq s)$. Hence finally $\psi \equiv \varphi_0 \in \mathcal{E}^0(\overline{O})$ which, by $0 = \delta_\chi(\psi) = cq^*(\varphi_0')$, yields $\psi \equiv 0$.

We have also

Theorem 2. *If G is nilpotent, δ_χ is an isomorphism for all coadjoint orbits and any complex polarization.*

Remark. 1) It is useful to note, that Theorem 1 retains its validity, if the reduced stabilizer is replaced by any of its open subgroups. 2) If we choose $(G_g)_0$, c can be assumed to be any complex number.

6. Global Darboux coordinates. The objective of this closing section is the following:

Theorem 3. *Suppose, that the coadjoint orbit O is simply connected. Then there are global Darboux coordinates on O.*

Proof: To establish our claim, it is obviously enough to show the following. There is a set of elements $(p_i, q_j\,; 1 \leq i,j \leq \sigma) \subset \mathcal{E}(O)$, where $2\sigma = d$, satisfying the following conditions: 1) We have $\{p_i, p_j\} \equiv 0 \equiv$

$\{q_i, q_j\}$, $\{p_i, q_j\} = \delta_{ij}$ $(1 \leq i, j \leq \sigma)$, 2) The map $g \mapsto (q_1(g), \ldots, q_\sigma(g), p_1(g), \ldots, p_\sigma(g))$ is a diffeomorphism of O onto $\mathbf{R}^d$.

(i) We note first, that by virtue of our assumption we have $(G_g)_0 \equiv G_g$ $(g \in G)$. Therefore we have also $\overline{E} = E$, etc. Let $\vartheta : G/E \to G$ be a global section modulo E, as in (ii) of the proof of Lemma 24. We shall use ϑ in place of γ in Section 3 and in Lemma 29 to define A. – Writing $\Theta = \vartheta(G/E)$, this is a closed submanifold of G. We have by Lemma 24 and 27: $GH = \Theta E_{\mathbf{C}} = \Theta \cdot H' \cdot H$ and thus $G/D = GH/H = \Theta \cdot H'$, or $G/D = \Theta \cdot H'$ setwise. Since we have also $E_{\mathbf{C}} = H' \cdot H$ setwise, we have a diffeomorphism $\Theta \simeq H'$. By transfer, we can use $(z_j\,;\, 1 \leq j \leq t)$ (as in Section 4) as holomorphic coordinates on H'.

(ii) We can assume in Section 3: $V = G/E$, and also $\tau(V) = \mathbf{R}^r$. Below we shall write $T(v)$ $(T = (t_1, \ldots, t_r))$ in place of $\tau(v)$ $(v \in G/E)$, and $T \to v(T)$ for the inverse map. – Let $(e_k)_{1 \leq k \leq r}$ be the standard basis in $\mathbf{R}^r$. Given k fix, we shall define a differential operator $D_k \in \mathrm{Diff}_1, (A(O/E))$ as follows. Given $g \in A(G/D)$, we set

$$(D_k g)(a \cdot h') = (d/d\tau) g(\vartheta(v(T + \tau e_k))h')|_{\tau=0} \quad (a = \vartheta(v(T))).$$

We write $\overline{D}_k = A^{-1} D_k A$ (cf. Lemma 29). Below we shall establish the following description of $\overline{D}_k$. We define a map $l_k : G/E \to g$ by:

$$-l_k(v(T)) = (d/d\tau)\vartheta(v(T + \tau e_k))(\vartheta(v(T)))^{-1}|_{\tau=0}.$$

Let $(l_j)_{1 \leq j \leq m}$ be a basis in $\mathfrak{g}$, and let us write:

$$l_k(v) = \sum_{j=1}^{m} a_j(v) l_j.$$

Then we shall show, that

$$\overline{D}_k = \sum_{j=1}^{m} \delta^*(a_j) d\gamma(l_j).$$

The role of these considerations in our construction of the global Darboux coordinates will be as follows. We assume Θ to be parametrized by $(t_1, \ldots, t_r) = T \in \mathbf{R}^r$ via $T \to \vartheta(v(T))$, and H', as already suggested, by $z_1, \ldots, z_t$. We have $D_k = \partial/\partial t_k$ $(1 \leq k \leq r)$. We shall define $(p_j)_{1 \leq j \leq r} \subset \mathcal{E}^1(O)$, by aid of Theorem 1 with $c = 1$, by stipulating $\delta_\chi(p_k) = \overline{D}_k$. The other variables will arise from $(z_k)_{1 \leq k \leq t}$, $(t_k)_{1 \leq k \leq r}$ by transfer. – To prove our claim above, we recall that $GH = \Theta \cdot H' \cdot H$ setwise. We have also with the notation of Section

4 and 5: $j(a \cdot h' \cdot h) = a \cdot h'$, $\sigma(a \cdot h') = a \cdot h' \cdot e$, and $h(a \cdot h' \cdot h) = h$. If $f \in \mathcal{E}(G, \chi)$, then $(Af)(a \cdot h') = F(a \cdot h')$ $(a \in \Theta, h' \in H')$. We have also

$$(D_k Af)(\vartheta(v)h') = (d/d\tau)F(\vartheta(T(v) + \tau e_k)h')|_{\tau=0}$$
$$=(d\gamma'(l_k(v))F)(\vartheta(v) \cdot h') = (Ad\gamma(l_k(v))f)(\vartheta(v) \cdot h')$$

and thus

$$(\overline{D}_k f)(x) = (d\gamma)(\delta * l_k)(x)(f(x)) = \sum_{j=1}^{m} (\delta * a_j)(x) d\gamma(l_j) f,$$

completing our proof. – We can remark at once, that if p_k is defined by $\delta_\chi(p_k) = \overline{D}_k$, we obtain $p_k = \sum_{j=1}^{m} \pi^* a_j \cdot \psi_{l_j}$.

(iii) We recall, that by Lemma 15 we have: $\{z_j, z_k\} \equiv 0 \equiv \{\overline{z}_j, \overline{z}_k\}$, and by Lemma 20: $\{z_j, \overline{z}_k\} \equiv (2/i)\delta_{jk}$ $(1 \le j, k \le t)$. Since $z_j = y_{r+j} + iy_{r+t+j}$, we can conclude, that $\{y_{r+j}, y_{r+k}\} \equiv 0 \equiv \{y_{r+t+j}, y_{r+t+k}\}$ and $\{y_{r+k}, y_{r+t+j}\} \equiv \delta_{jk}$. This being so, we propose the following definitions: $(p_k)_{1 \le k \le r}$ by $\delta_\chi(p_k) = \overline{D}_k$ via Theorem 1 with $c = 1$ (cf. Remark loc.cit.); then $p_{r+j} \equiv y_{r+j} (1 \le j \le t)$. In addition $q_k \equiv y_{s+k}$ $(1 \le k \le r)$ and $q_{r+j} \equiv y_{r+t+j}$ $(1 \le j \le t)$. By what we said in (ii) above, we have:

$$\left.\begin{array}{l}\{p_i, p_j\} \equiv 0 \equiv \{q_i, q_j\} \\ \{p_i, q_j\} \equiv \delta_{ij}\end{array}\right. \qquad (1 \le i, j \le r + t).$$

(iv) We complete our proof of Theorem 3 by showing, that $(p_i, q_j\,;\, 1 \le i, j \le r+t)$ constitute coordinates on O. As defined in Section 3 with $V = G/E$ loc.cit., $(y_j)_{1 \le j \le d}$ are cordinates on O. What we are proposing here is just to replace $(y_1, \ldots, y_r)$ by $(p_1, \ldots, p_r)$, while retaining all the remaining coordinates. Below we write $y = (y_{r+1}, \ldots, y_d) \in \mathbf{R}^s$. Since, by definition, $(p_j)_{1 \le j \le r}$ is in $\mathcal{E}^1(O)$, by Lemma 23 we can write:

$$p_j = a_{jo}(y) + \sum_{k=1}^{r} a_{jk}(y) y_k \qquad (1 \le j \le r).$$

Hence, in particular, p_j is linear in $y_1, \ldots, y_r$. Since $\{y_j, y_{s+k}\} = 0$ if $j > r$, $1 \le k \le r$, we obtain

$$\delta_{jk} = \{p_j, y_{r+k}\} = \sum_{u=1}^{r} a_{ju}(y)\{y_u, y_{r+k}\}.$$

To show, that the Jacobian of the transition is non zero, it is enough to note, that $\det(\{y_i, y_{r+k}\}, 1 \le i, k \le r) \neq 0$.

REMARK. By the Remark at the end of Section 5, we can also show the existence of global Darboux coordinates on the universal covering of any coadjoint orbit.

REFERENCES

[1] L.Auslander and B. Kostant, *Polarization and unitary representations of solvable Lie groups*, Invent.Math. **14** (1971), 255–354.

[2] B. Kostant, *Quantization and unitary representations*, in "Lecture Notes in Mathematics 170," Springer Verlag, Berlin, Heidelberg, New York, 1970, pp. 87–208.

[3] N. V. Pedersen, *On the symplectic structure of coadjoint orbits of (solvable) Lie groups and applications, I*, Math. Ann. **281** (1988), 633–669.

[4] L. Pukanszky, *Unitary representations of solvable Lie groups*, Ann. Sci. Ecole Norm. Sup **4** (1971), 457–608.

[5] L.Pukanszky, *Quantization and Hamiltonian foliations*, Trans. Amer. Math. Soc. **295** (1986), 811–847.

Department of Mathematics
University of Pennsylvania
Philadelphia, Pa. 19104
U.S.A.

The Poisson–Plancherel Formula for a Quasi-Algebraic Group with Abelian Radical and Reductive Generic Stabilizer

PIERRE TORASSO

I. INTRODUCTION AND GENERAL CONJECTURE

In this section we state following M. Vergne, as a conjecture, the Poisson-Plancherel formula for a unimodular quasi-algebraic group. So, first we give the notation and the material which is necessary to do that.

a) Quasi-algebraic groups.

A quasi-algebraic group is a triple $(G, j, \mathbf{G})$ where $\mathbf{G}$ is an algebraic group which is defined on the field of real numbers, G is a separable real Lie group and j is a Lie group morphism from G into $\mathbf{G}$, the image of which is an open subgroup of the group of the real points of $\mathbf{G}$ and the kernel of which is a discrete central subgroup of G.

Let $(G, j, \mathbf{G})$ be a quasi-algebraic group.

b) Very regular linear forms and Cartan-Duflo subalgebras.

Let $\mathfrak{g}$ denote the Lie algebra of G and let $f \in \mathfrak{g}^*$. Recall that if f is regular its stabilizer $\mathfrak{g}(f)$ is commutative. Then we denote by $\mathfrak{j}_f$ the unique reductive factor in $\mathfrak{g}(f)$, which is an algebraic torus.

Now we say that f is very regular if it is regular and futhermore if $\mathfrak{j}_f$ is of maximal dimension.

We denote by $\mathfrak{g}_r^*$ the set of very regular linear forms on $\mathfrak{g}$, which is a non empty G-invariant Zariski open subset of $\mathfrak{g}^*$, and by $\mathrm{car}(\mathfrak{g})$ the set of the torus $\mathfrak{j}_f$ for f in $\mathfrak{g}_r^*$.

The elements of $\mathrm{car}(\mathfrak{g})$ are the so-called Cartan-Duflo subalgebras of $\mathfrak{g}$; they all have same dimension. When $\mathfrak{g}$ is reductive $\mathrm{car}(\mathfrak{g})$ is nothing else but the set of Cartan subalgebras.

We denote by $\mathrm{Car}(G)$ a set of representatives of the G-conjugacy classes in $\mathrm{car}(\mathfrak{g})$. Then $\mathrm{Car}(G)$ is a finite set and if G is a complex Lie group the cardinalityit of $\mathrm{Car}(G)$ is one.

If $j \in \mathrm{car}(\mathfrak{g})$, let $\mathfrak{g}_{r,\mathfrak{j}}^*$ be the set of elements $f \in \mathfrak{g}_r^*$ such that $\mathfrak{j}_f$ is G-conjugated to $\mathfrak{j}$. Then $\mathfrak{g}_{r,\mathfrak{j}}^*$ is a G-invariant open set in $\mathfrak{g}^*$ and $\mathfrak{g}_r^*$ is the disjoint union of the $\mathfrak{g}_{r,\mathfrak{j}}^*$, $\mathfrak{j} \in \mathrm{Car}(G)$.

c) Invariant distributions.

Let $\mathfrak{j} \in \mathrm{car}(\mathfrak{g})$. We denote by $\mathfrak{h}_\mathfrak{j}$ its centralizer in $\mathfrak{g}$ and by $H_\mathfrak{j}$ (resp. $H_\mathfrak{j}'$) its centralizer (resp. normalizer) in G. Then $H_\mathfrak{j}$ and $H_\mathfrak{j}'$ are both closed Lie subgroups of G with Lie algebra $\mathfrak{h}_\mathfrak{j}$, and the group $W_\mathfrak{j} = H_\mathfrak{j}'/H_\mathfrak{j}$ is finite.

Since $\mathfrak{j}$ is a torus we have a unique $\mathfrak{j}$-invariant decomposition $\mathfrak{g} = \mathfrak{h}_{\mathfrak{j}} \oplus \mathfrak{q}_{\mathfrak{j}}$ to which corresponds the $\mathfrak{j}$-invariant decomposition of the dual space $\mathfrak{g}^* = \mathfrak{h}_{\mathfrak{j}} \oplus \mathfrak{q}_{\mathfrak{j}}^*$. We denote by $\mathfrak{h}_{\mathfrak{j},r}^*$ the intersection of $\mathfrak{h}_{\mathfrak{j}}^*$ with $\mathfrak{g}_r^*$.

Choose a volume form $\eta_{\mathfrak{j}}$ on the vector space $\mathfrak{q}_{\mathfrak{j}}$ and define a polynomial function $\pi_{\mathfrak{j}}$ on $\mathfrak{h}_{\mathfrak{j}}^*$ by the fact that for every $f \in \mathfrak{h}_{\mathfrak{j}}^*$, $\pi_{\mathfrak{j}}(f)$ is the Pfaffian with respect to $\eta_{\mathfrak{j}}$ of the restriction to $\mathfrak{q}_{\mathfrak{j}}$ of the alternate bilinear form B_f on $\mathfrak{g}$ defined by $B_f(X,Y) =< f, [X,Y] >$ for $X, Y \in \mathfrak{g}$. It turns out that this polynomial function is non zero. Its degree is the number $d = \frac{1}{2} \dim \mathfrak{q}_{\mathfrak{j}}$, which is independent of the choice of $\mathfrak{j}$.

If M is a manifold we denote by $\mathcal{D}(M)$ (resp. $\mathcal{D}'(M)$) the space of compactly supported C^∞ functions (resp. distributions) on M with its usual topology. If E is a set and Γ a group acting on E, we denote by E^Γ the set of Γ-invariants in E.

The volume form $\eta_{\mathfrak{j}}$ determines a unique quasi-invariant measure $d\dot{g}$ on the homogeneous space $G/H_{\mathfrak{j}}$. Now let dg be a Haar measure on G and dh the unique Haar measure on $H_{\mathfrak{j}}$ such that $dg = d\dot{g}\, dh$. These Haar measures determine Lebesque measures on $\mathfrak{g}$ and $\mathfrak{h}_{\mathfrak{j}}$, which, in turn, determine dual Lebesgue measures on $\mathfrak{g}^*$ and $\mathfrak{h}_{\mathfrak{j}}^*$, generically denoted df. To be more precise, if dx is a Lebesgue measure on the finite dimensional real vector space E, its dual Lebesgue measure is defined with respect to the following Fourier transform $\mathcal{F}\varphi$ of functions φ on E:

$$\mathcal{F}\varphi(f) = \int_E \varphi(x) e^{-i<f,x>} dx \quad \forall f \in E^*,$$

and is such that one has

$$\varphi(0) = \int_{E^*} \mathcal{F}\varphi(f) df .$$

Finally if E is a finite dimensional real vector space and if p is an element of the complex symmetric algebra $S(E)$ of E we denote by ∂_p the corresponding differential operator with constant coefficients on E.

LEMMA 1. *Keeping the notation from above we have*

(i) *The map* $\Omega \to \Omega \cap \mathfrak{h}_{\mathfrak{j},r}^*$ *is a bijection from* $\mathfrak{g}_{r,\mathfrak{j}}^*/G$ *on* $\mathfrak{h}_{\mathfrak{j},r}^*/H_{\mathfrak{j}}'$.

(ii) *The restriction map* $\varphi \to \varphi|_{\mathfrak{h}_{\mathfrak{j},r}^*}$ *from* $\mathcal{D}(\mathfrak{g}_{r,\mathfrak{j}}^*)^G$ *into* $\mathcal{D}(\mathfrak{h}_{r,\mathfrak{j}}^*)^{H_{\mathfrak{j}}'}$ *is an isomorphism of topological vector spaces and extends uniquely as a continuous map* $\theta \to \theta|_{\mathfrak{h}_{\mathfrak{j},r}^*}$ *from* $\mathcal{D}'(\mathfrak{g}_{r,\mathfrak{j}}^*)^G$ *into* $\mathcal{D}'(\mathfrak{h}_{r,\mathfrak{j}}^*)^{H_{\mathfrak{j}}'}$ *which is also an isomorphism and preserves Radon measures. Moreover one has for every* $\varphi \in \mathcal{D}(\mathfrak{g}_{r,\mathfrak{j}}^*)$ *and* $\theta \in \mathcal{D}'(\mathfrak{g}_{r,\mathfrak{j}}^*)^G$

(1)
$$\int_{\mathfrak{g}^*_{r,j}} \varphi(f)\theta(f)df =$$

$$(2\pi)^{-2d}[W_j]^{-1} \int_{G/H_j} |\det \operatorname{Ad} \dot{g}|^{-1} \left(\int_{\mathfrak{h}^*_{j,r}} \varphi(g.f)\theta|_{\mathfrak{h}^*_{j,r}}(f)\pi_j(f)^2 df \right) d\dot{g} .$$

(iii) *If $p \in S(\mathfrak{g}^*)^G$, ∂_p is a G-invariant differential operator on $\mathfrak{g}^*_{r,j}$ the radial part of which is $\pi_j^{-1} \circ \partial_{p|_{\mathfrak{h}_j}} \circ \pi_j$, that is, for every $\theta \in \mathcal{D}(\mathfrak{g}^*_{r,j})^G$ one has*

(2) $$(\partial_p\theta)_{|\mathfrak{h}^*_{j,r}} = (\pi_j^{-1} \circ \partial_{p|_{\mathfrak{h}_j}} \circ \pi_j)(\theta_{|\mathfrak{h}^*_{j,r}}) .$$

d) Admissible linear forms and Duflo's canonical measure.

Let $f \in \mathfrak{g}^*$. Then the alternate bilinear form B_f induces a symplectic structure on the quotient space $\mathfrak{g}/\mathfrak{g}(f)$. We denote by Sp^f the corresponding symplectic group and by Mp^f the corresponding metaplectic group, that is, its connected double covering group. Let $G(f)$ be the stabilizer of f in G; then $\operatorname{Ad} G(f)$ is a subgroup of Sp^f and we denote by $G(f)^f$ the subgroup of $G(f) \times Mp^f$, the elements of which are the pairs (g, m) such that g and m have the same image in Sp^f.

Then we denote by $X_G(f)$ the set of irreducible unitary representations τ of $G(f)^f$ which satisfy the following conditions:

(i) the restriction of τ to the neutral component of $G(f)^f$ is a multiple of the character, the differential of which is $if_{|\mathfrak{g}(f)}$,

(ii) if ε denotes the non trivial element in the kernel of the natural projection from $G(f)^f$ onto $G(f)$, then $\tau(\varepsilon) = -\operatorname{Id}$.

The subgroup $\operatorname{Ker} j$ of G is contained in $G(f)$ and admits a natural section in $G(f)^f$. If $\chi \in \operatorname{Ker} j^\wedge$, the unitary dual of $\operatorname{Ker} j$, we denote by $X_{G,\chi}(f)$ the subset of $X_G(f)$, the elements of which are the representations τ which, furthermore, satisfy the following condition:

(iii) $\tau(\gamma) = \chi(\gamma) \operatorname{Id} \qquad \forall \gamma \in \operatorname{Ker} j$.

We say that the linear form f is G-admissible (resp. G-χ-admissible) if $X_G(f)$ (resp. $X_{G,\chi}(f)$) is non empty. We denote by $\mathfrak{g}^*_G$ (resp. $\mathfrak{g}^*_{G,\chi}$) the set of very regular G-admissible (resp. G-χ-admissible) linear forms.

Let E_G (resp. $E_{\tilde{G}}$) be the set of elements T of $\mathfrak{g}$ such that $\exp_G T = 1$ (resp. $\exp_G T \in \operatorname{Ker} j$). Then E_G (resp. $E_{\tilde{G}}$) is a G-invariant subset of $\mathfrak{g}$, the elements of which are elliptic. We define a function $\tilde{1}_G$ on $\tilde{E}_G$, which is a generalization of the function ζ_ρ which was introduced in the reductive case in [Du-Ve], in the following way:

$$\tilde{1}_G(T) = \exp \frac{1}{2} \sum_{\substack{\lambda \in \mathrm{Spec}(\mathrm{ad}\, T) \\ i\lambda > 0}} m_\lambda \lambda \qquad \forall T \in \tilde{E}_G,$$

where for every $\lambda \in \mathrm{Spec}(\mathrm{ad}\, T)$, m_λ is its multiplicity.

Now if $\chi \in \mathrm{Ker}\, j^\wedge$ we put

$$\tilde{\chi}_G(T) = \tilde{1}_G(T)\chi(\exp_G T) \qquad \forall T \in \tilde{E}_G .$$

Let t be the anisotropic component of the algebraic torus $\mathfrak{j}$ and let $\mathfrak{t}(G)$ be the subspace of $\mathfrak{t}$ generated by $E_G \cap \mathfrak{j}$. Then $\tilde{E}_G \cap \mathfrak{j}$ (resp. $E_G \cap \mathfrak{j}$) is a lattice in $\mathfrak{t}$ (resp. $\mathfrak{t}(G)$) denoted by $\tilde{\mathfrak{t}}_G$ (resp. $\mathfrak{t}_G$). Moreover the restriction of the function $\tilde{\chi}_G$ (resp. $\tilde{1}_G$) to $\tilde{\mathfrak{t}}_G$ (resp. $\mathfrak{t}_G$) is a character. Now let $\tilde{\mathfrak{t}}^*_{G,\chi}$ (resp. $\mathfrak{t}^*_{G,1}$) be the set of elements $f \in \mathfrak{t}^*$ (resp. $\mathfrak{t}(G)^*$) such that

$$\tilde{\chi}_G(T) = e^{i<f,T>} \qquad \forall T \in \tilde{E}_G \cap \mathfrak{j}$$
$$(\text{resp. } \tilde{1}_G(T) = e^{i<f,T>} \qquad \forall T \in E_G \cap \mathfrak{j});$$

this is some translate of the dual lattice of $\tilde{\mathfrak{t}}_G$ (resp. $\mathfrak{t}_G$).

If E is a vector space, $F \subset E$ a vector subspace and $\mu \in F^*$ we denote by $F^\perp_\mu$ the subset of E^*, the elements of which are the linear forms λ such that $\lambda_{|F} = \mu$. With this notation, the set of G-χ-admissible (resp. G-admissible) elements of $\mathfrak{h}^*_\mathfrak{j}$ is

$$\bigcup_{\mu \in \tilde{\mathfrak{t}}^*_{G,\chi}} \mathfrak{t}^\perp_\mu \quad (\text{resp. } \bigcup_{\mu \in \mathfrak{t}^*_{G,1}} \mathfrak{t}(G)^\perp_\mu).$$

If A is a subset of $\mathfrak{g}^*$ we denote by A_r its intersection with $\mathfrak{g}^*_r$. Then it is clear that

$$\mathfrak{g}^*_{G,\chi} \cap \mathfrak{h}^*_\mathfrak{j} = \bigcup_{\mu \in \tilde{\mathfrak{t}}^*_{G,\chi}} (\mathfrak{t}^\perp_\mu)_r$$
$$(\text{resp. } \mathfrak{g}^*_G \cap \mathfrak{h}^*_\mathfrak{j} = \bigcup_{\mu \in \mathfrak{t}^*_{G,1}})\mathfrak{t}(G)^\perp_\mu)_r)$$

and that, for each $\mu \in \tilde{\mathfrak{t}}^*_{G,\chi}$ (resp. $\mathfrak{t}^*_{G,1}$), $(\mathfrak{t}^\perp_\mu)_r$ (resp. $(\mathfrak{t}(G)^\perp_\mu)_r$) is a Zariski open subset of the affine subspace $\mathfrak{t}^\perp_\mu$ (resp. $\mathfrak{t}(G)^\perp_\mu$).

Now it follows from the Poisson summation formula that the series

$$\sum_{T \in \mathfrak{i}_G} \tilde{\chi}_G(T) e^{i<f,T>} \quad (\text{resp. } \sum_{T \in \mathfrak{t}_G} \tilde{1}_G(T) e^{i<f,T>})$$

converges in the space of tempered distributions on $\mathfrak{h}_j^*$ towards a positive Radon measure which is denoted by $m_{G,\chi}^j$ (resp. m_G^j). The Radon measure $m_{G,\chi}^j$ (resp. m_G^j) is supported by the set of G-χ-admissible (resp. G-admissible) elements of $\mathfrak{h}_j^*$, and its restriction to each affine subspace $\mathfrak{t}_\mu^\perp$, $\mu \in \tilde{\mathfrak{t}}_{G,\chi}^*$ (resp. $\mathfrak{t}(G)_\mu^\perp$, $\mu \in \mathfrak{t}_{G,1}^*$) is a Lebesgue measure.

Then the restriction of $m_{G,\chi}^j$ (resp. m_G^j) to $\mathfrak{h}_{j,r}^*$ is a H_j'-invariant Radon measure and it follows from Lemma 1 that it extends, in a unique way, to a G-invariant Radon measure, $m_{G,\chi,j}$ (resp. $m_{G,j}$) on $\mathfrak{g}_{r,j}^*$.

Now we extend $m_{G,\chi,j}$ (resp. $m_{G,j}$) in a Borel measure on $\mathfrak{g}^*$, also denoted by $m_{G,\chi,j}$ (resp. $m_{G,j}$), such that $\mathfrak{g}^* - \mathfrak{g}_{r,j}^*$ is a set og measure zero. Then the measure $m_{G,\chi,j}$ (resp. $m_{G,j}$) depends only on the G-conjugacy class of j. So we define a positive G-invariant Borel measure on $\mathfrak{g}^*$ by putting

$$m_{G,\chi} = \sum_{j \in \mathrm{Car}(G)} m_{G,\chi,j} \quad (\text{resp. } m_G = \sum_{j \in \mathrm{Car}(G)} m_{G,j}).$$

These are the so called canonical measures on $\mathfrak{g}^*$, see [Du-2].

e) The functions $q_{G,\chi}$ and q_G.

Let Y_G be the set of pairs (f, τ) with $f \in \mathfrak{g}_G^*$ and $\tau \in X_G(f)$. Then M. Duflo has defined a function ζ_G on Y_G which serves to describe the Plancherel formula (see [Du-2] V.5, Theorem 40). Then, given $\chi \in \operatorname{Ker} j^\wedge$, we define a G-invariant function $q_{G,\chi}$ on $\mathfrak{g}_{G,\chi}^*$, by the formula

$$q_{G,\chi}(f) = [G(f) : \operatorname{Ker} j\overset{\circ}{G}(f)]^{-1} \sum_{\tau \in X_{G,\chi}(f)} (\dim \tau)^2 \zeta_G(f,\tau), \quad f \in \mathfrak{g}_{G,\chi}^*,$$

where $\overset{\circ}{G}(f)$ stands for the neutral component of the group $G(f)$.

Then if $f \in \mathfrak{g}_G^*$, there exists a character χ_f of $\operatorname{Ker} j$ such that $f \in \mathfrak{g}_{G,\chi_f}^*$ and moreover the set of elements $\chi \in \operatorname{Ker} j^\wedge$ which satisfy this condition is $\chi_f(\operatorname{Ker} j \cap \exp_G \mathfrak{t})^\perp$, where $(\operatorname{Ker} j \cap \exp_G \mathfrak{t})^\perp$ denotes the subgroup of $\operatorname{Ker} j^\wedge$ the elements of which are trivial on $\operatorname{Ker} j \cap \exp_G \mathfrak{t}$. It turns out that $(\operatorname{Ker} j \cap \exp_G \mathfrak{t})^\wedge$, is an abelian compact Lie group and we denote by $d\chi$ the normalized Haar measure on it. Now we define the G-invariant function q_G on $\mathfrak{g}_G^*$ by the formula

$$q_G(f) = \int_{(\operatorname{Ker} j \cap \exp_G \mathfrak{t})^\perp} q_{G,\chi_f\chi}(f) d\chi, \quad f \in \mathfrak{g}_G^*.$$

f) The orbital integrals on $\mathfrak{g}$.

Let T be an elliptic element of $\mathfrak{g}$, $\mathfrak{g}^T$ (resp. G^T) its centralizer in $\mathfrak{g}$ (resp. G) and $\mathfrak{q}_T$ the image of $\operatorname{ad} T$. One has $\mathfrak{g} = \mathfrak{g}^T \oplus \mathfrak{q}_T$ and consequently $\mathfrak{g}^* = (g^T)^* \oplus \mathfrak{q}_T^*$.

If $\lambda \in \operatorname{Spec}(\operatorname{ad} T)$ let us denote by $\mathfrak{g}_{\mathbf{C}}^{\lambda}$ the corresponding eigenspace in $\mathfrak{g}_{\mathbf{C}}$. Then let us define

$$\mathfrak{u}^+ = \bigoplus_{\substack{\lambda \in \operatorname{Spec}(\operatorname{ad} T) \\ i\lambda > 0}} \mathfrak{g}_{\mathbf{C}}^{\lambda} .$$

Clearly $\mathfrak{u}^+$ is a $\mathfrak{g}_{\mathbf{C}}^T$-invariant nilpotent subalgebra of $\mathfrak{g}_{\mathbf{C}}$. So we can define a polynomial function ω_T on $\mathfrak{g}_{\mathbf{C}}^T$ by putting

$$\omega_T(H) = \det{}_{\mathfrak{u}^+} \operatorname{ad} H \qquad \forall H \in \mathfrak{g}_{\mathbf{C}}^T .$$

Now let d_T be the dimension of $\mathfrak{u}^+$ and choose a basis $X_1, \ldots, X_{d_T}$ of $\mathfrak{u}^+$. Then $X_1, \ldots, X_{d_T}, \bar{X}_1, \ldots, \bar{X}_{d_T}$ is a basis of $\mathfrak{q}_{T,\mathbf{C}}$ and for every volume form η on $\mathfrak{q}_T$ the number $\eta(X_1, \ldots, X_{d_T}, \bar{X}_1, \ldots, \bar{X}_{d_T})$ is real and its sign does not depend on the choice of the basis $X_1, \ldots, X_{d_T}$ of $\mathfrak{u}^+$. So we can choose a volume form η_T on $\mathfrak{q}_T$ such that $\eta_T(X_1, \ldots, X_{d_T}, \bar{X}_1, \ldots, \bar{X}_{d_T}) > 0$.

With these data we define a polynomial function π_T on $(\mathfrak{g}^T)^*$ by the fact that for $f \in (\mathfrak{g}^T)^*$, $\pi_T(f)$ is the Pfaffian with respect to η_T of the alternate bilinear form $B_{f_{|\mathfrak{q}_T}}$. Now we have the

LEMMA 2. *With the same notations as above, π_T is non-zero if and only if there exists $\mathfrak{j} \in \operatorname{car}(\mathfrak{g})$ such that $T \in \mathfrak{j}$.*

We define, for every elliptic element T of $\mathfrak{g}$, a G-invariant distribution $M_{G,T}$ on $\mathfrak{g}$, which is supported in $G.T$ and which depends only on $G.T$, by the following formula

$$M_{G,T}(\varphi) = \left(\frac{i}{2\pi}\right)^{d_T} \int_{G/G^T} \partial\pi_T[\omega_T(H)\varphi(g.H)]|_{\substack{H = T \\ H \in \mathfrak{g}^T}} \, d\dot{g} ,$$

where $d\dot{g}$ stands for the quasi-invariant measure on G/G^T determined by the volume form η_T.

Suppose that G is a reductive group and let $\mathfrak{b}$ be a fundamental Cartan subalgebra of $\mathfrak{g}$. Then $M_{G,T}$ is $\operatorname{Card}[G.T \cap \mathfrak{b}]$-times the usual invariant tempered distribution obtained by derivation of the Harish-Chandra orbital integral for T (see [Du-Ve] for example).

One has the following result:

PROPOSITION 1. *For every elliptic element $T \in \mathfrak{g}$, the distribution $M_{G,T}$ is tempered.*

g) The general conjecture.

If θ is a tempered distribution on $\mathfrak{g}$ we denote by $\mathcal{F}\theta$ its Fourier transform.

Now we state the following conjecture, the main idea of which is due to M. Vergne [Ve-1].

CONJECTURE. *Let $(G, j, \mathbf{G})$ be a unimodular quasi-algebraic group and let $\chi \in \operatorname{Ker} j^{\wedge}$. Then*

(i) *$m_{G,\chi}$ and m_G are tempered positive Radon measures on $\mathfrak{g}^*$.*

(ii) a) *the distributional series*

$$\sum_{T \in E_{\tilde{G}}/G} \tilde{\chi}_G(T) M_{G,T} \quad \text{and} \quad \sum_{T \in E_G/G} \tilde{1}_G(T) M_{G,T} \tag{3}$$

converge in $\mathcal{S}'(\mathfrak{g})$ towards distributions which are respectively denoted $V_{G,\chi}$ and V_G.

b) *there exists an affine Zariski open set, $\mathcal{V}$, in $\mathfrak{g}^*$ such that the complement of $\mathcal{V} \cap \mathfrak{g}^*_{G,\chi}$ (resp. $\mathcal{V} \cap \mathfrak{g}^*_G$) is of measure zero with respect to $m_{G,\chi}$ (resp. m_G), on one hand, and the function $q_{G,\chi}$ (resp. q_G) is continuous on $\mathcal{V} \cap \mathfrak{g}^*_{G,\chi}$ (resp. $\mathcal{V} \cap \mathfrak{g}^*_G$), on the other hand*

c) *We have the following Poisson-Plancherel formulas for the group $(G, j, \mathbf{G})$*

$$\begin{aligned} \mathcal{F}_{\mathfrak{g}} V_{G,\chi} &= q_{G,\chi} dm_{G,\chi}, \\ \mathcal{F}_{\mathfrak{g}} V_G &= q_G dm_G \,. \end{aligned} \tag{4}$$

In particular $V_{G,\chi}$ and V_G are tempered distributions of positive type.

This conjecture was first proved by M. Vergne [Ve-2] for linear connected semi-simple Lie groups, then by P. Dourmashkin [Do] for connected groups of B_n-type and finally by M. Duflo and M. Vergne for connected semi-simple groups [Du-Ve]; their proof relies on a combinatorial result of Peterson-Vergne [Pe-Ve]. Moreover, M. Duflo gave a proof of a weaker form of this conjecture for complex algebraic groups; in particular he did not prove that the series (3) are convergent in $\mathcal{S}'(\mathfrak{g})$ (see [Du-1]). In the next section, we describe our proof of another case of the conjecture.

II. More precise results and some ideas on the proof in the handled case

We proved the conjecture for a quasi-algebraic group $(G, j, \mathbf{G})$ satisfying the following conditions

(i) the unipotent radical N of G is abelian,

(ii) the quotient group G/N (denoted by S) is semi-simple ,

(iii) if $\mathfrak{n}$ stands for the Lie algebra of N, then the isotropy subgroup in S for the elements of $\mathfrak{n}^*$ are generically reductive, or, equivalently, the S-orbits in $\mathfrak{n}^*$ are generically closed (see [Mu-Fo] Appendix to Chapter 1F).

This presents interesting new features. First of all, the value at a generic point f of $\mathfrak{g}^*$ of the Fourier transform of an orbital integral on $\mathfrak{g}$, is expressed by a very nice formula in terms of orbital integrals for a reductive factor of the stabilizer in G of $f_{|\mathfrak{n}}$ (see Theorem 1). Secondly we introduce new orbital integrals on $\mathfrak{g}^*$, which generalize the Harish-Chandra orbital integrals for the semi-simple case. But, in our case, these orbital integrals have weaker properties (Theorem 2), and for this reason, we have greater difficulties than in the semi-simple case, from an analysis point of view, in proving the conjecture. The complete proof is given in the long paper [To].

Now we describe our results and give some ideas about the proof of the conjecture in this case.

First of all we have the

Lemma 3. *There exists a non-zero S-invariant polynomial q defined on $\mathfrak{n}^*$ such that the complement V of its zero set has the following properties*

(i) *For all $u \in V$ the group $S(u)$ is reductive.*

(ii) *For all $u, v \in V$ if u and v belong to the same connected component of V then $S(u)$ and $S(v)$ are S-conjugated.*

Since V is a Zariski open set in the real vector space $\mathfrak{n}^*$ it has only a finite number of connected components. So, there are only a finite number of S-conjugacy classes for the stabilizers $S(u)$, $u \in V$.

Let ρ be the restriction map from $\mathfrak{g}^*$ on $\mathfrak{n}^*$ and let $\mathcal{V} = \mathfrak{g}^*_r \cap \rho^{-1}(V)$. If $f \in \mathfrak{g}^*$, we generically put $u = \rho(f)$.

For $\mathfrak{j} \in \operatorname{car}(\mathfrak{g})$ let $\mathcal{V}^{\mathfrak{j}} = \mathcal{V} \cap \mathfrak{h}_{\mathfrak{j}}$. Then we define an equivalence relation $\sim$ in $\mathcal{V}^{\mathfrak{j}}$ by deciding that for f, $f' \in \mathcal{V}^{\mathfrak{j}}$ one has $f \sim f'$ if and only if the stabilizers $S(u)$ and $S(u')$ are $S^{\mathfrak{j}}$-conjugated, where $S^{\mathfrak{j}}$ stands for the centralizer in S of the image in $\mathfrak{g}/\mathfrak{n}$ of $\mathfrak{j}$.

If f belongs to $\mathcal{V}$ let us choose a reductive factor $\mathbf{J}$ in $\mathbf{G}(f)$ and a reductive factor $\mathbf{R}$ in $\mathbf{G}(u)$ containing $\mathbf{J}$. Let us denote by J (resp. R)

the reductive factor in $G(f)$ (resp. $G(u)$) corresponding to $\mathbf{J}$ (resp. $\mathbf{R}$), that is its inverse image under j. Then R is naturally isomorphic to $S(u)$, and $(R, j, \mathbf{R})$ is a quasi-algebraic group. Let us denote by $\mathfrak{r}$ the Lie algebra of R, and by λ the restriction of f to $\mathfrak{r}$.

a) Factorization of the polynomial function $\pi_{\mathfrak{j}}$.

LEMMA 4. *Let $\mathfrak{j} \in \operatorname{car}(\mathfrak{g})$ and $f \in \mathcal{V}^{\mathfrak{j}}$. Then*

(i) $\mathfrak{j} \in \operatorname{car}(\mathfrak{r})$,

(ii) *the root system of the Cartan subalgebra $\mathfrak{j}_{\mathbf{C}}$ in $\mathfrak{r}_{\mathbf{C}}$ does not depend neither on the choice of $f \in \mathcal{V}^{\mathfrak{j}}$, nor on that of $\mathbf{R}$; we denote it by $\Delta_{\mathfrak{j},s}$,*

(iii) *for every $\alpha \in \Delta_{\mathfrak{j},s}$ let H_α be its coroot in the semi-simple algebra $[\mathfrak{r}, \mathfrak{r}]$; then this H_α does not depend on either of these choices,*

(iv) *if $\Delta^+_{\mathfrak{j},s}$ is a positive roots system in $\Delta_{\mathfrak{j},s}$ and if σ denotes the number of pure imginary roots in $\Delta^+_{\mathfrak{j},s}$ the polynomial function on $\mathfrak{h}^*_{\mathfrak{j}}$ defined by*

$$\pi_{\mathfrak{j},s} = i^{\sigma} \prod_{\alpha \in \Delta^+_{\mathfrak{j},s}} H_\alpha$$

is real on $\mathfrak{h}^*_{\mathfrak{j}}$ and divides $\pi_{\mathfrak{j}}$. Moreover the quotient $\pi_{\mathfrak{j},n} = \pi_{\mathfrak{j}}/\pi_{\mathfrak{j},s}$ is an element of $S(\mathfrak{n}^{\mathfrak{j}})$,

(v) *the intersection $\mathfrak{j}^{*\prime} = \mathfrak{j}^* \cap \mathfrak{r}^*_r$ is the complement in $\mathfrak{j}^*$ of the set of zeros of $\pi_{\mathfrak{j},s}$.*

In this lemma $\mathfrak{n}^{\mathfrak{j}}$ stands for the centralizer in $\mathfrak{n}$ of $\mathfrak{j}$.

b) A family of G-invariant polynomial functions on $\mathfrak{g}^*_{\mathbf{C}}$.

If q is a polynomial function as in Lemma 3 we denote by U the complement in $\mathfrak{g}^*_{\mathbf{C}}$ of its set of zeros. One has

PROPOSITION 2. *One can choose the polynomial function q of lemma 3 in such a way that*

(i) *U/S is a smooth amnifold,*

(ii) *for every element $u \in U$ one has*

$$\mathfrak{n}^*_{\mathbf{C}} = (\mathfrak{n}^*_{\mathbf{C}})^{S(u)} + \mathfrak{s}_{\mathbf{C}} \cdot u\,.$$

If these conditions are fullfilled, then

(iii) *if $\mathfrak{j} \in \operatorname{car}(\mathfrak{g}_{\mathbf{C}})$ and p is a polynomial function on $\mathfrak{j}^*$ which is invariant under the normalizer in $\mathbf{G}$ of $\mathfrak{j}$, then there exists a natural* *integer m and a G-invariant polynomial function r defined on $\mathfrak{g}^*_{\mathbf{C}}$, such that*

$$r(f) = p(f_{|\mathfrak{j}})q^m \circ \rho(f) \qquad \forall f \in \mathfrak{h}^*_{\mathfrak{j}}.$$

Moreover, r is uniquely determined by p and m.

In the sequel the polynomial q of Lemma 3 is supposed to satisfy conditions (i) and (ii) of Proposition 2.

COROLLARY. *There exists a natural integer k and a G-invariant polynomial function π_q on $\mathfrak{g}^*$ such that for every $\mathfrak{j} \in \mathrm{car}(\mathfrak{g})$ one has*

$$\pi_{q|\mathfrak{h}^*_{\mathfrak{j}}} = \pi^2_{\mathfrak{j},\mathfrak{s}} q^k \circ \rho_{|\mathfrak{h}^*_{\mathfrak{j}}}.$$

Moreover, one can deduce from the proof of Proposition 2 the following lemma, which will be useful later.

LEMMA 5. *If $\mathfrak{j} \in \mathrm{car}(\mathfrak{g})$, then the equivalence classes in $\mathcal{V}^{\mathfrak{j}}$ for the relation $\sim$ are open subsets of $\mathcal{V}^{\mathfrak{j}}$; in particular each of these is a union of some connected component of $\mathcal{V}^{\mathfrak{j}}$.*

c) Expression of the functions $q_{G,\chi}$ and q_G in terms of the corresponding functions for the groups R.

The notation being as above one has

LEMMA 6. *Let $\chi \in \mathrm{Ker}\, j^\wedge$ and let $f \in \mathcal{V}$. Then f belongs to $\mathfrak{g}^*_{G,\chi}$ (resp. $\mathfrak{g}^*_G$) if and only if λ belongs to $\mathfrak{r}^*_{\overset{\circ}{R},\chi_0}$ (resp. $\mathfrak{r}^*_{\overset{\circ}{R}}$), where χ_0 stands for the restriction of χ to $\mathrm{Ker}\, j \cap \overset{\circ}{R}$, and in this case one has*

$$q_{G,\chi}(f) = q_{\overset{\circ}{R},\chi_0}(\lambda) \quad (\textit{resp. } q_G(f) = q_{\overset{\circ}{R}}(\lambda)). \tag{5}$$

Moreover, let $\mathfrak{j}$ be an element of $\mathrm{car}(\mathfrak{g})$ and $\mathcal{W}$ an element of $\mathcal{V}^{\mathfrak{j}}/\sim$. Then, for $f \in \mathcal{W}$, the number $q_{G,\chi}(f)$ (resp. $q_G(f)$) depends only on the restriction of f to $\mathfrak{j}$.

d) The Fourier transform of the distribution $M_{G,T}$.

Let us denote by $\Theta_{G,T}$ the Fourier transform of $M_{G,T}$. We have

THEOREM 1. *Let T be an elliptic element of $\mathfrak{g}$. Then the distribution $\Theta_{G,T}$ is an essentially bounded measurable function on $\mathfrak{g}^*$, which is analytic on the open subset $\mathcal{V}$. Moreover one has*

$$\Theta_{G,T}(f) = \sum_{T' \in (G.T \cap \mathfrak{t})/R} \Theta_{R,T}(\lambda) \qquad \forall f \in \mathcal{V}. \tag{6}$$

REMARK: If T is an elliptic element of $\mathfrak{g}$ denote by $\mathrm{Car}_T(G)$ the set of elements of $\mathrm{Car}(G)$, the image of which in $\mathfrak{s} = \mathfrak{g}/\mathfrak{n}$ is S-conjugated, for a certain $u \in V^T$, to a Cartan subalgebra of $\mathfrak{s}(u)$. Then the support of $\Theta_{G,T}$ is contained in

$$\bigcup_{\mathrm{j} \in \mathrm{Car}_T(G)} \overline{\mathfrak{g}^*_{r,\mathrm{j}}},$$

and moreover it follows from the properties of the function $\Theta_{R,T}$ when G is reductive (properties due to Harish-Chandra) that $\lim_{s \to 0, \neq} \Theta_{G,sT}$ exists and is a step function which restricted to each set $\mathfrak{g}^*_{r,\mathrm{j}}$, $\mathrm{j} \in \mathrm{Car}_T(G)$, is a non zero multiple of its characteristic function. There are examples for which this function is non-constant, unlike the case when G is reductive.

e) The orbital integrals for $\mathfrak{g}^*$.

If $\mathrm{j} \in \mathrm{car}(\mathfrak{g})$ and if $\mathcal{W}$ is a subset of $\mathcal{V}^{\mathrm{j}}$ we put

$$\tilde{\mathcal{W}} = \mathrm{j}^\perp \cap \mathfrak{h}_\mathrm{j}^*(\rho^{-1}(\rho(\mathcal{W}))).$$

Now let $\mathcal{W} \in \mathcal{V}^{\mathrm{j}}/\sim$. Then for every element φ of $\mathcal{S}(\mathfrak{g}^*)$ we define a function $F^{\mathcal{W}}_{\mathrm{j},\varphi}$ on $\mathrm{j}^{*'}$ (see Lemma 4) by the following formula

(7)

$$F^{\mathcal{W}}_{\mathrm{j},\varphi}(f) = \pi_{\mathrm{j},\mathfrak{s}}(f) \int_{G/H_\mathrm{j}} \left(\int_{\tilde{\mathcal{W}}} \varphi(g.(f+u))\pi_{\mathrm{j},\mathfrak{n}}(u)^2 du \right) d\dot{g} \qquad \forall f \in \mathrm{j}^{*'}.$$

When G is semi-simple, that is N is trivial, $\mathcal{W}$ is always equal to $\mathrm{j}^{*'}$ and the functions $F^{\mathcal{W}}_{\mathrm{j},\varphi}$ are nothing else but Harish-Chandra's orbital integrals. For this reason we call the functions $F^{\mathcal{W}}_{\mathrm{j},\varphi}$ the orbital integrals on $\mathfrak{g}^*$ with respect to the group G.

Now it is a consequence of Proposition 2 (i) that $\tilde{\mathcal{V}}^{\mathrm{j}}/H_\mathrm{j}$ is a smooth manifold. The group W_j acts on this manifold and it is possible to choose on $\tilde{\mathcal{V}}^{\mathrm{j}}/H_\mathrm{j}$, a W_j-invariant odd volume form γ_j, which defines a positive Radon measure $d\mu_{\gamma_\mathrm{j}}$. Now if $\Omega \in \tilde{\mathcal{V}}^{\mathrm{j}}/H_\mathrm{j}$ let γ^j_Ω be the odd volume form which is the quotient along Ω of the Lebesgue measure df on $\tilde{\mathcal{V}}^{\mathrm{j}}$ by γ_j. Then one has for every function $\varphi \in \mathcal{D}(\tilde{\mathcal{V}}^{\mathrm{j}})$:

$$(8) \qquad \int_{\tilde{\mathcal{V}}^{\mathrm{j}}} \varphi(f)df = \int_{\tilde{\mathcal{V}}^{\mathrm{j}}/H_\mathrm{j}} \left(\int_\Omega \varphi(f) d\mu_{\gamma^\mathrm{j}_\Omega} \right) d\mu_{\gamma_\mathrm{j}}(\Omega).$$

With these definitions we define for every $\varphi \in \mathcal{S}(\mathfrak{g}^*)$ the function $\tilde{F}_{\mathrm{j},\varphi}$ on the space $\mathrm{j}^{*'} \times \tilde{\mathcal{V}}^{\mathrm{j}}/H_\mathrm{j}$ by the following formula

$$(9) \quad \tilde{F}_{\mathrm{j},\varphi}(f) = \pi_{\mathrm{j},\mathfrak{s}}(f) \int_{G/H_\mathrm{j}} \left(\int_\Omega \varphi(g.(f+u))\pi_{\mathrm{j},\mathfrak{n}}(u)^2 d\mu_{\gamma^\mathrm{j}_\Omega}(u) \right) d\dot{g},$$

where $f \in \mathfrak{j}^{*'}$ and $\Omega \in \tilde{\mathcal{V}}^{\mathfrak{j}}/H_{\mathfrak{j}}$. Then using (8) it is easy to see that we have

$$F^{\mathcal{W}}_{\mathfrak{j},\varphi}(f) = \int_{\tilde{\mathcal{W}}/H_{\mathfrak{j}}} \tilde{F}_{\mathfrak{j},\varphi}(f,\Omega)d\mu_{\gamma_{\mathfrak{j}}}(\Omega)\,.$$

f) Some functional spaces and the regularity properties of the orbital integrals on $\mathfrak{g}^*$.

Let E be a finite dimensional real vector space, π an element of $S(E^*_{\mathbf{C}})$ which is a product of linear forms, and (Y,μ) a σ-finite measurable space. In E we consider an open cone Γ, for which there exists an open cone Γ' with a finite number of connected components, all of them being convex, such that $\Gamma' \subset \Gamma \subset \overline{\Gamma'}$. Moreover, let $S \subset S(E)$ be a subalgebra containing the element 1 and such that $S(E)$ is a finitely generated S-module.

We denote by $L^1_S(\Gamma \times Y, |\pi|dxd\mu)$, where dx stands for a Lebesgue measure on E, the space of functions φ which are defined on $\Gamma \times Y$ and satisfy the following conditions

(i) φ is $|\pi|dxd\mu$-measurable,

(ii) for μ a.a. $y \in Y$ the function $\varphi_y(x) = \varphi(x,y)$ is of C^∞-class on Γ,

(iii) for every $p \in S$, $\partial_p\varphi \in L^1(\Gamma \times Y, |\pi|dxd\mu)$,

and we endow $L^1_S(\Gamma \times Y, |\pi|dxd\mu)$ with the topology defined by the semi-norms

$$\nu^\pi_p(\varphi) = \int_{\Gamma\times Y} |\partial_p\varphi(x,y)|\, |\pi(x)|dxd\mu(y), \quad p \in S\,.$$

If $S = S(E)$ the space $L^1_S(\Gamma\times Y, |\pi|dx\,d\mu)$ is denoted $L^1_\infty(\Gamma\times Y, |\pi|dx\,d\mu)$. We define also the spaces $L^1_S(\Gamma, |\pi|dx)$ or $L^1_\infty(\Gamma, |\pi|dx)$ by saying that they are the corresponding spaces above in case (Y,μ) is the one point measurable space and μ is the measure of total mass 1.

Now the natural projection $\mathfrak{g} \to \mathfrak{s} = \mathfrak{g}/\mathfrak{n}$ induces an injection $S(\mathfrak{s}^*) \hookrightarrow S(\mathfrak{g}^*)$ and one has $S(\mathfrak{s}^*)^S \subset S(\mathfrak{g}^*)^G$.

If $p \in S(\mathfrak{g}^*)$ and $\mathfrak{j} \in \mathrm{car}(\mathfrak{g})$ we denote by $p_{\mathfrak{j}}$ the restriction of p to $\mathfrak{j}$. Let $S_{\mathfrak{j}} \subset S(\mathfrak{j}^*)$ be the subalgebra the elements of which are the $p_{\mathfrak{j}}$ with $p \in S(\mathfrak{s}^*)^S$. Then it turns out that $S(\mathfrak{j}^*)$ is a finitely generated module over $S_{\mathfrak{j}}$.

We have the following

THEOREM 2. *Let $\mathfrak{j} \in \mathrm{car}(\mathfrak{g})$ and $\mathcal{W} \in \mathcal{V}^{\mathfrak{j}}/\sim$. Then for every $\varphi \in \mathcal{S}(\mathfrak{g}^*)$ and $f \in \mathfrak{j}^{*'}$ the orbital integral $F^{\mathcal{W}}_{\mathfrak{j},\varphi}(f)$ is absolutely convergent. Moreover the map $\varphi \to F^{\mathcal{W}}_{\mathfrak{j},\varphi}$ is a continuous homomorphism from $\mathcal{S}(\mathfrak{g}^*)$ into $L^1_\infty(\mathfrak{j}^{*'}, |\pi_{\mathfrak{j},\mathfrak{s}}|df)$ and if $p \in S(\mathfrak{s}^*)^S$ and $\varphi \in \mathcal{S}(\mathfrak{g}^*)$ one has*

$$F^{\mathcal{W}}_{\mathfrak{j},\partial_p\varphi} = \partial_{p_{\mathfrak{j}}}(F^{\mathcal{W}}_{\mathfrak{j},\varphi})\,.$$

To prove this result, first we establish the following result:

PROPOSITION 3. *Let* $\mathfrak{j} \in \operatorname{car}(\mathfrak{g})$. *Then for every* $\varphi \in \mathcal{D}(\mathfrak{g}^*)$ *and* $(f, \Omega) \in \mathfrak{j}^{*'} \times \tilde{\mathcal{V}}^{\mathfrak{j}}/H_{\mathfrak{j}}$ *the integral (9) is absolutely convergent.*

Moreover, the function $\tilde{F}_{\mathfrak{j},\varphi}$ *is of* C^∞*-class and for every* $p \in S(\mathfrak{s}^*)^S$, *one has*

$$\tilde{F}_{\mathfrak{j},\partial_p\varphi} = \partial_{p|_{\mathfrak{j}}}(\tilde{F}_{\mathfrak{j},\varphi}).$$

Finally the map $\varphi \to \tilde{F}_{\mathfrak{j},\varphi}$ *is a continuous morphism from the space* $\mathcal{D}(\mathfrak{g}^*)$ *endowed with the topology induced by* $\mathcal{S}(\mathfrak{g}^*)$ *into the space* $L^1_{S_{\mathfrak{j}}}(\mathfrak{j}^{*'} \times \tilde{\mathcal{V}}^{\mathfrak{j}}/H_{\mathfrak{j}}, |\pi_{\mathfrak{j},\mathfrak{s}}| df\, d\mu_{\gamma_{\mathfrak{j}}})$.

The proof of this proposition rely on the existence of the polynomial π_q (see the corollary of Proposition 2) and is very similar to the proof of the regularity of Harish-Chandra's orbital integrals for a reductive Lie algebra given by Varadarajan in [Va].

Now to deduce Theorem 2 from Proposition 3 we use a result concerning the spaces $L^1_S(\Gamma, |\pi| dx)$ introduced above.

Before stating theorem 3 we need more notation. We denote by $L^\infty_\infty(\Gamma)$ the space of functions of C^∞-class on Γ which are, together all their derivatives, uniformly bounded on Γ, endowed with the natural topology.

THEOREM 3. *With notations and assumptions as above, we have*

(i) *the space* $L^1_S(\Gamma, |\pi| dx)$ *is contained in* $L^\infty_\infty(\Gamma)$, *the injection being continuous,*

(ii) *the space* $L^1_S(\Gamma, |\pi| dx)$ *is a Frechet space,*

(iii) *the spaces* $L^1_S(\Gamma, |\pi| dx)$ *and* $L^1_\infty(\Gamma, |\pi| dx)$ *are equal and the topology of the latter is in fact the one defined by the semi-norms* ν^π_p, $p \in S$.

To prove this theorem, it suffices to handle the case where Γ is a convex open cone in $\mathbf{R}^n$ containing the canonical basis of $\mathbf{R}^n$. To do that we use an elementary solution of a sufficiently great power of the differential operator $(1 - \frac{\partial}{\partial x_1}) \dots (1 - \frac{\partial}{\partial x_n})$ such that this elementary solution is supported in the closed cone $\bar{C} = \{x \in \mathbf{R}^n \mid x_1 \leq 0, \dots, x_n \leq 0\}$. These results contain, for a cone Γ as considered, the results of Varadarajan in Part I, Appendix 1, 2 and 3 of [Va], and we believe that our proof is simpler.

g) A sketch of the proof of the Poisson-Plancehrel formula.

Using Theorem 1 formula (6), Lemma 6 formula (5), and the integration formula (1) in Lemma 1, we prove that the Poisson-Plancherel formula (4) with respect the character χ of $\operatorname{Ker} j$ reduces to the following assertion:

If $\mathfrak{j} \in \text{car}(\mathfrak{g})$, $f \in \mathcal{V}^{\mathfrak{j}}$ and R is a reductive factor in $G(u)$, then for every element λ of $\mathfrak{j}^{*'}$, we have

$$(10) \qquad \pi_{\mathfrak{j},s}(\lambda) \sum_{T \in \tilde{E}_R/R} \tilde{\chi}_R(T)\Theta_{R,T}(\lambda) = \pi_{\mathfrak{j},s}(\lambda) q_{R,\chi}(\lambda) dm^{\mathfrak{j}}_{R,\chi}(\lambda)$$

as an equality between linear functionals on the space $L^1_\infty(|\pi_{\mathfrak{j},s}|df)$.

In fact we have to prove that the two members of (10) applied to an orbital integral $F^{\mathcal{W}}_{\mathfrak{j},\varphi}$ give the same result.

The equality (10), understood as an equality between functionals on the space $\mathcal{S}(\mathfrak{j}^{*'})$ (obvious definition), was proved by M. Duflo and M. Vergne in [Du-Ve]. But, as we saw it by computing an example, the properties of the orbital integrals on $\mathfrak{g}^*$ established in Theorem 2 are the best ones. So an important piece of our work is devoted to proving that equality (10) is still true in the more general case we indicated.

Finally the proof relies on the classical Poisson summation formula for a sufficiently large class of functions, as stated below.

Let U be the open subset of $\mathbf{R}^n$, the elements of which are the $x \in \mathbf{R}^n$ such that $x_1 \dots x_n \neq 0$ and let $\mathcal{E}_n$ be the space of continuous functions φ on $\mathbf{R}^n$ which are such that

(i) $\varphi_{|U}$ is of C^∞-class

(ii) $\nu_p(\varphi) = \int_U |\partial_p \varphi(x)| dx < +\infty$ for all $p \in S(\mathbf{R}^n)$.

We endow $\mathcal{E}_n$ with the topology defined by the semi-norms ν_P, $p \in S(\mathbf{R}^n)$.

Recall that $\mathcal{F}_{\mathbf{R}^n}$ stands for the Fourier transform on $\mathbf{R}^n$. One has

THEOREM 4. *For every $\varphi \in \mathcal{E}_n$ the series*

$$\sup_{x \in \mathbf{R}^n} \sum_{k \in \mathbf{Z}^n} |\varphi(x+k)| \quad \textit{and} \quad \sup_{x \in \mathbf{R}^n} \sum_{k \in \mathbf{Z}^n} |\mathcal{F}_{\mathbf{R}^n}\varphi(x+2\pi k)|$$

are finite and define continuous semi-norms on $\mathcal{E}_n$.

Moreover one has the following Poisson summation formula which is valid for every $x, y \in \mathbf{R}^n$

$$\sum_{k \in \mathbf{Z}^n} \varphi(x+k) e^{-i<x+k,y>} = \sum_{k \in \mathbf{Z}^n} \mathcal{F}_{\mathbf{R}^n}\varphi(y+2\pi k) e^{i<2\pi k,x>} .$$

REFERENCES

[Do] P. Dourmashkin, *A Poisson-Plancherel formula for groups of type B_n*, Thése M.I.T. (1984); a paraître.

[Du-1] M. Duflo, *Représentations unitaires des groupes de Lie et méthode des orbites*, in "G.M.E.L.," Bordas, Paris, 1982.

[Du-2] M. Duflo, *On the Plancherel formula for almost algebraic real Lie groups*, in "Lie Groups Representations III;" Lecture Notes in Mathematics 1077, Springer Verlag, Berlin, Heidelberg, New York, Tokyo, 1984.

[Du-Ve] M. Duflo et M. Vergne, *La formule de Plancherel des groupes de Lie semi-simples réels*, in "Representations of Lie Groups;" Kyoto, Hiroshima (1986), Advanced Studies in Pure Mathematics 14, 1988.

[Mu-Fo] D. Munford and J. Fogarty, *Geometric Invariant Theory*, in "A Series of Modern Surveys in Mathematics," Springer Verlag, Berlin, Heidelberg, New York, Tokyo, 1982.

[Pe-Ve] D. Peterson and M. Vergne, *Recurrence relations for Plancherel functions*, in ";" Lecture Notes in Mathematics 1243, Springer Verlag, Berlin, Heidelberg, New York, Tokyo, 1987, pp. 240-261.

[To] P. Torasso, *La formule de Poisson-Plancherel pour un groupe presque algébrique á radical abélien: cas oú le stabilisateur générique est réductif*, Prépublication de Département de Mathématiques de l'Université de Poitiers No. 40 (212p.); (to appear in Mémoires de la S.M.F.).

[Va] V.S. Varadarajan, *Harmonic analysis on real reductive groups*; Lecture Notes in Mathematics 1576, Springer Verlag, Berlin, Heidelberg, New York, Tokyo.

[Ve-1] M. Vergne, *A Poisson-Plancherel formula without group representations*, in "O.A.G.R. Conference," INCREST, Bucarest, Roumania, 1980.

[Ve-2] M. Vergne, *A Poisson-Plancherel formula for semi-simple Lie groups*, Ann. of Math 115 (1982), 639–666.

Laboratoire de Mathématiques
URA CNRS D 1322 "*Groupes de Lie et Géométrie*"
Université de Poitiers
40, Avenue du Recteur Pineau
F-86022 – Poitiers Cedex
FRANCE

Progress in Mathematics

Edited by:

J. Oesterlé
Département de Mathématiques
Université de Paris VI
4, Place Jussieu
75230 Paris Cedex 05
France

A. Weinstein
Department of Mathematics
University of California
Berkeley, CA 94720
U.S.A.

Progress in Mathematics is a series of books intended for professional mathematicians and scientists, encompassing all areas of pure mathematics. This distinguished series, which began in 1979, includes authored monographs and edited collections of papers on important research developments as well as expositions of particular subject areas.

All books in the series are "camera-ready", that is they are photographically reproduced and printed directly from a final-edited manuscript that has been prepared by the author. Manuscripts should be no less than 100 and preferably no more than 500 pages.

Proposals should be sent directly to the editors or to: Birkhäuser Boston, 675 Massachusetts Avenue, Suite 601, Cambridge, MA 02139, U.S.A.

A complete list of titles in this series is available from the publisher.

21 KATOK. Ergodic Theory and Dynamical Systems II
22 BERTIN. Séminaire de Théorie des Nombres, Paris 1980–81
23 WEIL. Adeles and Algebraic Groups
24 LE BARZ/HERVIER. Enumerative Geometry and Classical Algebraic Geometry
25 GRIFFITHS. Exterior Differential Systems and the Calculus of Variations
26 KOBLITZ. Number Theory Related to Fermat's Last Theorem
27 BROCKETT/MILLMAN/SUSSMAN. Differential Geometric Control Theory
28 MUMFORD. Tata Lectures on Theta I
29 FRIEDMAN/MORRISON. Birational Geometry of Degenerations
30 YANO/KON. CR Submanifolds of Kaehlerian and Sasakian Manifolds
31 BERTRAND/WALDSCHMIDT. Approximations Diophantiennes et Nombres Transcendants
32 BOOKS/GRAY/REINHART. Differential Geometry
33 ZUILY. Uniqueness and Non-Uniqueness in the Cauchy Problem
34 KASHIWARA. Systems of Microdifferential Equations
35 ARTIN/TATE. Arithmetic and Geometry: Papers Dedicated to I.R. Shafarevich on the Occasion of His Sixtieth Birthday, Vol. 1
36 ARTIN/TATE. Arithmetic and Geometry: Papers Dedicated to I.R. Shafarevich on the Occasion of His Sixtieth Birthday. Vol. II
37 BOUTET DE MONVEL. Mathématique et Physique
38 BERTIN. Séminaire de Théorie des Nombres, Paris 1981–82
39 UENO. Classification of Algebraic and Analytic Manifolds
40 TROMBI. Representation Theory of Reductive Groups
41 STANELY. Combinatorics and Commutative Algebra
42 JOUANOLOU. Théorèmes de Bertini et Applications
43 MUMFORD. Tata Lectures on Theta II
44 KAC. Infinite Dimensional Lie Algebras

45 BISMUT. Large Deviations and the Malliavin Calculus
46 SATAKE/MORITA. Automorphic Forms of Several Variables, Taniguchi Symposium, Katata, 1983
47 TATE. Les Conjectures de Stark sur les Fonctions L d'Artin en s = 0
48 FRÖHLICH. Classgroups and Hermitian Modules
49 SCHLICHTKRULL. Hyperfunctions and Harmonic Analysis on Symmetric Spaces
50 BOREL, ET AL. Intersection Cohomology
51 BERTIN/GOLDSTEIN. Séminaire de Théorie des Nombres, Paris 1982–83
52 GASQUI/GOLDSCHMIDT. Déformations Infinitésimales des Structures Conformes Plates
53 LAURENT. Théorie de la Deuxième Microlocalisation dans le Domaine Complexe
54 VERDIER/LE POTIER. Module des Fibres Stables sur les Courbes Algébriques, Notes de l'Ecole Normale Supérieure, Printemps, 1983
55 EICHLER/ZAGIER. The Theory of Jacobi Forms
56 SHIFFMAN/SOMMESE. Vanishing Theorems on Complex Manifolds
57 RIESEL. Prime Numbers and Computer Methods for Factorization
58 HELFFER/NOURRIGAT. Hypoellipticité Maximale pour des Opérateurs Polynomes de Champs de Vecteurs
59 GOLDSTEIN. Séminaire de Théorie des Nombres, Paris 1983–84
60 PROCESI. Geometry Today: Giornate Di Geometria, Roma. 1984
61 BALLMANN/GROMOV/SCHROEDER. Manifolds of Nonpositive Curvature
62 GUILLOU/MARIN. A la Recherche de la Topologie Perdue
63 GOLDSTEIN. Séminaire de Théorie des Nombres, Paris 1984–85
64 MYUNG. Malcev-Admissible Algebras
65 GRUBB. Functional Calculus of Pseudo-Differential Boundary Problems
66 CASSOU-NOGUÈS/TAYLOR. Elliptic Functions and Rings and Integers
67 HOWE. Discrete Groups in Geometry and Analysis: Papers in Honor of G.D. Mostow on His Sixtieth Birthday
68 ROBERT. Autour de L'Approximation Semi-Classique
69 FARAUT/HARZALLAH. Deux Cours d'Analyse Harmonique
70 ADOLPHSON/CONREY/GHOSH/YAGER. Number Theory and Diophantine Problems: Proceedings of a Conference at Oklahoma State University
71 GOLDSTEIN. Séminaire de Théorie des Nombres, Paris 1985–1986
72 VAISMAN. Symplectic Geometry and Secondary Characteristic Classes
73 MOLINO. Riemannian Foliations
74 HENKIN/LEITERER. Andreotti–Grauert Theory by Integral Formulas
75 GOLDSTEIN. Séminaire de Théorie des Nombres, Paris 1986–87
76 COSSEC/DOLGACHEV. Enriques Surfaces I
77 REYSSAT. Quelques Aspects des Surfaces de Riemann
78 BORHO/BRYLINSKI/MACPHERSON. Nilpotent Orbits, Primitive Ideals, and Characteristic Classes
79 MCKENZIE/VALERIOTE. The Structure of Decidable Locally Finite Varieties
80 KRAFT/PETRIE/SCHWARZ. Topological Methods in Algebraic Transformation Groups
81 GOLDSTEIN. Séminaire de Théorie des Nombres, Paris 1987–1988
82 DUFLO/PEDERSEN/VERGNE. The Orbit Method in Representation Theory
83 GHYS/DE LA HARPE. Sur les Groupes Hyperboliques d'après Mikhael Gromov